W9-AWR-813

METEOROLOGY: AN INTRODUCTION

METEOROLOGY: AN INTRODUCTION

BY CHRISTOPHER LAMPTON

A GROLIER COMPANY

FRANKLIN WATTS

NEW YORK | LONDON | TORONTO | SYDNEY | 1981

AN IMPACT BOOK

FRONTISPIECE: A U.S. SATELLITE PHOTOGRAPH
SHOWS THE CLOUD COVER OVER THE CONTINENTAL
UNITED STATES ON JUNE 2, 1980.

Drawings by Vantage Art, Inc.

Photographs courtesy of the
U.S. Department of Commerce/NOAA:
pp. ii and 72; and NASA: p. 87.

Library of Congress Cataloging in Publication Data

Lampton, Christopher
Meteorology, an introduction.

(An Impact book)
Bibliography: p.
Includes index.
SUMMARY: Discusses the various components
of weather, constantly changing phenomena
which get their energy from the sun.
1. Meteorology—Juvenile literature.
[1.   Meteorology] I. Title.
QC863.5.L35   551.5   80–22688
ISBN 0-531-04260-X

CONTENTS

CHAPTER
1

Weather controls much of your life.

You don't agree? Maybe you just haven't thought about it. How many times have you missed school or work because of snow? How many trips and picnics have you canceled because of rain? How many colds have you caught when the temperature suddenly changed? And how many times have you stayed indoors and watched television just because it was too cold or too hot to go out?

Weather affects your life in subtler ways, too. Certain foods grow only in certain seasons. Many sports that are played in winter aren't played in summer, and many sports that are played in summer aren't played in winter. When the weather is nice, you feel nice. When the weather is dreary, you feel dreary. Hot, dry days make you feel restless. Hot, damp days make you feel tired.

Weather can take its toll—in money and in lives. If it were not for the cold of winter, there would be no heating bills. If not for the heat of summer, there would be no air-conditioning bills. If the weather never changed, we could live in less expensive houses, drive less expensive cars, wear less expensive clothes. There would be fewer automobile accidents, fewer illnesses, fewer flooded basements. We would never have to fear lightning, tornadoes, or hurricanes.

If you've ever been through a violent storm, you don't need to be told that weather exacts a price. In the ferocious winds of a hurricane, hundreds of thousands of people can lose their lives in a few hours, and billions of dollars' worth of damage can be wreaked. A tornado can reduce a house to splinters or a building to rubble. Even the common thunderstorm can pelt the earth with hailstones—balls of ice as big as eggs, as destructive as bullets.

For better or worse, weather is always with us, so we've learned to live with it—learned to live with it so well that we sometimes forget that it's there.

But weather is a constant phenomenon—constantly changing, constantly moving. The atmosphere of our planet is like a great engine, always operating, its cogs and gears forever turning and grinding.

What *are* the gears of this engine? What is its fuel? For what purpose does it run? Why does it keep churning, day in, day out, year after year?

To understand the weather, it might help to step back a distance, to look at the earth as it might be seen from outer space. From a few hundred thousand miles away, the planet seems like a tiny ball, the atmosphere around it (where the process of weather takes place) hardly thicker than a coat of paint.

How insignificant the earth seems at this distance. How fragile the atmosphere. And yet we live inside that shallow atmosphere, spend our lives within it.

Anything that affects the atmosphere affects us. A

mere ripple in that wisp of air can wipe out cities, create storms, floods, tidal waves.

From our distant perspective the earth seems calm and serene. The space around it is vast and empty. What could possibly affect that atmosphere? What could send ripples of weather through that thin shell of gases?

Only one thing. The sun.

From 93,000,000 miles (149,665,000 km) away, the sun bombards our planet with heat. The atmosphere grows warm under this bombardment—and begins to move.

Because of the way the earth is positioned in space, some parts of its surface receive more of this heat than other parts. We know, of course, that it is warmer at the equator than at the poles. This is because sunlight falls on the poles at a low angle, spreading a small amount of heat over a wide area and thus decreasing the sun's overall effect.

But heat tends to travel. The solar heat that falls on the equator doesn't stay at the equator. Instead, it begins making its way north and south, toward the poles, to correct this temperature imbalance.

The process by which that heat works its way from the equator to the poles is called . . . weather.

And that process gets its energy from the sun. . . .

CHAPTER
2

Anybody who has read a newspaper or listened to the news lately has some idea of the importance of energy. We are running out of it, the headlines scream. We are in the middle of an energy crisis, say the news magazines. Yet how many people know just what energy is?

Ask any physicist, and you'll probably be told that energy is the ability to perform work. But that definition may leave you more confused than ever. Certainly it takes energy to split logs or dig ditches, and surely no one will argue that splitting logs isn't work. But does that mean that it doesn't take energy to play games or engage in sports?

Of course not. A game of touch football uses up a lot more energy than an hour spent solving math problems, but you may be tempted to call the math "work" and the football "fun." To a scientist, however, the term *work* has a particular meaning. It is "the application of a force over a distance."

That, however, may leave you even more confused than before. If so, consider that if you apply enough force to something, whether it's a book or a bicycle or an old car, it will eventually move. And when it moves, you can say that you have applied "a force over a distance," the distance being how far the object moved. For our purposes, then, we can say that work is "making something move," and that energy is "the ability to make something move." A scientist might quibble with this definition, but for our purposes it is good enough. It is safe to say that if you see something move, energy is involved.

Energy comes in a number of different forms, and it can change suddenly from one form to another. If an object is moving, for instance, we say that it has *kinetic* energy, from a Greek word meaning "to move." A baseball that has just been batted toward the outfield would be an example of an object with kinetic energy.

Suppose, however, that you are simply holding a baseball in your hand. The baseball would appear to have no energy at all, because it is not moving. But if you let go of the ball, it will move suddenly toward the ground. Where did this kinetic energy come from?

It is one of the basic understandings of science that energy cannot be created from nothing. (This is called *conservation of energy* and is usually expressed as follows: "Energy can neither be created nor destroyed.") Therefore, we say that the ball, while you held it in your hand, had *potential energy,* or *energy of position.* All you had to do to convert this potential energy into kinetic energy was to drop the ball.

Another form of energy is heat. You probably have a good idea as to what heat is. It's what you feel when you lie in the sun on a summer day or when you stand next to a radiator or hot stove. But, once again, the term *heat* has a special meaning to scientists.

All matter is made up of tiny particles called *atoms,* which group together into collections of particles called *molecules.* When molecules are in motion (and they are always in motion), we say that they have *heat energy.* The faster the molecules move, the hotter we say they are. The heat that you feel when you stand next to a hot stove is simply the way your nervous system interprets this molecular motion. In other words, you are feeling the molecules move. Because heat involves the motion of molecules, we can say that heat energy is a form of kinetic energy on the molecular level.

To remove all molecular motion in a piece of matter, you would have to lower the temperature of the matter to −459° F (−273° C, or 0° Kelvin), a temperature that scientists call *absolute zero.* And even then an infinitesimal amount of molecular motion would remain, no matter how hard you tried to eliminate it.

As any boy or girl scout knows, you can convert kinetic energy into heat energy by rubbing two sticks together. The friction of wood against wood "excites" the motion of the molecules in the sticks until a spark is ignited and a fire starts. (Actually, in almost any transaction involving energy, a certain amount of heat energy is produced, though the amount may be too small to be noticed.) There are any number of devices for converting heat energy back into kinetic energy: the steam engine, for instance, or gunpowder.

Another form of energy is *electricity.* By running an electric current through the filament of a light bulb, you can convert electricity into heat energy, which in turn produces yet another kind of energy, *light.*

But what, you might ask, does all this have to do with weather?

Well, it's not hard to see that weather involves a

great deal of energy. Rains fall, winds blow, clouds scud across the sky—all of this calling for kinetic energy. Air temperatures go up and down, so heat energy must be involved. Thunderstorms generate electricity in the form of lightning, which in turn produces dazzling flashes of light. That's two more forms of energy to be accounted for. And clouds must contain potential energy, because they produce raindrops that fall great distances to the ground, just as the baseball we discussed earlier falls when you release it from your hand.

In fact, the amount of energy required to keep the weather process in motion is quite huge, by human standards. An average hurricane uses enough energy to supply electric power to the entire United States for six months. A summer thunderstorm packs the same energy wallop as a small atomic bomb.

And all this energy has to come from someplace. But where? Where does it come from?

From the greatest source of energy this planet has ever known—the sun.

The sun has been producing huge amounts of energy for five billion years and will continue producing energy for at least five billion more, through a process known as *nuclear fusion.* The amount of energy that the sun produces is so great that the human mind has great difficulty comprehending it. We can measure it, however, with a unit called the *calorie*—the amount of heat energy required to raise the temperature of one gram of water by 1° C. The sun produces roughly 5,600,000,000,000,000,000,000,000,000 calories *every minute!*

Throughout history, almost all of the energy needs of the human race have been met, either directly or indirectly, by the sun. It is the sun (as we shall see) that makes the wind blow, to turn windmills. It is the sun that makes trees grow, trees that in turn supply wood for burning. It is the sun that makes food grow, food that in turn supplies energy to the muscles of our bodies. It is the sun that evaporates water and recycles it in the form of rain, rain that makes rivers flow and turn the great turbines in hydroelectric dams. It is the sun that is the initial source of the energy in the fossil fuels, such as natural gas, oil, and coal, that run our cars and factories and heat our homes.

And it is the sun that provides the energy for the weather process.

But how does that energy cross 93,000,000 miles (149,665,000 km) of empty space to reach the earth from the sun?

To answer that question we'll have to examine a particular form of energy—light.

Drop a stone into a perfectly still pool of water and what happens? Ripples spread outward in a circular pattern from the point where the stone breaks the surface. We call these ripples *waves.*

Let's take a closer look at these waves. Even though they are easy to see and can have a definite effect on any object floating in the water, the waves themselves have no substance. They are simply a pattern of motion in the water. Each wave has a high point, which we can call the *peak,* and a low point, which we can call the *trough.* The distance from peak to peak (or trough to trough) is called the *wavelength,* because it represents the length of a single wave.

Waves don't just travel through water. They can also travel through air. If you dropped the stone onto a wooden floor instead of into the water, it would create waves in the air, and when these waves reached your ear, you would hear them as sound. In this case, you would hear a clatter as the stone bounced across the wood. Any movement that creates sound (the beating of a drum, the vibrating of your vocal cords, an explosion) sets up similar waves in the air, which in turn cause vibrations inside your ear. The nature of the sound you hear depends on the length of the wave. If the wavelength is long, the sound will be low-pitched and rumbling. If the wavelength is short, the sound will be high-pitched and shrill.

Light is also a wave—sort of.

When you turn on the burner of an electric stove, for instance, it will begin to glow. We can describe that glow as a kind of wave, though not a wave like the ones we just discussed.

You might ask what would cause a hot burner to create a wave. We saw earlier that all matter is made up of small particles called atoms. Atoms, however, are made up of even smaller particles. One of these particles is the *electron.* Electrons orbit around the outer part of the atom in somewhat the same way as the moon orbits around the earth. When an atom becomes hot, these electrons begin to oscillate, or move in and out of their orbits. The oscillating electron emits a kind of wave we call *electromagnetic radiation.* Depending on its wavelength, electromagnetic radiation can take many forms: X rays, gamma rays, radio waves, microwaves—and light.

When we turn on the electric burner, the coil becomes hot and its electrons start to oscillate. You can see this oscillation as a dull red light. This, in fact, is why we refer to such burners as being "red hot."

It is important to note, however, that there are differences between light waves and sound or water waves. Water waves have no substance of their own;

neither do sound waves. They are simply a recurring motion in some other substance—either air or water. Without air, for instance, there would be no sound, because there would be nothing for the sound to travel through.

This is not true of electromagnetic radiation. Light —or radio waves, microwaves, and so forth—needs no substance through which to travel. The light from our sun, for instance, travels through the vacuum of space with no trouble at all. This is because, in a sense, light carries its own substance with it in the form of particles called *photons.* (This is a simplification of what actually happens but, once again, it is good enough for our purposes.)

Just as your ear is attuned to detecting air waves as sound, your eye will detect light waves as color. The particular color that you see depends (as you may have guessed) on the wavelength of the light. Light of a fairly long wavelength will appear to your eyes as red, and light of a fairly short wavelength will appear blue. There are many wavelengths of light that cannot be detected by the human eye. For that reason, we refer to the wavelengths that can be seen (those between red and blue) as *visible light.*

The nonvisible forms of light are grouped into two categories: *infrared* and *ultraviolet.* The infrared ("beyond red") wavelengths are longer than those of visible light, and the ultraviolet ("beyond violet") wavelengths are shorter.

The sun is very hot. In fact, the temperatures at its core may be higher than 59,000,000° F (15,000,000° C)! The sun's surface, at roughly 9,964° F (5,500° C), is much cooler, but overall the sun is hot enough to produce light at all of the visible wavelengths, including the infrared and the ultraviolet. It is the energy in this light (or, actually, in the extremely small percentage of the sun's total light that is intercepted by the earth) that serves as the fuel for the weather process.

Once the light leaves the sun, it takes only eight minutes to cover the vast distance between the sun and the earth. In order to have a significant effect on the weather process, however, light not only has to reach the earth, but it also has to penetrate the earth's atmosphere so that the heat energy it contains can be absorbed by the ground. This isn't nearly as simple as it sounds. In fact, only about half of the light that reaches the outer portion of our atmosphere gets all the way to the earth's surface.

What stands in the way? Clouds, for one thing. Clouds are very white and shiny. Like any bright object, they reflect light. In fact, clouds can reflect as much as 75 percent of the light that falls on them. The reflected light simply shoots back out into space and never reaches the ground.

(Interestingly enough, this is why the planet Venus seems so bright in the early evening or morning sky—because Venus's surface is almost completely wrapped in thick clouds.)

Snow also reflects light. Snow-covered ground can throw off as much as 90 percent of the sunlight that falls on it, so that only 10 percent is actually absorbed by the earth. Because of this, the presence of snow can, in a sense, create cold weather. So much heat energy is lost by reflection that the air cannot become warm enough to melt the snow—a vicious circle!

Water can also reflect sunlight, but only if the sun is very low in the sky. Just before sunset, when light comes skimming in over the water at a low angle, as much as 80 percent will bounce back into space.

All in all, the earth reflects back into space about

35 percent of the sun's radiation that reaches it. (We say, then, that the earth has an *albedo* [al-BEE-doe] of 35 percent, the albedo being the percentage of light that an object reflects.) However, we said before that only 50 percent of the sun's light goes all the way to the ground. What happens to the remaining 15 percent?

You'll recall that heat energy causes electrons to oscillate and that this oscillation produces light. Under certain conditions this process can happen in reverse. Light will be *absorbed* by electrons and converted back into heat. This happens, for instance, when the ultraviolet in sunlight meets a gaseous substance in the earth's atmosphere called *ozone.* The ultraviolet is absorbed by the ozone, and the ozone in turn becomes warm.

This has important consequences for those of us who live on the earth's surface. Ultraviolet light is very harmful to living creatures. If the full ultraviolet radiation of the sun were allowed to reach the ground, life as we know it could not exist.

The absorption of ultraviolet light also has a significant effect on the temperature of the atmosphere. We'll discuss this in a later chapter.

We should note that a certain percentage of the sun's light is neither reflected nor absorbed but is instead scattered by the earth's atmosphere. Light striking tiny particles in the air is knocked aside. Some of this scattered light is simply reflected back into space. The rest finds its way to the ground. If it were not for this scattering of sunlight, the sky would appear to be pitch black, even in the middle of the day. The only light we would see would be that coming directly from the sun. Because the short blue wavelengths of light are scattered more easily than the long red wavelengths (which can "slip" around the atmospheric particles), we see the sky as blue.

About 50 percent of the light that strikes the atmosphere, then, makes it to the ground. Except for the ozone, this light has no heating effect at all on the air that it passes through. When it reaches the ground, however, it is absorbed and converted back into heat.

Where does this heat go? Into the air, of course. And at that point the weather process begins.

CHAPTER
3

We live at the bottom of a sea of air, but what is that air made of? We know it exists, because we breathe it, fly kites and airplanes in it, and feel it brush across our skin every time the wind blows.

Yet what is air?

It is a mixture of gases, actually, made up mostly of nitrogen (its most abundant component) and oxygen. Other gases in air include carbon dioxide, neon, krypton, xenon, and methane, though for the most part these gases exist in insignificant amounts.

We refer to this sea of air that blankets our planet as the *atmosphere.* The atmosphere reaches a height of quite a few miles. If sunshine is the fuel of the weather engine, this atmosphere acts as its gears and pistons. Without air, there would be no weather. The weather process takes place entirely within this fragile shell of gases.

And so, in order to understand the weather process, we must know something about how gases behave.

In 1660, a scientist named Robert Boyle demonstrated that gases could be compressed.

What does this mean? Well, suppose you have a small chamber (a metal box, perhaps) filled with nothing but air, and you drive a piston into this chamber in such a way that no air can escape. What will happen to the air in the chamber as the piston presses against it?

Boyle showed that as the piston took up more and more space in the chamber, the air would take up less and less, even though no air was allowed to leave the chamber. The piston would crowd the air into an increasingly smaller area. The air would be *compressed*—that is, the space between the molecules of air would grow smaller, allowing the air to take up less room.

However, Boyle also showed that air resists this sort of compression. As the piston pushes against the air, the air pushes back against the piston. And the more compressed the air becomes, the harder it pushes back. When the air is compressed to half its original size, it pushes back twice as hard as it did while still uncompressed. Not only does the air push back against the piston, but it pushes back against the sides of the chamber also. In fact, if the piston pushes too hard against the air, the chamber might well explode.

We refer to this tendency of air to push back when compressed as *air pressure*. The more compressed air becomes, the greater the air pressure will be. This relationship, which is true of all gases (not just air), is called *Boyle's Law*.

What does this have to do with the atmosphere?

The air around us is constantly being compressed by the weight of the air above it.

Perhaps this will be easier to picture if we compare it to something more familiar. Imagine that we have a pile of mattresses. Under normal conditions each mattress is 1 foot (30 cm) thick. However, in this pile there are a thousand mattresses, one on top of another. The bottom mattress is so crushed by the weight of the 999 mattresses on top of it that it is only 1 inch (2.5 cm) thick. The second mattress from the bottom has 998 mattresses on top of it and is also crushed, but not quite as much as the one on the bottom. If we were able to measure it, we would find that each mattress was a little less crushed than the one beneath it. The mattress at the top would not be crushed at all.

The same is true of the air in our atmosphere. The air at the bottom—that is, the air that we breathe—is very dense and crushed. As we travel up through the atmosphere we find that the air grows less and less compressed, just as the mattresses from the bottom to the top grow progressively less crushed. The air at the top of the atmosphere is very thin, its molecules quite far apart from one another. In fact, it is so thin that it is difficult to say just where the earth's atmosphere ends and the vacuum of space begins. (Even space, which is supposedly airless, contains some molecules of gas.)

We can say, then, that air pressure decreases with height, as the air becomes less compressed.

The ocean of air is very heavy at the bottom as a result of all the compression. So heavy is it, in fact, that we would be crushed beneath it ourselves were it not for the air inside our bodies (which is also highly compressed) pushing back against the air outside, just as the air in the chamber pushes back against the piston. If you took an airtight box and pumped all of the air out of it through a specially drilled hole (so that there would be no air left inside to push against the air outside), the box would be crushed by the weight of the atmosphere. This, essentially, is the principle of the suction cup. When you squeeze the air out of the rubber "cup," the weight of the atmosphere will cause it to stick to any smooth surface. If there were air inside the cup, it would push back and the cup would not stick.

Suppose, now, that we have a parcel of air under pressure. (Meteorologists like to speak of air as coming in *parcels.* A parcel is simply a quantity of air of more or less uniform temperature and pressure, existing alongside the rest of the air in the atmosphere but not mixing with it. This is not to imply, however, that there is any physical barrier between the air inside the parcel and the air outside of it.) When the pressure is removed, the parcel will expand; that is, it will become decompressed. However, in order for the air to expand, a certain amount of energy is required. The air obtains this energy by converting its heat energy into *energy of expansion.* Because its heat energy has been changed into something else, the air will grow colder as it expands. The amount of energy in the air, however, will remain the same. When air loses heat without losing energy, we say that it has *cooled adiabatically* (aah-dee-aah-BAA-tik-lee).

The same thing happens, in reverse, when air is compressed. Energy of expansion is converted back into heat energy, and the air grows warmer. In this case, we say that the air has *warmed adiabatically.*

As a parcel of air rises through the atmosphere, the pressure of the air around it will drop. The parcel, therefore, will expand. As it expands, its heat energy changes form, and the air becomes cooler.

Most of the heat in our atmosphere comes from the ground and is carried upward by rising air. Yet, we

(17

just saw how rising air loses its heat energy through expansion. As a result of this, the temperature of our atmosphere tends to drop with height.

This is an important concept. To put it another way, air pressure decreases with height. Therefore air expands with height, and as it expands, it cools. Therefore, air cools with height.

If you were to rise from the earth's surface (in the gondola of a balloon, perhaps), you would find, under normal circumstances, that the air temperature dropped steadily as you went up. The rate at which the temperature drops varies according to where you are and what the weather is like, but ordinarily, by the time you reached a height of about 9 miles (14.4 km), you would find air temperatures between 50 and 80 degrees below zero, Fahrenheit (45 and 62 degrees below zero, Centigrade).

Oddly, at this point, the temperature stops going down. If you kept going upward, the temperature would begin rising again. This gives meteorologists a convenient way of dividing up the atmosphere and describing its parts. Because most weather phenomena take place in the first 9 miles (14.4 km) of our atmosphere (for reasons that will be obvious later), this region is called the *troposphere,* from Greek words meaning "sphere of movement." The point at which the temperature stops plunging is called the *tropopause,* meaning "end of movement." The region where the temperature starts to rise again is called the *stratosphere,* meaning "sphere of layers" (because of certain layers of gas found within it).

Why does the temperature start going up again in the stratosphere? Because this is where the ozone is found. You'll recall from the previous chapter that ozone absorbs ultraviolet light from the sun and con-

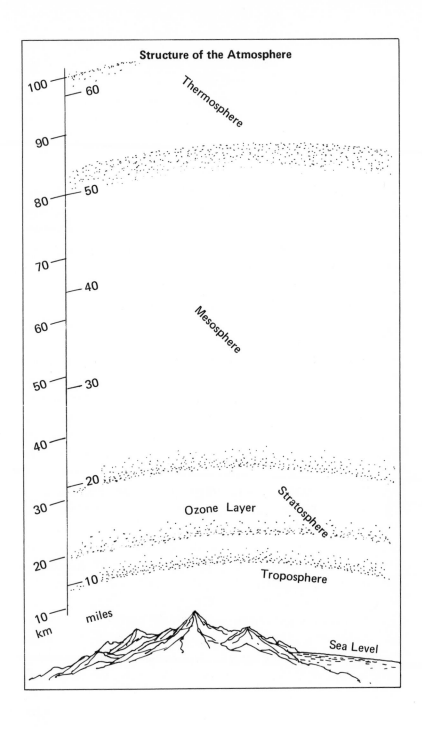

Structure of the Atmosphere

Thermosphere

Mesosphere

Stratosphere

Ozone Layer

Troposphere

Sea Level

km
miles

verts it into heat. This heat warms the air in the stratosphere, so that temperatures here are higher than those in the upper troposphere. But 20 miles (32 km) up (at the *stratopause,* which marks the end of the stratosphere), the temperature starts going down again. This region of falling temperature is called the *mesosphere* and extends to a height of about 53 miles (84.8 km), where the temperature (at the *mesopause*) can go as low as −165° F (−110° C). Above the mesopause is the *thermosphere* ("sphere of heat"), where the temperature rises yet again. In the upper reaches of the thermosphere temperatures soar as high as 1,000° F (540° C).

This sounds more than a little incredible. One thousand degrees Fahrenheit? The oven in your kitchen probably won't produce temperatures that high. Yet we have sent astronauts to the moon, and not one was cooked on his way through the thermosphere.

Of course the space capsule, with its protective skin, would have prevented the astronauts from being burned. The fact is, however, that if you were to expose your bare skin to the air in the thermosphere, it wouldn't feel hot. It would feel extremely cold. Why is this?

In the dense atmosphere at the earth's surface, we are accustomed to thinking of temperature and heat as being the same thing. When we speak of the air as having a high temperature, we mean that it is very hot, and vice versa. Yet heat and high temperature are not the same.

In Chapter 2 we saw that heat is simply the motion of molecules. When a molecule contains heat energy, it moves. The more heat energy it has, the faster it moves. The speed at which a molecule moves is, in simple terms, its temperature. The higher its temperature, the faster it moves. The lower its temperature, the slower it moves.

It follows that the temperature of a parcel of air can be defined as the average speed of its molecules. When you heat a parcel of air, the movement of its molecules increases—and its temperature goes up. The amount of heat required to raise a parcel of air to a certain temperature depends on how many molecules there are in the parcel. The more molecules there are, the more heat is required. When we say that air "feels warm," we are referring to the overall heat content of the air, not to its temperature. In very thin air, only a small amount of heat is needed to raise the temperature. Thus, the thin air of the thermosphere can have a very low heat content (and feel very cold) while still having a high temperature (because its molecules are moving quickly).

There is another way to look at it. On the earth's surface we are constantly being bombarded by billions of molecules, and our senses interpret this bombardment as heat. In the upper reaches of the thermosphere, however, we would receive only a weak bombardment by a mere few thousand molecules. Even though these molecules are moving fast, and therefore are "hot," there are so few that the total effect would be negligible, and we would feel almost no heat.

This is why the astronauts could pass safely through the thermosphere. There is not enough actual *heat* in this region to kindle even the smallest fire.

In the last chapter we saw how light makes its way through the earth's atmosphere to be absorbed by the ground. This light is then re-radiated by the ground as heat. Where does the heat go from there?

It enters the air and becomes the main source of

heat for the troposphere. However, it only enters the air close to the ground. How then does it spread to the air above?

There are two ways in which heat can travel through the atmosphere: *conduction* and *convection*.

Conduction is the transfer of heat from molecule to molecule. When air molecules are heated, they move quickly. Sometimes these hot, fast-moving molecules bump into cold, slow-moving molecules. These collisions cause the slow-moving molecules to move faster, just as one billiard ball starts another billiard ball moving by running into it. In this way heat, which is nothing more than molecular motion, spreads. The same thing happens when you put a pan of water on a hot burner. The fast-moving molecules of the burner collide with the slow-moving molecules in the pan, causing the pan molecules to move more quickly. Once the pan molecules are moving quickly, they collide with the molecules in the water, and the water molecules are speeded up in turn.

In air, however, conduction is a slow way of spreading heat. Because air molecules are rather widely spaced (even in the dense air at the bottom of the troposphere), collisions between air molecules are rare, no matter what the temperature. Mostly, heat spreads through air by convection.

Convection is the transfer of heat by the movement of entire parcels of air, rather than by the movement of individual air molecules.

When the molecules in a parcel of air are heated, and therefore moving at high speeds, they spread out over a wider area than when they are cold. The air becomes thinner as a result. Thin air is lighter than dense air because there are fewer molecules in a given amount of space. Therefore, warm air is lighter than cold air.

When a parcel of air is warmed by heat from the ground, it becomes lighter than the cooler air above it. The heavier, cooler air sinks, forcing the lighter, warmer air in the parcel to move upward. Because of this, we say that *warm air rises.* As it rises, it cools, because its heat energy is being converted into energy of expansion. When it is no longer warmer than the air around it, the parcel will stop rising. (We'll study this process in more detail in a later chapter.)

In this way, hot air is set into motion. As it moves upward, it spreads its heat, even though much of that heat is lost through expansion. This rising of air is extremely important to the weather process, especially when the air contains water, as we shall see in the next chapter.

CHAPTER
4

"Water, water, everywhere, nor any drop to drink."

In this famous quotation from "The Rime of the Ancient Mariner," Samuel Taylor Coleridge's mariner is referring to the water in the ocean, which is so salty that a sailor would die from drinking it. But he might as well have been referring to air.

At any given time, as much as 4 percent of our air may be made up of water. This water is in the form of a gas, called *water vapor,* which finds its way into the atmosphere through *evaporation.*

Evaporation is the process by which water turns into a gas. For evaporation to take place, energy is required. As with so many other parts of the weather process, this energy is supplied in the form of heat.

As we saw a moment ago, the hotter a molecule is —that is, the more heat energy it contains—the faster it moves. Water molecules are no exception to this rule.

In a given quantity of water, however, not all molecules will be moving at the same speed. Suppose we have a pan of water sitting quietly at room temperature —say, 70° F (21° C). Some of the water molecules will be moving quickly—that is, they will possess a great deal of heat energy—and some will be moving slowly. As with air molecules (see Chapter 3), when we speak of the temperature of the water, we mean the average temperature (or speed) of the water molecules in the pan. In this case, the average water molecule will be moving at a speed represented by a temperature of 70° F (21° C).

Occasionally, one of the fast-moving molecules will attain such a high speed that it will shoot out of the pan into the air, where it will remain. We say that this molecule has evaporated—that is, it has become part of the air as water vapor. As other fast-moving molecules do the same, the amount of water in the pan will become smaller. Only the slower-moving, cooler molecules will remain behind.

It is important to note what happens to the water that remains in the pan. Because evaporation removes the warmer molecules from the pan and leaves the cooler molecules behind, the temperature of the water in the pan goes down. Evaporation actually cools the water, even though, ironically, it is heat that causes the evaporation in the first place.

We can generalize, therefore, by saying that *evaporation cools.*

You have probably noticed that right after you step out of a shower or swimming pool, you feel quite cool. This is because the heat of your body (and of the sun, in daytime) provides the energy to evaporate the water on your skin. The evaporation, in turn, carries your body heat away into the air, literally leaving you cold. Of course, if the wind is blowing you will become even colder, because moving air increases the rate of evaporation.

(25

How does it do this?

Air can hold only so much water at a given time. If, for instance, the air around our imaginary pan of water is standing still, it will eventually become so full of water vapor that as much water will be moving from the air to the pan as is moving from the pan to the air. At this point it is impossible to increase the amount of water in the air through further evaporation, so we say that the air has become *saturated.*

The evaporation process takes place more rapidly in air that is completely dry—that is, air that does not already contain evaporated water. As the air grows closer to saturation, the process takes place more and more slowly.

However, if the air is moving (as when the wind is blowing), the wet air is carried away before it can become saturated. It is replaced, in turn, by drier air, which can absorb water vapor more quickly than the wet air. Therefore, a breeze actually speeds up the evaporation process. And that is why the wind blowing across your wet body makes you feel cold.

(It might be added here that this is one of the reasons why you perspire on a hot day. As the perspiration evaporates from your skin, it carries heat away with it. Therefore, sweat keeps you cool, especially when there's a breeze blowing.)

What happens to the heat that is carried away by evaporation? Remember that energy can be neither created nor destroyed. It can only change form. Therefore, the heat lost through evaporation has to go somewhere.

In fact, it enters the air, though you cannot feel it, in the form of *latent heat.* This is a kind of potential energy (see Chapter 2). This latent heat will remain locked away in the molecules of air until the water vapor is converted back into liquid form, at which point it will become *sensible heat* (heat that you *can* feel, or sense). We shall see the importance of this in the next chapter.

The amount of water a given parcel of air can hold depends on the temperature of the parcel. Warm air can hold more water vapor than cold air. It is useful for meteorologists to note the amount of water that a given parcel of air can hold and compare it with the amount of water it actually does hold. This comparison is expressed as a percentage and is called the *relative humidity.* For instance, if a given parcel of air contains 5 ounces (140 g) of water but could potentially hold 10 ounces (280 g) of water, we would say that the parcel is halfway to being saturated. Another way of expressing this is to say that the parcel has a relative humidity of 50 percent. If that same parcel contained 7½ ounces (210 g) of water out of a potential 10 (280), we would say that it had a relative humidity of 75 percent. And if it was completely saturated, we would say that it had a relative humidity of 100 percent. If the humidity goes *above* 100 percent, we say that the air is *supersaturated.* We will see what happens to such supersaturated air later in this chapter.

The relative humidity of a parcel of air depends on its temperature. Lowering the temperature of a parcel of air will actually *increase* its relative humidity, even though the amount of water vapor it contains remains the same. This is because the potential for holding water decreases as the temperature goes down. It takes less water to saturate a cold parcel of air than it takes to saturate a warm one. Therefore, the same amount of water represents a larger percentage (or relative humidity) in cold air than in warm air.

When air becomes supersaturated, the water vapor in the air starts turning back into liquid water. This process, the opposite of evaporation, is called *condensation.*

However, supersaturation in itself is not enough to cause condensation. In perfectly clean air, for instance, supersaturations of more than 400 percent have occurred.

This is because water vapor requires a surface on which to condense, and in perfectly clean air no surfaces are available. Under normal conditions, however, air is never perfectly clean. It is filled with tiny particles, bits of matter too small for the eye to see. These particles are specks of dust carried aloft by the wind, salt tossed from the crests of waves, soot sent up from trash fires and smokestacks or from the cones of active volcanoes, pollen ejected by flowers, and micrometeorites that have fallen from outer space.

Almost all air contains water vapor. We saw in the previous chapter that the temperature of air goes down as the air rises. As the temperature of the air drops, its ability to hold water decreases and its relative humidity goes up. Therefore, when the air reaches a certain height (and its temperature falls to a certain point), it becomes supersaturated. The water vapor in the air begins to condense back into liquid water.

This condensation takes place on the tiny particles of dust and soot and pollen we just mentioned. For that reason, we refer to these particles as *condensation nuclei.* (The word *nucleus*—singular of nuclei—comes from a Latin word meaning "nut," or "kernel." We use it to mean the hard center around which a larger object, or cluster of objects, forms. ·A condensation nucleus is, therefore, the center around which condensation takes place.)

Not surprisingly, condensation will take place more easily on particles that are somewhat water-absorbent. Such particles are called *hygroscopic* particles, meaning that they have an unusual attraction for water.

The temperature at which condensation begins for a given parcel of air is called the *dew point,* because when this temperature is reached at ground level it causes a form of condensation known as *dew.* When rising air reaches this temperature and condensation nuclei are present, as they always are, condensation will take place. All at once the air will be full of tiny droplets of water, millions of them, each formed around a single nucleus. Some of these droplets will be less than 1 micrometer in diameter. (A micrometer is equal to one-thousandth of a millimeter, or one-millionth of a meter.)

When enough of these droplets gather together to be visible from the ground, we refer to the resulting conglomeration as a *cloud.* The droplets themselves are called *cloud droplets.*

The average cloud droplet is less than 10 micrometers in diameter. Because of their small size, they can remain in the air for long periods of time. Gravity has very little effect on these droplets and cannot overcome the force of the rising air that is carrying them upward.

Even if a cloud droplet falls back to earth, it will evaporate before it reaches the ground. The rising air will then carry it upward until it condenses and once again becomes part of the cloud.

If cloud droplets cannot fall, where does rain come from?

A raindrop obviously has to be larger than a

cloud droplet. It has to be heavy enough to resist the force of the rising air and to fall back to earth before evaporation can take its toll. No matter how fast it falls, however, it is going to lose water to evaporation, so it must be large enough to survive that loss.

Given this information, we can calculate that a raindrop must have a volume roughly 30,000 times greater than a cloud droplet. How does a droplet this size form?

It would make sense to suppose that raindrops are overgrown cloud droplets—but it doesn't work that way. Even the largest cloud droplets are only about 20 micrometers in diameter. The smallest raindrops are about 500 micrometers in diameter—a significant difference! For reasons that we will not go into here, it is impossible for a cloud droplet to grow any larger than 20 micrometers through condensation. If it does grow any larger, it becomes unstable and immediately begins shrinking again.

In order to bridge the gap between cloud droplet size and raindrop size, the droplet must grow very quickly, before it has time to shrink again. Condensation is just too slow a process to produce a raindrop.

Nevertheless, meteorologists believe that raindrops form from cloud droplets and that there are two ways in which this can happen. The first is called *collision and coalescence.*

We saw earlier that cloud droplets are light enough to be carried upward on the rising air, but the rate at which they rise depends on how big they are. Cloud droplets of different sizes will move upward at different speeds. It follows that droplets moving at different speeds will sometimes run into one another. When two or more droplets collide, they can coalesce (join together) to form a larger droplet. The larger droplet may be so large and heavy that it will begin falling back to earth. As it falls, it will collide with still more droplets, growing larger with every collision.

Eventually, after thousands of such collisions, the droplet will be large enough to reach the ground without evaporating. The cloud droplets will have coalesced into a raindrop.

But not all clouds produce rain. Why should collision and coalescence take place in some clouds but not in others?

Well, for droplets to run into one another it is necessary for them to be of differing sizes. (Droplets of the same size, which move at the same speed, will not run into each other.) Yet condensation is a very even process and tends to produce cloud droplets all of the same size.

You'll remember, though, that certain nuclei have a stronger attraction for water than others. Hygroscopic particles can form droplets much larger than those formed around ordinary particles. Salt is very hygroscopic. For obvious reasons, salt particles are more common in clouds formed over the ocean. Therefore, the process of collision and coalescence is more likely to take place in clouds over the sea than in clouds over land.

It is also more likely to take place in tall clouds than in short ones. Raindrops need room in which to form. If they fall out of the bottom of the cloud before they reach their full size, they simply evaporate before they reach the ground and the process has to begin all over again. Therefore, the more cloud there is for a coalescing droplet to fall through, the more likely the droplet is to become a raindrop.

Meteorologists also suspect that electricity may play a role in the collision-coalescence process. Under normal conditions, colliding cloud droplets tend to bounce off one another instead of coalescing. A strong electrical field may help to break down this resistance, so that larger droplets can form. Such fields occur in certain kinds of clouds, such as those that generate thunderstorms.

Not all rain is formed by collision and coalescence. In fact, meteorologists suspect that the *Bergeron process*—named after the scientist who proposed it—may be even more important in the creation of rainstorms.

As strange as it sounds, air that would be merely saturated in the presence of liquid water can become supersaturated in the presence of ice, even though no change in temperature is involved. When air becomes saturated over ice, the water vapor in the air will *sublime*—change directly into ice—instead of condensing into water.

When ice finds its way into a cloud, water will tend to sublime on the ice before it will condense on the cloud droplets that are forming. This, in turn, lowers the amount of water in the air, which causes the cloud droplets to evaporate. The water vapor created by this evaporation will then sublime on the ice particles.

In this way, the ice particles steal the water from the cloud droplets. The amount of ice in the cloud increases, and the amount of water in the droplets decreases. The ice particles, which can grow to larger sizes than cloud droplets can, form into *snow crystals.*

Far heavier than cloud droplets, snow crystals (also called *snowflakes*) glide downward toward the ground without evaporating. If the air near the ground is cold enough, they remain in frozen form until they have finished falling, and the result is a snowstorm. If the air near the ground is warm, however, the snowflakes will melt, and the result will be rain.

(32

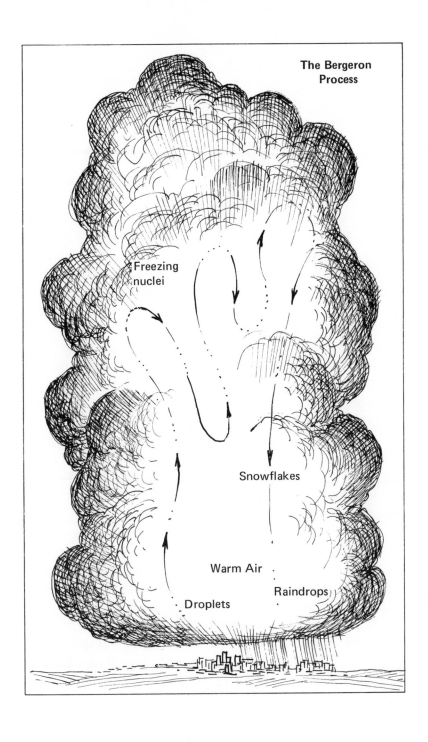

For the Bergeron process to work, however, there must be ice in the cloud. You might ask where this ice comes from.

Actually, this question is not as simple as it sounds. In the average cloud, temperatures are well below freezing, so that the water in the cloud droplets might be expected to turn to ice.

Yet this is not so. Water can be exposed to temperatures as low as −40° F (−40° C) without freezing. In order for water to turn to ice, certain particles must be present. As you might expect, these particles are called *freezing nuclei.*

The best freezing nuclei are particles with a composition similar to that of ice—certain kinds of crystal, for instance. In fact, the best freezing nuclei of all are particles of ice. If particles of ice (or silver iodide, which resembles ice in its basic structure) are dumped into a cloud from above (out of an airplane, for instance), snow crystals will begin to form. This process is called *cloud seeding* and can be used to create rainstorms.

However, under ordinary conditions clouds seed themselves. When particles of ice form inside a cloud, they can act as freezing nuclei for the formation of still more particles of ice. And as these particles grow, they become snow crystals.

Rain and snow are not, of course, the only kinds of precipitation produced by clouds. In order for rain to form by the Bergeron process, the snow crystals must melt on their way to the ground. But what if, after melting, the raindrops pass through a layer of cold air? The falling water can become *supercooled*—that is, it will become freezing cold without actually turning to ice; for ice to form there must always be freezing nuclei available. When this supercooled water strikes the ground, it freezes on contact. (In a sense, the ground serves as a freezing nucleus for the super-cooled water.) This kind of precipitation is called *freezing rain,* and it can create hazardous conditions. When the rain freezes around electric lines, for instance, the lines can break and endanger the lives of persons on the ground. Ice on highways means dangerous driving, of course, and slippery sidewalks can be hazardous to pedestrians.

What if the falling flakes are only partially melted before they pass through that layer of cold air? The still-frozen portion of the crystal will serve as a freezing nucleus for the liquid portion of the crystal, and pellets of ice will develop. These pellets are called *sleet.*

The most dangerous form of frozen precipitation is *hail.* When very strong updrafts are present within a thunderstorm, particles of ice that would normally be heavy enough to fall to the ground can become suspended within the cloud for an unusual length of time. When this happens, the particles can continue growing and growing until, under certain conditions, they become as large as golf balls. These *hailstones* can destroy crops and sometimes take lives when they strike the ground. Methods of controlling the development of hail—by firing cannons and rockets into clouds, for instance—have been developed, but no one is sure how effective they are.

CHAPTER
5

We have seen that a large part of the weather process depends on the upward motion of air. It would be a good idea at this point to look at some of the conditions under which air tends to rise and some of the conditions under which air tends *not* to rise.

In Chapter 3 we saw that hot air rises because it is lighter than cold air. The heavier cold air sinks, displacing the hot air and buoying it upward. We also saw that air temperature goes down as height increases. By putting these facts together, we might guess that air is in constant motion. As the air near the ground becomes hot, it rises. The cold air from above sinks to take its place. As the cold air falls, its energy of expansion is converted back into heat energy, and it becomes warmer. The warm air, on the other hand, expands as it rises, and it becomes cooler. Eventually, the warm air becomes cold air, the cold air becomes warm air, and the process starts all over again. In fact, this process can keep starting over and over again indefinitely.

You would expect, therefore, that air in the tropo-sphere would be constantly rising and falling.

This is true—some of the time. But a lot depends on how fast the temperature of the rising air drops in relation to the temperature of the air around it.

Temperatures in the troposphere drop as you move upward. The rate at which they drop is called the *lapse rate.* The lapse rate varies according to condi-tions, but on the average the temperature drops about 43.7° F for every half mile (6.5° C for every kilometer) that you go upward. Therefore, we can say that the average lapse rate is 43.7° F per half mile (6.5° C per kilometer). This is true until you reach the tropopause. In the stratosphere, of course, the temperature begins to rise again, and the lapse rate reverses.

The temperature of a parcel of air also drops as the parcel rises. The rate at which it drops is called the *adiabatic rate.* (An adiabatic temperature change is any change in temperature that takes place without an overall loss or gain of energy, such as the falling temperature in rising air.) For dry air—that is, air in which condensation is not taking place—the adiabatic rate is 50° F per half mile (10° C per kilometer). In other words, for every kilometer that a parcel of air rises, its temperature drops by 50° F (10° C). Almost all air contains water vapor, though, and when the tem-perature falls below a certain level, condensation be-gins. And then the adiabatic rate will change.

Why should this happen?

In Chapter 4, we saw that evaporation requires heat energy, which enters the air in the form of latent heat. When condensation takes place, this latent heat is released as sensible heat.

As the parcel of air rises and condensation takes place, the release of latent heat causes the parcel to grow warmer. At the same time, expansion causes the air to grow cooler. Overall, the air continues growing cooler, but more slowly than before. The rate at which air cools while rising *and* condensing is called the *wet adiabatic rate.* (We can call the rate at which air cools while rising and *not* condensing the *dry adiabatic rate.*)

The wet adiabatic rate varies according to how much water is in the air. For very dry air it can be as high as 9° C per kilometer. For very wet air, it can be as low as 5° C per kilometer.

We know that a parcel of air will rise if it is warmer than the surrounding air. We also know that it will cool as it rises and that it will keep rising until it is the same temperature as the surrounding air. But the surrounding air is also getting cooler as the parcel rises, according to the lapse rate prevailing in that area. If the temperature of the rising parcel goes down more quickly than the temperature of the surrounding air, the two temperatures will eventually be the same. At this point—because the air is at the same temperature as its environment—it will stop rising. If the temperature of the rising parcel does *not* go down faster than the temperature of the surrounding air, the parcel will simply keep rising until it reaches the tropopause, where the lapse rate reverses and the temperature of the surrounding air starts going up again. (This explains why weather in the troposphere rarely invades the stratosphere. Rising air almost always stops at the tropopause.)

We can express this in fewer words by saying that the parcel of air will stop rising if the adiabatic rate (the rate at which the temperature of the parcel goes down) is greater than the lapse rate (the rate at which the temperature of the surrounding air goes down).

(38

If the adiabatic rate is *not* greater than the lapse rate, however, the air will keep rising, because the air parcel will always be warmer than its environment.

In the first of these two situations, where the tendency of the air is to seek its own level and not rise, we say that the air is *stable,* by which we mean that up-and-down motion is unlikely. In the second situation, where the tendency of the air is to keep rising, we say that the air is *unstable,* by which we mean that up-and-down motion is very likely.

The fact is, though, that in many situations the air is neither stable nor unstable. You'll recall that air in which condensation is taking place (which we can call *wet air*) has a lower adiabatic rate than air in which condensation is not taking place (*dry air*). The lapse rate of the surrounding air will often fall between these two adiabatic rates. Therefore, if the air is dry, the adiabatic rate will be greater than the lapse rate. The air will tend not to rise and will therefore be stable. But if the air is wet (condensing), its adiabatic rate will be *less* than the lapse rate. The air *will* tend to rise and will be unstable. Because the air can therefore be both stable and unstable, depending on its condition, we say that it is *conditionally stable.*

The importance of air stability is obvious. Unstable air, because it tends to rise, will produce clouds and rain. Stable air, because it tends not to rise, is usually a source of fair weather.

Usually, but not always. Sometimes air, stable or unstable, is forced to rise—to get over the top of a mountain, for instance. This is called *mechanical lifting* of air. When air is lifted mechanically, condensation will take place just as though it had been lifted by other means. For this reason, rain is common on the windward slopes (slopes toward the wind) of mountains. As the air climbs the mountain, water vapor condenses, clouds form, and rain falls. Once over the top, the air will sink again, but now it is drier than before, having lost much of its water to precipitation. For this reason, deserts are common on the lee sides (sides away from the wind) of mountains.

Often, air of one temperature will be lifted over air of another temperature, through a process called *frontal lifting.* We'll take a closer look at this form of mechanical lifting in Chapter 8.

When air *converges* (comes together from several directions at once) it can be forced upward. Over the state of Florida, for instance, wind blows east from the Gulf of Mexico and west from the Atlantic Ocean. When this air meets over land, it has nowhere to go but up. Because this rising air tends to precipitate, rainstorms are common in Florida.

When stable air is forced to rise, the clouds that form are flat and wide, and the precipitation they produce is a thin drizzle, though this drizzle can last for long periods. Unstable air, on the other hand, will form tall, bulky clouds that can produce devastating, but usually brief, storms.

When conditionally stable air is forced to rise, it can become unstable. As water vapor begins to condense, the adiabatic rate changes. The release of latent heat buoys the air up. Towering clouds begin to form.

If changes in the adiabatic rate can alter the stability of air, so can changes in the lapse rate. The steeper the lapse rate—that is, the faster the atmospheric temperature drops with increasing height —the more likely that a parcel of air will be unstable. (A parcel becomes unstable, remember, when its adiabatic rate is lower than the lapse rate. The steeper the

(40

lapse rate, the more likely it is to be greater than the adiabatic rate.)

Anything that warms the air close to the ground without warming the air above will increase or steepen the lapse rate—and increase instability. What warms air near the ground without warming air higher up? Sunlight, of course.

Because the lapse rate is steepened by intense sunlight and because the steepened lapse rate encourages instability, intense storms caused by rising air (such as thunderstorms and hurricanes) are more common in summer than in winter.

Other conditions can also have an effect on stability. When cold air moves over a warm surface, the lapse rate will be steepened (because the air near the ground will be warmed), and the air will become unstable. When warm air moves over a cold surface, the lapse rate will decrease, and the air will become more stable.

CHAPTER
6

Anyone who has ever listened to a weather forecast has heard of *barometric pressure.* Usually the forecaster says either that the barometer is rising or it is falling—that is, that the barometric pressure is going up or going down. If you've listened carefully to these forecasts, you know that a rising barometer means that there is clear weather ahead, whereas a falling barometer indicates an upcoming storm. You may even have a barometer in your home, so that you can check it for yourself. But have you ever wondered exactly what the barometer measures?

In Chapter 3 we saw that air exerts pressure. This pressure is created by the weight of the atmosphere. In 1643, a scientist named Evangelista Torricelli invented a device for measuring this weight.

Torricelli filled a long glass tube with mercury, a heavy substance that is liquid at room temperature. (Water was first used, but mercury was substituted soon after because it is heavier than water and thus works better.) He then put a stopper in the mouth of the tube and turned the tube upside down in a dish also filled with mercury, so that the mouth was completely submerged in the liquid. Finally he removed the stopper from the tube. Some of the mercury flowed out, but not all of it. Almost 30 inches (75 cm) remained in the tube, above the level of mercury in the dish.

Torricelli reasoned that it was the weight of the air that held the column of mercury in place. The air pressed down on the mercury in the dish, forcing it up into the tube. If there had been air above the column of mercury in the tube, it would have counterbalanced this pressure, just as the air inside your body counterbalances the pressure of the air outside (see Chapter 3). However, when the mercury flowed out, it left a vacuum behind it in the tube, and so the air outside pushed the mercury back up into the tube. The 30 inches (75 cm) of mercury was just enough to counterbalance the weight of the air in the atmosphere.

You might look at this device of Torricelli's as a scale for weighing air. Today we call it a *mercury barometer* and use it to measure air pressure. There are several different kinds of barometers. The one in your home is probably an *aneroid barometer.*

Since at sea level the height to which the column rises in a mercury barometer is, on the average, 29.92 inches (just under 75 cm), we say that the standard atmospheric pressure is 29.29 inches of mercury. Another measure of air pressure, more commonly used by meteorologists, is the *millibar.* The millibar is a measure of the *magnitude* of air pressure. The higher the millibar reading, the higher the pressure. The average atmospheric pressure at sea level is 1,013.25 millibars. (We have to specify that these readings were taken at sea level because pressure drops with height. It's also necessary to specify that this is an *average* reading, because air pressure varies from place to place and time to time, according to conditions that we'll discuss in a moment.)

What makes the wind blow?

Imagine for a moment that you are trying to move a stalled automobile. The road is flat, and the car is free to move in whichever direction you push it. You and your friends decide to push the car forward, but another group has decided to push the car backward. If both groups push in opposite directions at the same time, which way will the car roll?

Let's assume that you and your friends push with enough force so that, if no one were pushing the other way, the car would speed up to 20 miles (32 km) per hour within 15 seconds. Furthermore, let's assume that the other group pushes with enough force that, if you weren't pushing against them, the car would speed up (backward) to 15 miles (24 km) per hour within 15 seconds.

Within 15 seconds, then, the car (with both groups pushing simultaneously from opposite directions) will move forward at 5 miles (8 km) per hour, because you and your friends are pushing with greater force. The speed at which the car moves is the difference between the force with which you are pushing and the force with which the others are pushing—in this case a difference of 5 miles (8 km) per hour (20 minus 15).

Another way of putting this is to say that the car will move in the direction of the least pressure.

The same thing happens when a parcel of air is caught between two regions of different air pressure. (Remember, when air is compressed from above, it will push outward in all directions.) The parcel moves in the direction of the least pressure. The air, like the car, is being pushed in two directions at the same time, so it moves in the direction in which it is being pushed the hardest. (Actually, the pressure on the parcel will be coming from more than two directions, but the principle is the same.) And, as with the car, the speed at which the parcel moves will be equivalent to the difference in pressure. If the difference is small, the parcel will move slowly. If the difference is large, the parcel will move quickly.

This motion of air is called *wind.*

In order to predict what direction the wind will blow, meteorologists draw maps with air pressure readings marked on them in millibars. They then connect areas of equal pressure on these maps with lines called *isobars.* (The prefix *iso-* means "equal.")

The result is a map that looks similar to contour maps that show elevations on the earth's surface, except that the "mountains" on this map are areas of high pressure and the "valleys" and "craters" are areas of low pressure. In fact, the resemblance to a contour map is so strong that meteorologists refer to the area over which the pressure goes up or down as the *pressure gradient* (where gradient means "slope," as in the slope of a hill). And, because parcels of air move from regions of high pressure to regions of low pressure, we say that the wind blows along the pressure gradient.

It might help if we compare the air to, say, a rubber ball. Just as a rubber ball will roll down a smooth hillside, the air "rolls" down the pressure gradient. And the steeper the gradient (the more suddenly the pressure drops), the faster the air moves, just as a rubber ball will roll down a steep hill faster than it will roll down a gentle incline.

The differences in pressure that we are talking

A Typical Weather Map

Showers

Millibar

1,008

29.75
(barometer reading)

80°

80°

Isobar

80°

80°

80°

60°

70°

60°

50°

40°

30°

40°

50°

60°

70°

80°

H

H

L

about are actually rather small, but air is quite light and can be pushed to high speeds by pressure differences of only a few millibars.

This tendency of air to follow the pressure gradient has a powerful—and complicated—effect on large-scale weather patterns. Suppose that we have an area of high pressure, roughly circular in shape. We can call this area a *high pressure cell.* (Once again, when we speak of high as opposed to low pressure cells, we are speaking only relatively. The difference between a high pressure cell and a low pressure cell may be only a few millibars.) You can spot this high pressure area easily on a pressure map, because it will be surrounded by circular isobars, showing that the pressure drops in all directions as you move away from the middle of the cell. In fact, the region around the high pressure area will resemble (on the map) the kind of circular target that is used in archery and darts, with the middle of the cell as the bull's-eye. (The isobars will not be *perfectly* circular, of course, but the resemblance is still striking.)

Wind, you'll remember, blows *away* from an area of high pressure. If you take a pencil and draw lines to represent the flow of air on this map, you will find that the wind blows outward in all directions from the middle of the high pressure cell the way spokes radiate outward from the center of a wheel.

To help imagine this, pretend that the high pressure region is a cone-shaped mountain (which it resembles on the map) with perfectly smooth slopes. If you dumped hundreds of rubber balls over the peak of this mountain, they would roll down the slopes in all directions, just as wind "rolls" down the pressure gradient away from the high pressure area. When air moves away in all directions from a single point, we say that it is *diverging.*

What will happen to the high pressure zone when all of the air has blown outward along the pressure gradient? Will there be nothing left but airless vacuum? We'll come back to this question in a moment.

Suppose now that we have an area of low pressure at another point on our map. Like the high pressure cell, this *low pressure cell* is roughly circular in shape. It is surrounded by isobars and also looks something like an archery target.

The difference, however, is that while air is flowing *away* from the high pressure zone, it is flowing *toward* the low pressure zone. We might imagine that the region of low pressure is a deep crater, and the wind is hundreds of rubber balls rolling down its slopes (the pressure gradient) from all sides, heading toward the center. When air comes together in this manner, we say that it is *converging* (see Chapter 5).

A few moments ago we saw that when air comes together (converges), it is forced to move upward. This is what happens in the low pressure region. Exactly the opposite happens in the high pressure zone. As the wind diverges from the center, air is forced to move downward from above to take its place.

At this point we can see how a direct relationship could form between the two cells. As the air rises in the low pressure cell, it has to go *somewhere.* Otherwise it would begin to pile up in the upper atmosphere. Similarly, the air in the high pressure zone can sink for only so long before there is no air left to sink.

But what if the rising air in the low pressure cell were to travel to the high pressure cell and replace the sinking air?

This is pretty much what happens, in a simplified

sort of way. Actually, the rising air in the low pressure cell diverges (moves outward) in all directions once it has risen far enough above the ground (the opposite of what it was doing nearer to the ground). And the air above the high pressure region converges from all directions, replacing the sinking air. In this way, our two pressure cells are connected not only with each other, but with high and low pressure cells all around the world. The air rising in all low pressure cells eventually becomes the sinking air in all high pressure cells and vice versa. Because the pressure gradient reverses at a certain distance above the ground, however, the wind blows in one direction on the ground and in the opposite direction aloft.

We'll look at this process in more detail later. First, however, let's examine another factor that affects the motion of the wind.

So far, we've assumed that the wind blows along the pressure gradient in a straight line. Alas, this isn't true. Earlier, we compared the divergence of air to the spokes of a wheel, radiating outward from the center. But this would be an accurate image only if the wheel were turning at the time.

Turning? Why would a high pressure region be turning?

Let's perform another imaginary experiment. Suppose that you are standing at the north pole. In this experiment, you are very, very tall, so tall that you can see all the way around the curve of the earth to the equator.

You have a baseball in your hand. You toss the baseball toward Brazil, a country whose northern part is located on the equator. The ball flies due south at, let us say, 2,000 miles (3,200 km) per hour. (In this thought-experiment, you obviously have a very powerful pitching arm.)

It takes the ball roughly three hours to reach the equator at that speed. Yet when it gets there, Brazil is gone. The ball drops into the ocean. What happened?

Remember that the earth rotates around its axis once every twenty-four hours. Because the north pole (where you are standing) is one of the two points around which it spins, the entire planet will turn around you once a day. In the three hours that it took the ball to reach the equator, Brazil had already gone one-eighth of the way around the earth.

To a viewer in North America, watching the ball fly past overhead, it would seem as though the ball were curving toward the southwest, even though you pitched it due south. The viewer might assume that some mysterious force was deflecting it to the right of its path. Actually, the viewer would be moving eastward underneath the ball. It would merely *seem* as though the ball were moving westward over the viewer. There would be no mysterious force involved.

There might as well be, however. When we fire a missile at a distant target, for instance, we must take the earth's rotation into account, because while the missile is in flight, the target can rotate over a considerable distance, thus making it seem as though the missile has been deflected to the right of its path. Similarly, when the wind blows from north to south (or from east to west, for that matter), the earth will turn underneath it. To those on the earth's surface, however, it seems as though the wind is being deflected to the right of its path (that is, to the right of the pressure gradient).

So important is this deflection that it even has a name—the *Coriolis effect*.

The Coriolis Effect

How does the Coriolis effect influence the wind around pressure cells? We said earlier that wind diverges (moves outward) around a high pressure cell. We now know that it is also deflected to the right by the Coriolis effect. As a result of all this, the wind is made to move around the high pressure cell in a clockwise direction. We call this clockwise motion of air an *anticyclone.* However, wind converging (moving inward around a low pressure cell) and being deflected (to the right) also by the Coriolis effect, is made to move around the cell in a counterclockwise direction. We call this counterclockwise motion of air a *cyclone.*

(The term *cyclone* is sometimes used to mean a hurricane or tornado. Although these *are* kinds of cyclones, the term actually refers to any counterclockwise motion of air.)

Before we temporarily leave the subject of cyclones and anticyclones, let's consider what effect they have on the weather. When air spirals upward in a low pressure cell, it cools, and condensation begins. As in most rising air, clouds form and rain often follows. On the other hand, when air falls in a high pressure cell, it is warmed by compression and cannot become saturated. Therefore, condensation does not occur, and there are no clouds or rain.

As a rule, low pressure systems mean clouds and precipitation, whereas high pressure systems mean clear skies and mild weather. And that, to answer the question we began this chapter with, is how you can tell weather with a barometer. If the air pressure is falling, a low pressure system is on the way, and a storm may be coming with it. If the air pressure is rising, a high pressure system is on the way, and you should expect blue skies and calm weather.

CHAPTER
7

Why should the air pressure be high in some places and low in others?

Once again, the answer is heat. Heat affects the molecules of the atmosphere and therefore affects air pressure. Because some areas are warmer than others, this effect varies from place to place. And so does air pressure.

To see how this effect works, let's review the way in which the air is heated by the sun.

Sunlight has no direct warming effect on the air in the troposphere. Rather, the air is warmed when the ground absorbs sunlight and re-radiates it as heat. The amount of heat in the air, therefore, depends on how much sunlight the ground absorbs.

Land—especially land that is dark in color—absorbs sunlight rapidly and easily. Water, on the other hand, absorbs sunlight very slowly, and therefore takes a long time to become warm. If you go to the beach often, you've probably noticed that the water stays cold through spring and early summer, only becoming warm in late July and August.

However, because water takes so long to become warm, it *stays* warm for long periods of time, even in the absence of sunlight. Land, on the other hand, starts losing its heat the moment the sun goes down.

What does this have to do with differences in air pressure? Imagine an ocean beach at sunrise. The land will have lost most of its heat during the night, so that now, at sunrise, it will be cold. The water, on the other hand, will have held on to much of its heat through the night. At sunrise, therefore, it is still warm. Once the sun has risen, however, the situation reverses. The land heats up more quickly than the water.

Now the beach is warmer than the ocean. The air above the land too has become warmer than the air above the water, and its molecules—which have begun moving more quickly than the molecules of air over the ocean—will spread out over a wider area. For this reason, the atmosphere over the land extends upward for a greater distance than the atmosphere over the sea. To put this another way, the column of air over the beach is taller than the column of air over the ocean.

You'll remember that air pressure results from the weight of the air above pressing down on the air below. In this case, the amount of air over the land is the same as the amount of air over the sea, even though it extends over a wider area. Therefore the pressure, at ground level, is the same. However, as you move upward, the pressure drops more quickly over the ocean than over the beach.

Why? The air over the water is denser than the air over the land—that is, the molecules of air are more tightly compressed. Therefore, in a given amount of space, there will be more molecules of air over the ocean than over the land.

Suppose that you climb into the gondola of a balloon and float straight upward above the point where the beach meets the ocean. Let's further suppose that you have excellent vision—so excellent that you can count molecules in the air. At a height of 100 feet (30 m), you look down, begin counting, and find that there are more molecules below you on the ocean side than on the beach side. This makes sense, because the air is denser on that side, and there are more air molecules packed into that 100 feet (30 m).

However, this also means that *above* you there are more air molecules on the *beach* side, because these molecules are spread along a taller column of air. More molecules above you means a greater weight of air—and greater air pressure. Therefore, the air pressure at 100 feet (30 m) is greater over land than over water. The wind, flowing along the pressure gradient, blows from the land out over the sea.

This flow of air, however, increases the amount of air over the water and decreases the total amount of air above the beach. At ground level, therefore, the weight of air becomes greater over the ocean. This increases the ground level air pressure, so that the pressure over the sea is greater than over land—a reverse of conditions aloft. We can say, therefore, that the pressure gradient runs in one direction on the ground and in the opposite direction aloft, which explains why the wind can blow in two different directions, according to its height above the ground.

Now, with the ground level pressure greater over the sea than over the land, the wind blows from the sea onto the beach. This kind of wind is called a *sea breeze.* Higher up, where the gradient turns around, the wind blows the other way. When the sun goes down, however, the sea breeze reverses itself. The land loses its heat more quickly and becomes cooler than the water. The pressure over the land rises, becoming higher than the pressure over the sea. The wind then begins to blow back out to the ocean. This reversible wind is familiar to anyone who lives near the seashore.

In countries near the equator, where the sunshine is intense, the land becomes so hot during a summer day that it will not cool down sufficiently at night to reverse the pressure difference. The wind will blow from water to land constantly, day and night. During the winter, however, the sea keeps its summer heat and remains warmer than the land for months at a time. Then the wind blows constantly from land to water. This kind of seasonal sea breeze, blowing for months in one direction then for months in the other, is called a *monsoon,* from an Arabic word meaning "season."

In certain parts of the world, the monsoons are of tremendous importance. In southern Asia, for instance, the monsoon blows out to sea for much of the year. There is little moisture in this air, and it brings little precipitation. Then in the summer, the air, heavy with water vapor, comes in from the sea. When this air meets the Asian mountains and is lifted mechanically, precipitation begins. Rain cascades down onto mountainside villages. In Cherrapunji (Assam, India) alone, there is an annual rainfall of 36 feet (11 m), most of it during a few months in summer.

The agriculture of this region depends on this rain. If the monsoon comes late, crops may fail, and thousands of people will starve. The monsoon is very important to the people of southern Asia.

We've seen how local differences in air pressure can affect the weather process. Now let's take a look at how large-scale differences in air pressure can affect the weather of the entire world.

Look at almost any map of the earth. You will see lines running across the earth's face from north to south and from east to west. These lines do not actually exist on the earth, of course. On the map, they help you determine where you are. The *lines of latitude,* which are the ones we are concerned with here, run from east to west—or from right to left, as you look at the map. If you are looking at a globe rather than a map, the lines of latitude will wrap around the earth in circles that grow increasingly smaller as they approach the poles. Each line has a number, given in degrees. For instance, the line that wraps around the equator is numbered 0° (zero degrees) latitude. The line that passes through Denver, Colorado, and Ankara, Turkey, is marked 40° (forty degrees) latitude, and so forth.

Because the equator is at 0°, the number of degrees increases as you go north from the equator *and* as you go south. To tell these numbers apart, we refer to the lines north of the equator as "degrees north latitude" and the lines south of the equator as "degrees south latitude." By this system, Denver is at 40° north latitude. So is Ankara. But Melbourne, Australia, is at 40° *south* latitude.

Because of the earth's position in space, more sunlight falls on the equator than anywhere else on the planet. The earth is slightly tilted, however, so that the area of maximum sunlight wanders periodically to the north and south. For instance, on the first day of summer in the northern hemisphere, the sun is directly above 23½° north latitude. On the first day of winter in

(57

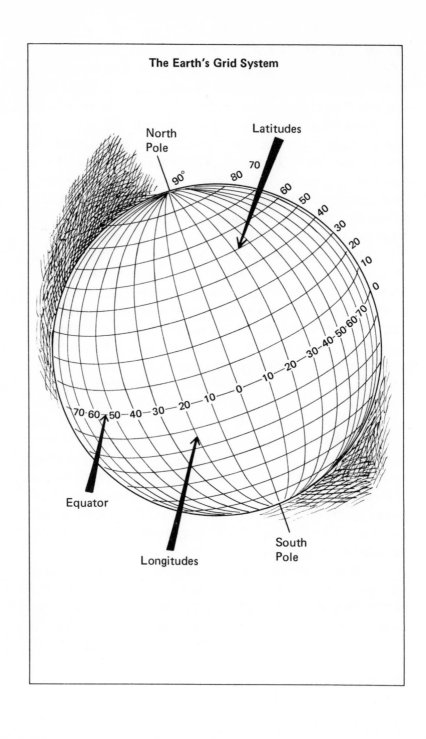

The Earth's Grid System

the northern hemisphere (which is also the first day of summer in the southern hemisphere), the sun shines directly above 23½° *south* latitude. On the average, though, throughout the year the largest amount of sunlight falls on the equator.

As a result, there is a zone of low pressure at the equator that stretches around the world. Air converges on this low pressure cell from the north and south, forcing the air at the equator upward. Aloft, the pressure gradient reverses. The air diverges, half heading north and half heading south.

Where does this air go?

The regions on earth that receive the least heat energy from the sun are the north and south poles. In fact, because of the way the earth is tilted in space, both poles go without sun for months at a time. Furthermore, both polar regions have a permanent snow cover called *permafrost,* which reflects back into space most of what little sunlight is received. Thus, the poles are cells of very high pressure.

It follows that the air at the poles must be sinking. Where does the air come from to replace this sinking air? We saw in the last chapter that the sinking air in a high pressure cell is replaced by the rising air in a low pressure cell. We know that there is a very large high pressure cell at the equator. It would make sense, then, to assume that the air diverging aloft over the equator simply travels north and south until it reaches the poles, where it sinks and returns to the equator again.

This is the model of air circulation developed in 1735 by the scientist George Hadley. Unfortunately, it doesn't agree with our observations of the way air actually moves. Hadley failed to take sufficient account of the Coriolis effect. The only part of the world where the air circulates as Hadley believed it did is between the equator and 30° north and south latitude. In his honor, this region is called the *Hadley cell.*

To see where Hadley went wrong, let's examine the circulation of air in the northern hemisphere. (The circulation in the southern hemisphere is a mirror image of that to the north.) As the air heads north from the equator, the earth rotates underneath it. The Coriolis effect seems to deflect this air toward the east. By the time the air reaches 30° north latitude, it is moving due eastward. In fact, it practically doubles back on itself. The air begins to pile up, the way cars on a turnpike will pile up if one stops too suddenly. By this time, the air is much cooler than it was when it was at the equator. Therefore, it moves in the only direction it can —down. It sinks back to the ground, creating a region of high pressure around 30° north latitude.

Once back at ground level, the air diverges both north and south. The portion of the air that returns to the equator is deflected again by the Coriolis effect, this time to the west. When it reaches the equator it rises again, having come full circle, and begins diverging northward again.

The portion of the air that does not return to the equator heads north instead. While this air is headed north, cold air from the high pressure region at the north pole is diverging to the south. What happens when these two currents meet?

They collide at 60° north latitude—the region known as the *polar front.* Because the two currents are converging, they are forced upward. Aloft, the air diverges north and south once again.

By now this picture of the general circulation of the atmosphere may have you thoroughly confused, so let's back up and see what we've learned.

We can divide the earth's air circulation into six cells, three in the northern hemisphere and three in the southern hemisphere. Let's take a look at the three in the northern hemisphere.

The first is the *equatorial cell* (or *Hadley cell*), which stretches from the equator to 30° north latitude. The second, from 30° north latitude to 60° north latitude, is the *midlatitude cell*. The third, above 60° north, is the *polar cell.*

In the equatorial cell, the air on the ground flows southward toward the equator, while the air aloft flows northward. The rising air at the equator creates a zone of low pressure, with frequent rain and gentle, monotonous winds. For centuries, sailors have referred to this region as the *doldrums.* Even today, when life seems monotonous and unexciting, we say that we are "in the doldrums."

The sinking air at 30° north creates a low pressure system of calm weather and almost nonexistent winds. Sailing vessels are often becalmed in this region. Spanish sailors in the seventeenth century dubbed the area the *horse latitudes* because (legend has it) they were often forced to throw their horses overboard to conserve food and water.

The winds that blow southward in this cell, given a westward twist by the Coriolis effect, become the *trade winds.* In the sixteenth century these constant and reliable currents of air were used by explorers and traders traveling to the New World from Europe. It is on these winds that Columbus made his first voyage to the Americas in 1492.

Between 30° north and 60° north, in the midlatitude cell, the flow of air is opposite to that in the Hadley cell. The wind on the ground blows northward (with an eastward twist from the Coriolis effect). These

(61

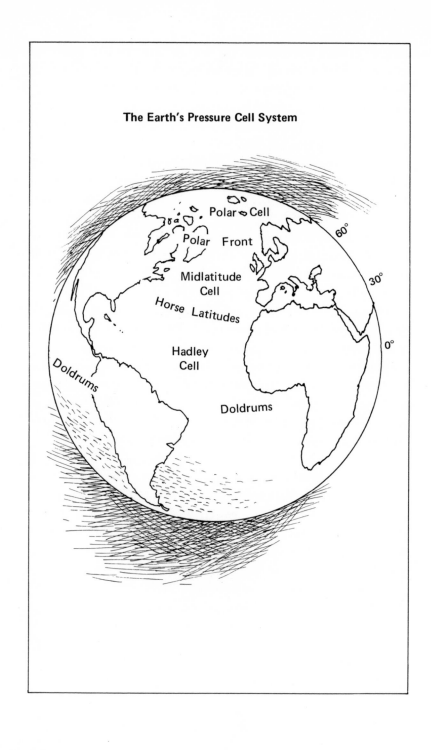

The Earth's Pressure Cell System

winds, called the *prevailing westerlies* (terms such as *westerly* and *easterly* refer to the direction from which the wind is coming, not the direction in which it is blowing), were used by sailors returning from the New World to Europe and are not as constant and reliable as the trade winds to the south.

At 60°, where the polar cell begins, the rising air creates a region of low pressure and frequent storms. At the pole itself, however, the weather is very calm and precipitation rare. (This may seem odd, since the ground at the pole is always covered with snow. Though precipitation *is* rare at the pole, what little snow does fall tends to remain, because of the extremely low temperatures.)

Names have also been given to the high and low pressure systems between the cells. At 0°, for instance, is the *equatorial low* (also called the *intertropical convergence zone*). Between the equatorial cell and the midlatitude cell is the *subtropical high.* (Here, around 30° north latitude, are found most of the world's great deserts.) At the polar front (60° north) is the *subpolar low,* and at the pole, of course, is the *polar high.*

You'll recall that the air pressure drops more quickly with altitude in a high pressure cell than in a low pressure cell, so that the pressure gradient reverses aloft. Not only does it reverse, but the difference in pressure increases as you go upward, so that the gradient keeps growing steeper, right up to the tropopause. And as the gradient grows steeper, the winds blow faster.

This is true of the global pressure systems we just described. As a result, there are winds in the upper troposphere that reach speeds of greater than 250 mph (400 kmph)! These winds, found between 35,000 and 40,000 feet (10,500 and 12,000 m) above the ground,

The General Circulation of the Atmosphere

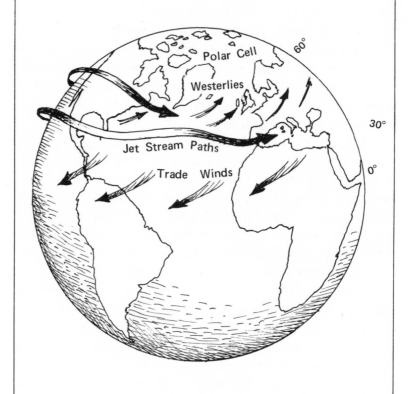

are called the *jet streams.* They were discovered by high-flying airmen during World War II, when pilots found themselves being blown violently off course or pushed forward at twice the speed at which they were supposed to be flying. At least one hapless pilot found himself flying directly into an oncoming jet stream— and standing still in midair!

So far, we have treated these belts of high and low pressure as though they stretch evenly around the entire planet, like the colored rings around a beach ball. Unfortunately, this isn't true. If the surface of the earth were entirely water, or entirely land, this might be an accurate picture. But we saw earlier in this chapter that water and land respond differently to the sun's heat. Therefore, pressure tends to be higher over water than over land.

When we speak of the subtropical high, for instance, we are actually referring to a series of high pressure cells centered above the oceans at roughly 30° latitude. To see how these cells affect the weather, let's take a closer look at one in the northern hemisphere, the one over the Atlantic Ocean.

Because it is a high pressure region, the air within the cell is sinking. The diverging air on the ground is given a twist to the right by the Coriolis effect, creating a clockwise anticyclone (see Chapter 6). To the left of this anticyclone is the southeastern United States. To the right is northern Africa. If you picture this clockwise motion in your mind, you'll see that the air on the left side of the anticyclone is moving from south to north. These southerly (from the south) winds carry warm, moist air from the tropics northward into the southwestern United States, creating humid conditions and frequent rainfall in states such as Florida,

Georgia, and Louisiana. On the left side of the anticyclone, however, the air moves from north to south, bringing dry air southward into Africa. Precipitation in this air is infrequent, and so northern Africa is mostly desert. A similar high pressure system over the Pacific helps to create the deserts of the southwestern United States.

What we have just described can be termed the *general circulation of the atmosphere,* because it represents the ways in which the wind *tends* to travel around the world. Local conditions also have an effect on wind tendencies. Sea breezes and monsoons are two examples of *local winds.* Let's look at some others.

There is a kind of wind found in mountain regions called a *foehn.* In Chapter 5 we saw how air can be lifted over a mountain and forced to condense. This rising air is cooled at the wet adiabatic rate while it is condensing. On its way down the other side of the mountain, however, it warms at the faster, dry adiabatic rate. When the air reaches the ground again, it is warmer than when it started up the mountain originally. It is also very dry.

To someone living on the lee side of the mountain (the side away from the wind), it would seem as though a hot, dry wind had descended mysteriously from the sky. These winds can raise the temperature by as much as 30° F in a few hours. If there is snow on the ground, it will be melted before the day is out. Certain Indian tribes in the American West have dubbed these winds "snow-eaters."

The term *foehn* originated in the Alps, where these hot winds are common. In the United States, they are sometimes called *chinooks.* Perhaps the most famous chinook is the *Santa Ana.* Named after the Santa

Ana Canyon in southern California, these winds help create the conditions of extreme dryness that lead to forest fires in this region.

Another form of local wind is the *katabatic wind,* or *fall wind.* Cold air on high, snow-covered plateaus will sometimes fall over the edge and descend to the ground below. Though the air is warmed adiabatically as it descends, it remains far colder than the surrounding air. The effect is the reverse of the foehn—a cold wind blowing down a mountainside. Two famous katabatic winds are the *bora,* in the Austrian Alps, and the *mistral,* along the French coast of the Mediterranean.

Winds can be created in valleys too, where the mountains are heated all day by the sun but the air above the valley remains cool. The difference in temperature creates a pressure difference—low over the mountain and high over the valley. The wind blows along the pressure gradient, from the valley up the side of the mountain. This is called a *valley wind.* At night the mountain cools, and the process reverses itself. The wind blows back down the mountain into the valley. This is called a *mountain wind.*

CHAPTER
8

We've talked a lot about differences in air. Differences in pressure make the wind blow. Differences in temperature create differences in pressure. Differences in humidity cause differences in precipitation. Obviously these differences are very important to the weather process.

Are there times, though, when there simply *aren't* any differences? Times when you can travel for thousands of miles without noticing any significant change in air pressure or temperature or humidity?

Of course. Whenever there is a large body of air all pretty much the same throughout in terms of temperature, pressure, and humidity, that body is called an *air mass.*

(When we say that the temperature and pressure are the same throughout the air mass, of course, we mean in a horizontal—sideways—sense. As you move upward—vertically—through an air mass, the temperature and pressure *always* change.)

What are the conditions under which such an air

mass will form? We saw earlier that air temperature (and therefore pressure) varies according to how easily the ground absorbs sunlight. The air near the ground also varies in moisture content, the moisture coming from lakes, streams, and oceans. So in order for air temperature and humidity to become constant over a large area, ground conditions must be constant over a large area. Otherwise, the ground will be warmer and wetter in some areas than in others, and the temperature and humidity of the air will vary accordingly. Where can we find conditions that are this uniform over thousands of miles?

Not in the United States or Europe, certainly. Not only do ground conditions vary too much on these continents, but the low pressure cells at these latitudes stir the air too much, giving air masses no chance to form.

On the other hand, over the oceans conditions are perfect, especially in the subtropic highs, where the weather is usually calm and still. The ice fields of the arctic are also ideal for the formation of air masses.

Once formed, these air masses can move into other regions and affect weather conditions there. Naturally, an air mass will have characteristics typical of the area where it was created.

For instance, air from the polar regions and the arctic will be very cold as well as very dry. Air from the tropical oceans will be wet and warm.

You can often tell where an air mass was formed by measuring its temperature and humidity. This gives meteorologists a method of classifying air masses.

A cold air mass will be referred to as either *polar* or *arctic*, depending on how cold it is. (Polar air is colder than arctic air.) Warm air will be either *tropical* or *equatorial*. If an air mass is wet it is referred to as *maritime* (meaning that it formed over the sea), and if it is dry it is referred to as *continental* (meaning that it formed over land).

As these air masses wander from their *source regions* (the areas where they were formed), they have varying effects on the weather. Continental polar air is very stable, because the air next to the ground is colder than the air aloft. Maritime tropical air masses, on the other hand, are usually unstable and are frequently a source of rain and other precipitation.

In the northern hemisphere, cold air masses tend to form to the north, in the polar regions. Warm air masses tend to form to the south, in the tropical regions. If you live in the middle latitudes—that is, in the United States or Europe—your weather is as likely to be dominated by cold air from the north as by warm air from the south.

What would happen if two air masses, moving in opposite directions, collided with one another? Would they simply mix together, blending into a single mass with characteristics somewhere between those of the two colliding masses?

Not really. Air masses don't like to mix. Even when they collide, they tend to remain separate, as though there were an invisible wall between them.

The point at which two air masses meet is called a *front.* These fronts are extremely important in determining weather conditions.

Imagine, for instance, that we have a cold air mass. A warm air mass, moving northward, bumps into this cold mass and starts pushing it along. The two masses both move to the north. Because the warm mass is pushing the cold mass ahead of it, we refer to this as a *warm front.*

By the opposite token, if the cold air mass is pushing the warm air mass, we refer to the system as a *cold front.*

Let's take a closer look at the warm front. We said before that when the two masses collide, they refuse to mix. Does this mean that the warm air mass simply flattens out against the cold mass, forming a perfectly straight "wall"?

No. Cold air is denser and heavier than warm air. It usually remains closer to the ground. When the warm air, which rises easily, strikes the cold air mass, it actually begins to ride up over the top of the cold air! The heavier cold air, in turn, slides underneath the warm air. There is still an invisible wall between the two masses, but it is tilted at an angle to the ground. In fact, the wall between the two masses is tilted at such an angle that it is practically parallel to the ground. In this way, the warm air can flow over the top of the cold air for thousands of kilometers, slowly climbing to great heights.

And what happens to the warm air when it does this? Precisely the same thing that happens to air that is forced to rise over a mountain. If there is enough moisture in the air, it begins to condense. Clouds form. Precipitation takes place.

We can say, then, that this *frontal lifting* (the lifting of air over a front) is a form of mechanical lifting. The amount of condensation and precipitation that occurs as the warm air is lifted over the front depends on the air's stability. If the air is stable, flat clouds and drizzly rains will result. If the air is unstable or conditionally stable, large clouds and torrential rains will develop.

You can tell when a warm front is headed in your direction by watching the sky. When the front is a thousand kilometers or so distant, wispy white clouds

(71

will appear high overhead. These clouds, looking almost like wispy feathers floating in the air, are called *cirrus clouds.* They are in the warm part of the front, which at this point is still far above the ground. As the front grows closer and the warm air gets nearer to the ground, the cirrus clouds will flatten out and become *stratus clouds*—thin, shallow sheets of cloud that seem to stretch across the entire sky. The moon, when seen through a particularly thin cirrus cloud, will appear to have a ring around it. Such rings have long been regarded as omens of upcoming storms.

About 300 miles (480 km) ahead of the front, rain or snow will begin to fall. If the air in the warm part of the front is stable, this precipitation will be mild. If not, *cumulus clouds* will form. These are bulky, clumpy clouds that may produce a thunderstorm or blizzard.

(Meteorologists like to put the suffix *-nimbus* after the name of the cloud producing precipitation. For instance, a stratus cloud that is producing rain or snow is called a *stratonimbus* cloud. A precipitating cumulus becomes a *cumulonimbus* cloud.)

In this January 1980 satellite photograph, clouds can clearly be seen covering almost the entire continental United States. Low clouds cloak the east from the mid-Atlantic region to the southeast, multilayered clouds extend into New Mexico and Texas, and middle-level clouds stretch into the northwest. Snow is also visible from the plains to the Ohio Valley.

A process similar to the one just described for a warm front takes place with a cold front also, but everything happens more rapidly and with consequently greater violence.

In a warm front, the warm air rises slowly, gently, over the cold air. In a cold front, the warm air also rises above the cold, but the angle of the front is much steeper, and the air rises more quickly. The precipitation, therefore, tends to be more extreme. When you see a group of cumulonimbus clouds (called *thunderheads)* clustering in the distance, it may mean that a cold front is on the way. But usually such fronts give little advance warning of their approach.

Some of the most spectacular weather occurs when two air masses pass side by side rather than colliding. Imagine, for instance, that we have a continental arctic air mass moving down from the north and a maritime tropical air mass moving up from the south. As you face northward, the arctic air mass will be on the left, the tropical air mass on the right. The point where they touch is called a *stationary front,* because it is not moving in the direction of either air mass.

So far, we have spoken of fronts as though they are points where one air mass ends and another suddenly begins. Actually, there *is* some overlap between two air masses when they meet at a front. We call this overlap the *transition zone.* The greater the difference in temperature between the air masses, the smaller this zone will be. However, we can expect it to be at least a few hundred feet wide.

As our two air masses move past each other, one heading north and the other south, there will be a certain amount of friction between them. The air molecules in the transition zone are moving in two different directions. It follows that these molecules will frequently collide. This, in turn, puts a drag on the motion of the two air masses.

To understand what happens next, let's perform a simple experiment. Place a stick or a pencil between the open palms of both your hands. Move your right hand forward and your left hand backward. The pencil will begin to rotate in a counterclockwise direction.

The same thing happens to the air in the transition zone. It receives a counterclockwise twist from the two passing air masses. A mini-cyclone of sorts develops. Some of the air from the tropical air mass is yanked by this cyclone into the colder arctic mass.

Think of how this would look on an isobar map. A segment of the warm air from the tropical mass has invaded the arctic mass. It is now surrounded on three sides by cold air. The air pressure is naturally lower in the warm air than in the cold. We have, therefore, an area of low pressure surrounded by an area of higher pressure. On an isobar chart, you would see a series of nearly circular isobars wrapped around the area of low pressure. The same thing happens here as happens in the low-pressure system we described in Chapter 6. The wind, flowing along the pressure gradient, blows inward toward the low pressure cell. The Coriolis effect gives it a counterclockwise twist (reinforcing the counterclockwise motion created by the friction of the two air masses), and a full-scale cyclone develops. In fact, meteorologists refer to such a situation as a *wave cyclone.*

Inside the wave cyclone we have two simultaneous fronts. The warm air from the tropical mass rides up over the top of the cold air in the arctic mass, creating a warm front. The arctic air, however, has been given a strong counterclockwise twist by the cyclone. As the warm front forms, the arctic air wraps com-

(75

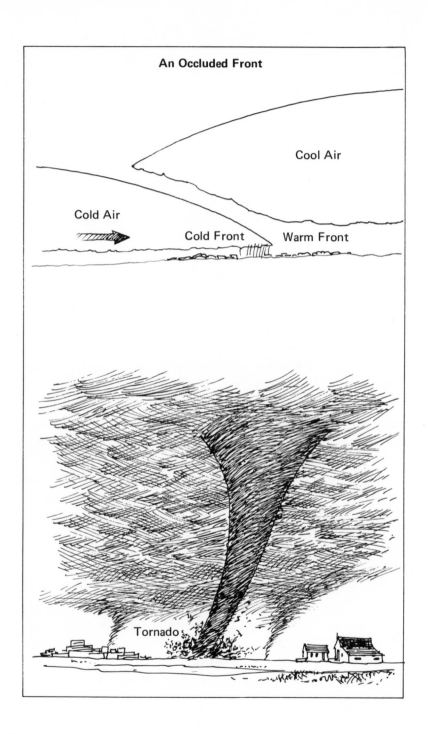

An Occluded Front

Cool Air

Cold Air

Cold Front   Warm Front

Tornado

pletely around the segment of warm air and comes racing up *behind* it. Sliding under the warm air, the arctic air creates a cold front as well. This cold front continues moving in a counterclockwise direction until it has gone full circle and catches up once again with the warm front. At this point, the warm air is lifted right off the ground, squeezed from both front and rear by cold air. The warm air is now completely separate from the tropical mass of which it once was a part. When two fronts collide in this manner, we call it an *occluded front.*

You can well imagine, from everything we've talked about so far, that violent weather will accompany this wave cyclone, especially if the warm air is unstable. For one thing, there is a sudden lifting of air. This means condensation (clouds) and precipitation (rain, snow, sleet, or hail). The more sudden the lifting, the more violent the precipitation. In the cyclone itself there are very strong winds. The greater the difference in temperature (and therefore pressure) between the two air masses, the stronger these winds will be. Occasionally in the springtime (and especially in the midwestern United States), two air masses of such differing temperatures will come together that a sort of supercyclone occurs. Winds start rotating at extremely fast speeds, and the air pressure at the center of the cyclone plunges as low as 100 millibars below the pressure outside. The resulting storm is called a *tornado.*

The winds in a tornado can move at speeds of over 650 kilometers per hour! The tornado itself can travel at speeds of perhaps 45 kilometers per hour. Though such storms are no more than 600 meters wide (usually much less), they can wreak staggering amounts of damage in property and lives. In a storm that devastated an area from Canada to Georgia in 1974, three hundred people were killed, making it the deadliest storm in half a century. In 1979, a series of tornadoes roared along the Texas-Oklahoma border, taking a hundred lives and causing more than half a billion dollars worth of damage.

Much of the devastation from a tornado comes not from the speed of its winds but from the tremendous pressure differences. When a tornado passes near a building, for instance, the pressure outside will drop so suddenly that the building will literally explode, because the air pressure inside the building is no longer counterbalanced by the pressure outside. In the event of a tornado, it is a good idea to open doors and windows in your home, so that the pressure will equalize in a less violent way.

Roughly seven hundred tornadoes strike the United States every year. Unfortunately, these storms are difficult to predict; meteorologists can only guess when conditions are ripe for one to materialize. When a tornado seems imminent, a *tornado watch* is declared, meaning that a tornado could hit the area within one to seven hours. If a tornado should actually be sighted, a *tornado warning* is issued. In the midwestern United States, a network of radar antennae, capable of detecting cumulus clouds 300 kilometers away, keeps constant watch for these storms.

CHAPTER
9

Probably most of you reading this book have been through a thunderstorm. Among common midlatitude storms, these are probably the most dramatic. The sudden darkening of the sky as the cumulonimbus cloud rolls overhead, the sharp wind, the crackle of lightning—all conspire to make the thunderstorm seem sinister and rather frightening. Even so, we sometimes forget about the damage that these storms can leave in their wake. Lightning can start fires and take lives; hail (often generated by thunderstorms) can destroy crops and kill cattle; squall lines can produce destructive winds.

Although thunderstorms can be produced as part of a front or a wave cyclone, all that is really needed is an abundant supply of warm, moist air. Because this is usually available only in spring and summer, it is during these times of the year that the storms normally occur. Heavy sunshine causes moisture-laden air to rise, and a cumulus cloud begins to form.

A full-fledged thunderhead (cumulonimbus) is very

large, stretching to heights of well over 10,000 meters. Because these clouds tend to re-evaporate into the summer air given half a chance, the thunderhead must form quickly, and it must be supplied with moisture continuously. Therefore, air conditions must be highly unstable. Strong updrafts are also necessary to keep the cloud supplied with moisture and to support droplets within the cloud as the Bergeron process operates to create rain. If the updrafts are strong enough, hailstones will form (see Chapter 4).

Once the cloud has formed and the raindrops begin to fall, a strong downdraft is created—literally pulled out of the cloud by the falling rain. This downdraft exists side by side with the updrafts that form the cloud. If you observe from the ground the beginning of a thunderstorm, you will notice that these downdrafts arrive before the rain, stirring the grass and cooling the air.

Without a continuous supply of moist air, the thunderstorm will soon die. The area over which the thunderhead formed is quickly depleted of its moisture. Thus once the storm has begun, it has to keep moving in order to find new areas to supply it with "fuel."

Eventually, however, the downdrafts inside the cloud become more powerful than the updrafts, and the cloud dissipates. The storm ends.

We can divide thunderstorms roughly into three types: those that are caused by unstable tropical air, those that are caused by mechanical lifting, and those that form in squall lines.

Storms of the first kind usually occur in tropical air from the Gulf of Mexico. Because maximum solar heating is also necessary, these storms usually take place in the afternoon. And, because the instability of the air varies widely from place to place, these storms are also usually scattered—that is, they form randomly over a wide area, with rarely more than one storm passing over a single spot.

Thunderstorms are common at the equator because of the convergence of air in the Hadley cell. They are common also on peninsulas—such as the state of Florida—because of the convergence of air from the ocean.

And, of course, thunderstorms are often formed in low-pressure cyclones, where convergence and the rising of air are constant.

The second type of thunderstorm occurs when conditionally stable air is forced to rise mechanically and becomes unstable through condensation. Cold fronts are particularly fertile sources of this kind of storm, because air is lifted more sharply and rapidly in a cold front than in a warm front, thereby creating the proper conditions for rapid formation of a cumulonimbus cloud.

Frontal lifting of warm air will, in turn, bring about the third type of thunderstorm, that generated by a squall line. Squall lines occur at cold fronts. You read a moment ago that a thunderstorm generates strong downdrafts. The air in these downdrafts is quite cold. Sometimes they create a front of their own. The cold air from the downdraft races ahead of the thunderstorm and lifts the warm air up in front of it. This lifting causes condensation and creates still more thunderstorms. These thunderstorms, in turn, create more downdrafts, which cause more lifting, and so on. In this fashion, a whole line of thunderstorms—called a *squall line*—can spring up in front of a cold front.

The thunderstorm gets its name, of course, from the characteristic thunderclaps that follow the bolts of lightning the storm produces. These bolts are a form of energy created by differences in what scientists call *electric potential.* If one part of the cloud develops a negative potential and another part develops a positive potential, streams of electrons will race from one area to the other to balance out the difference, just as wind blows from areas of high pressure to areas of low pressure. No one is quite sure, however, how these differences in potential come about.

Sometimes these streams of electrons jump out of the cloud and race downward to the ground. As they pass through the atmosphere, the air becomes hot and begins to glow. It is this glow, not the stream of electrons itself, that we see as the lightning bolt. The hot air expands very quickly. In fact, it expands with explosive pressure, and we hear that explosion as thunder.

In the Philippines they are called *baguios;* off Australia, they are *willy-willies;* near Japan they are *typhoons.* In the West Indies and in North America, they are *hurricanes.*

Meteorologists call these storms *tropical cyclones.* Tropical cyclones are the most violent storm systems, in terms of potential for destruction, generated by the weather process. They are born in the warm waters of the Hadley cell, far enough south (when they occur in the northern hemisphere) to thrive on the intense energy of the equatorial sun, but far enough from the equator to be driven by the Coriolis effect. (The Coriolis effect weakens toward the equator.) In the summertime, the heat becomes so great in the tropical oceans that it produces moist updrafts and towering cumulonimbus clouds that reach nearly to the tropopause. If a jet stream should be passing over this low-pressure cyclone, its racing winds will pump

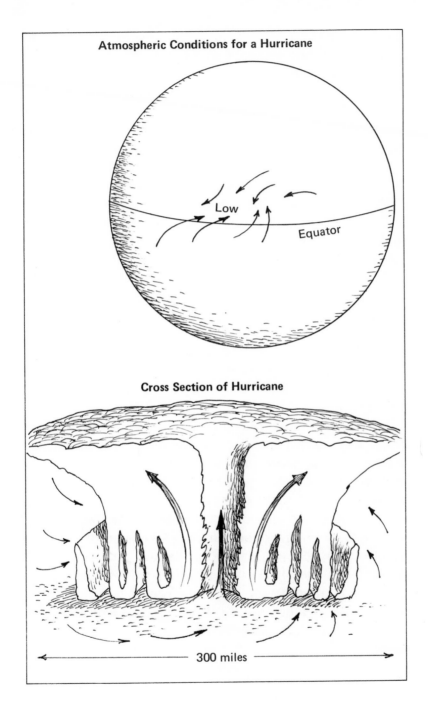

Atmospheric Conditions for a Hurricane

Low

Equator

Cross Section of Hurricane

300 miles

the air out of the cumulonimbus. More air will rise (very rapidly) to replace it, and the pressure will plunge even lower. This, in turn, causes the cyclonic (counterclockwise) winds of the low-pressure system to spin even faster. As the moist air rises in the cyclone, condensation takes place. This releases latent heat, which causes the cyclone to spin faster still.

The storm officially becomes a hurricane when the winds begin rotating at speeds greater than 72 miles (115 km) per hour. This, however, is only the lower limit for hurricane winds. They have been known to go as fast as 200 miles (320 km) per hour. These winds can generate waves 50 feet (15 m) high. Actually, the wind speed of a hurricane may not sound terribly impressive when compared to that of a tornado, but consider that the average hurricane has a diameter of nearly 400 miles (640 km)! Only in the center of these swirling winds (an area referred to as the *eye* of the storm) is the wind calm. The pressure in this area, however, can be as much as 60 millibars below that outside the storm.

The real damage from a hurricane takes place when the storm hits a populated area. Waves will be driven against the shore with such incredible speed that they will actually pile up, and a giant wave called a *storm surge* may develop. This wave can race through an entire city, destroying buildings and drowning anyone unlucky enough not to have fled at the hurricane's approach. In 1900, a storm surge passed through Galveston, Texas, killing five thousand people. One man who survived the surge said that the water level in his home rose from ankle level to chest level—in less time than it took him to turn around!

Once the hurricane moves inland, it dies. The power of a hurricane comes from warm water, and once the storm loses its source of water it begins to fade, until it is little more than an ordinary low-pressure cyclone.

Casualties from hurricanes have been reduced in recent years through improved methods of prediction. The storms can be photographed by space satellites orbiting the earth hundreds, even thousands, of miles out in space. Meteorologists can watch the distinctive circular motion of the clouds in these tropical cyclones and make educated guesses as to the direction in which the storm will move, so that endangered areas can be evacuated hours or days in advance. Still, the movements of these storms are difficult to predict, and they must be carefully monitored. In some areas of the world, a last-minute evacuation is not feasible. In 1970, for instance, 200,000 lives were lost when a hurricane struck Bangladesh.

Ironically, though, if it were not for hurricanes, still more lives might be lost as a result of droughts in certain parts of the world where hurricanes supply much of the annual precipitation. It's even conceivable that, looked at in this way, hurricanes save more lives than they take.

EPILOGUE

In the first chapter we saw that the process called weather is nothing more than the movement of heat from the equator to the poles. Nothing more.

Obviously, this process isn't terribly successful. The poles are still cold. The equator is still warm. Actually, before the heat can reach the poles, most of it is re-radiated back into space. But it *tries* to get there.

Along the way the radiant heat is absorbed by the ground and given off as sensible heat, deflected by the Coriolis effect, converted into latent heat by evaporation, converted back into sensible heat by condensation, removed from the air by expansion, and restored to the air by compression. This heat triggers hurricanes in the tropics, tornadoes in the midlatitudes, and blizzards at the poles. It causes air to rise and fall and spin clockwise and counterclockwise. It helps sailors cross the sea, turns windmills and weathervanes, and fills riverbeds with flowing water.

From a few hundred thousand miles away, out where the earth looks like a tiny ball, this weather process seems like a trivial thing. Trivial. Insignificant.

But to the tinier creatures who live on that ball, the ones who call themselves human beings, that process is one of the most important forces in the universe.

After all, it controls so much of their lives.

(86

SEPTEMBER 3, 1974

A 1974 NASA satellite was used to track Hurricane Carmen (arrow) as it moved northward from Mexico's Yucatan Peninsula toward the U.S. Gulf Coast.

RECOMMENDED READING

Chandler, T. J. *The Air Around Us: Man Looks at his Atmosphere.* New York: Doubleday & Co., Inc., 1969.

*Kirk, Ruth. *Snow.* New York: William Morrow & Co., Inc., 1978.

*Lutgens, Frederick K. and Tarbuck, Edward J. *The Atmosphere: An Introduction to Meteorology.* New Jersey: Prentice-Hall, Inc., 1979.

May, Julian. *Rainbows, Frost, and Foggy Dew.* Minn.: Creative Educational Inc., 1972.

Milgrom, Harry. *Understanding Weather.* New York: Macmillan, Inc., 1970.

Spar, Jerome. *The Way of the Weather.* Minn.: Creative Educational Inc., 1967.

*Thompson, Philip D., and O'Brien, Robert. *Weather.* Virginia: Time-Life Books, 1968.

*Young, Louise B. *Earth's Aura.* New York: Alfred A. Knopf, Inc., 1977 (paperback by Avon, 1979).

* Indicates written for adults. These books have been included because they are highly readable or popular treatments of the subject.

INDEX

PRAISE FOR *CAT'S EYE*
MARGARET ATWOOD'S MASTERPIECE

"Atwood [is] writing at the top of her energy. . . . *Cat's Eye* is so fine that simply to observe how it works is the best praise."
—*The Philadelphia Inquirer*

"Brilliant . . . Atwood is a gifted and complex writer."—*Ms.*

"A daring piece of work . . . With each small scene offering traditional reminiscence, Atwood ups the aesthetic ante, placing the mundane in a speculative framework where nothing can be taken for what it seems to be, in art or life, in past or present, in time or out of it."—*San Francisco Chronicle Book Review*

"*Cat's Eye* is a story about the power of secrets. . . . Atwood has accomplished something rare and valuable. . . . Her depictions of little-girl life (not just the details, but the mood and the feelings) are among the most powerful I can remember."
—*Mademoiselle*

"A literary event!"—*Toronto Star*

"The power of *Cat's Eye* . . . is in the stunning portrait of pain inflicted, and of the disability that freezes the protagonist's life ever after. . . . [The book] refuses to console or be consoled; its bleakness is contagious. Its joy—also contagious—is in the flash of language. . . . Atwood conceives Elaine with a poet's transforming fire; and delivers her to us that way, a flame inside an icicle."—*Los Angeles Times*

"A work of intensity and tart wit . . . All the better Atwood trademarks are here—wry humor, unforgiving detailed observation—and likely to attract a wide audience."—*Kirkus Reviews*

MARGARET ATWOOD
CAT'S EYE

"A rich novel . . . The freshness of Atwood's images should earn sustained applause from readers. . . . With amazing deftness, she rubs together strong personalities and dozens of incidents and combines them to produce a sympathetic, sometimes acerbic and hard woman who is more than the sum of her experiences."—*Newsday*

"Beautifully sustained . . . Endlessly surprising."—*Quill and Quire*

"As fine a novel as you are likely to find this year—or maybe any year . . . Atwood's portrait of the unblinking cruelty of children is remarkable, and it shades naturally into an exploration of adult friendships and alliances."—*Cincinnati Post*

"Truly brilliant . . . Atwood's most empathetic work."—*Calgary Herald*

MARGARET ATWOOD
CAT'S EYE

"A beautifully written, complexly engineered novel."—*St. Louis Post-Dispatch*

"Atwood's portrayal of . . . friendship gives the novel its fraught and mysterious center."—*Publishers Weekly*

"Atwood writes about childhood as if she had just lived it yesterday. The rituals, the secrets, the best friends, the desperate need to belong—they are all here in heartbreaking vividness. . . . *Cat's Eye* is a book to linger over."—*Inside Books*

BANTAM BOOKS
NEW YORK · TORONTO · LONDON · SYDNEY · AUCKLAND

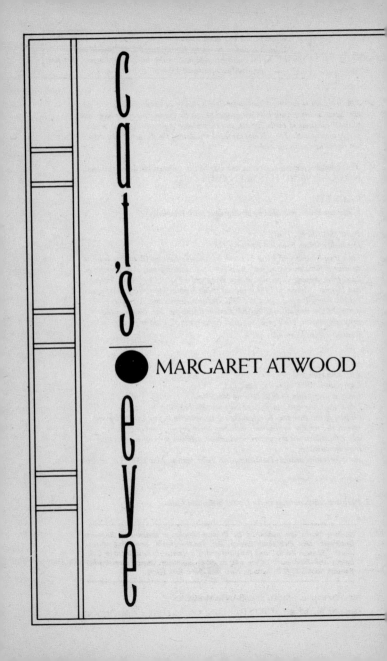

cat's eye

● MARGARET ATWOOD

This is a work of fiction. Although its form is that of an autobiography, it is not one. Space and time have been rearranged to suit the convenience of the book, and, with the exception of public figures, any resemblance to persons living or dead is purely coincidental. The opinions expressed are those of the characters and should not be confused with the author's.

This edition contains the complete text of the original hardcover edition.
NOT ONE WORD HAS BEEN OMITTED.

CAT'S EYE

A Bantam Book / published by arrangement with Doubleday

PUBLISHING HISTORY
Doubleday edition published February 1989

Lines from Memory of Fire: Genesis *by Eduardo Galeano (Pantheon Books, a division of Random House, Inc., New York: 1985) are reproduced by permission. Lines from the song, "Coming in on a Wing and a Prayer," by Jimmy McHugh and Harold Adamson © 1943 (renewed 1970) Robbins Music Corp. All rights of Robbins Music Corp. assigned to SBK Catalogue Partnership. All rights controlled and administered by SBK Robbins Catalogue, Inc. International copyright secured. Made in U.S.A. All rights reserved. Used by permission.*

Bantam edition / December 1989

This book is for S.

Contents

The paintings and other modern works of art in this book do not exist. Nevertheless, they have been influenced by visual artists Joyce Wieland, Jack Chambers, Charles Pachter, Erica Heron, Gail Geltner, Dennis Burton, Louis de Niverville, Heather Cooper, William Kurelek, Greg Curnoe, and pop-surreal potter Lenore M. Atwood, among others; and by the Isaacs Gallery, the old original.

The physics and cosmology sideswiped herein are indebted to Paul Davies, Carl Sagan, John Gribbin, and Stephen W. Hawking, for their entrancing books on these subjects, and to my nephew, David Atwood, for his enlightening remarks about strings.

Many thanks to Graeme Gibson, for undergoing this novel; to my agent, Phoebe Larmore; to my English agents, Vivienne Schuster and Vanessa Holt; to my editors and publishers, Nan Talese, Nancy Evans, Ellen Seligman, Adrienne Clarkson, Avie Bennett, Liz Calder, and Anna Porter; and to my indefatigable assistant, Melanie Dugan; as well as to Donya Peroff, Michael Bradley, Alison Parker, Gary Foster, Cathy Gill, Kathy Minialoff, Fanny Silberman, James Polk, Coleen Quinn, Rosie Abella, C. M. Sanders, Gene Goldberg, John Gallagher, and Dorothy Goulbourne.

When the Tukanas cut off her head, the old woman collected her own blood in her hands and blew it towards the sun.

"My soul enters you, too!" she shouted.

Since then anyone who kills receives in his body, without wanting or knowing it, the soul of his victim.

—EDUARDO GALEANO
Memory of Fire: Genesis

Why do we remember the past, and not the future?

—STEPHEN W. HAWKING
A Brief History of Time

IRON LUNG

1

Time is not a line but a dimension, like the dimensions of space. If you can bend space you can bend time also, and if you knew enough and could move faster than light you could travel backward in time and exist in two places at once.

It was my brother Stephen who told me that, when he wore his raveling maroon sweater to study in and spent a lot of time standing on his head so that the blood would run down into his brain and nourish it. I didn't understand what he meant, but maybe he didn't explain it very well. He was already moving away from the imprecision of words.

But I began then to think of time as having a shape, something you could see, like a series of liquid transparencies, one laid on top of another. You don't look back along time but down through it, like water. Sometimes this comes to the surface, sometimes that, sometimes nothing. Nothing goes away.

?

"Stephen says time is not a line," I say. Cordelia rolls her eyes, as I knew she would.

"So?" she says. This answer pleases both of us. It puts the nature of time in its place, and also Stephen, who calls us "the teenagers," as if he himself is not one.

Cordelia and I are riding on the streetcar, going downtown, as we do on winter Saturdays. The streetcar is muggy with twice-breathed air and the smell of wool. Cordelia sits with nonchalance, nudging me with her elbow now and then, staring blankly at the other people with her gray-green eyes, opaque and glinting as metal. She can outstare anyone, and I am almost as good. We're impervious, we scintillate, we are thirteen.

We wear long wool coats with tie belts, the collars turned up to look like those of movie stars, and rubber boots with the tops folded down and men's work socks inside. In our pockets are stuffed the kerchiefs our mothers make us wear but that we take off as soon as we're out of their sight. We scorn head coverings. Our mouths are tough, crayon-red, shiny as nails. We think we are friends.

On the streetcars there are always old ladies, or we think of them as old. They're of various kinds. Some are respectably dressed, in tailored Harris tweed coats and matching gloves and tidy no-nonsense hats with small brisk feathers jauntily at one side. Others are poorer and foreign-looking and have dark shawls wound over their heads and around their shoulders. Others are bulgy, dumpy, with clamped self-righteous mouths, their arms festooned with shopping bags; these we associate with sales, with bargain basements. Cordelia can tell cheap cloth at a glance. "Gabardine," she says. "Ticky-tack."

Then there are the ones who have not resigned themselves, who still try for an effect of glamour. There aren't many of these, but they stand out. They wear scarlet outfits or purple ones, and dangly earrings, and hats that look like stage props. Their slips show at the bottoms of their skirts, slips of unusual, suggestive colors. Anything other than white is suggestive. They have hair dyed straw-blond or baby-blue, or, even more startling against their papery skins, a lusterless old-fur-coat black. Their lipstick mouths are too big around their mouths, their rouge blotchy, their eyes drawn screw-jiggy around their real eyes. These are the ones most likely to talk to themselves. There's one who says "mutton, mutton," over and over again like a song, another who pokes at our legs with her umbrella and says "bare naked."

This is the kind we like best. They have a certain gaiety to them, a power of invention, they don't care what people think. They have escaped, though what it is they've escaped from isn't clear to us. We think that their bizarre costumes, their verbal tics, are chosen, and that when the time comes we also will be free to choose.

"That's what I'm going to be like," says Cordelia. "Only I'm going to have a yappy Pekinese, and chase kids off my lawn. I'm going to have a shepherd's crook."

"I'm going to have a pet iguana," I say, "and wear nothing but cerise." It's a word I have recently learned.

Now I think, what if they just couldn't see what they looked like? Maybe it was as simple as that: eye problems. I'm having that trouble myself now: too close to the mirror and I'm a blur, too far back and I can't see the details. Who knows what faces I'm making, what kind of modern art I'm drawing onto myself? Even when I've got the distance adjusted, I vary. I am transitional; some days I look like a worn-out thirty-five, others like a sprightly fifty. So much depends on the light, and the way you squint.

I eat in pink restaurants, which are better for the skin. Yellow ones turn you yellow. I actually spend time thinking about this. Vanity is becoming a nuisance; I can see why women give it up, eventually. But I'm not ready for that yet.

Lately I've caught myself humming out loud, or walking along the street with my mouth slightly open, drooling a little. Only a little; but it may be the thin edge of the wedge, the crack in the wall that will open, later, onto what? What vistas of shining eccentricity, or madness?

There is no one I would ever tell this to, except Cordelia. But which Cordelia? The one I have conjured up, the one with the rolltop boots and the turned-up collar, or the one before, or the one after? There is never only one, of anyone.

If I were to meet Cordelia again, what would I tell her about myself? The truth, or whatever would make me look good?

Probably the latter. I still have that need.

I haven't seen her for a long time. I wasn't expecting to see her. But now that I'm back here I can hardly walk down a street without a glimpse of her, turning a corner, entering a door. It goes without saying that these fragments of her—a shoulder, beige, camel's-hair, the side of a face, the back of a leg—belong to women who, seen whole, are not Cordelia.

I have no idea what she would look like now. Is she fat, have her breasts sagged, does she have little gray hairs at the corners of her mouth? Unlikely: she would pull them out. Does she wear glasses with fashionable frames, has she had her lids lifted, does she streak or tint? All of these things are possible: we've both reached that borderline age, that buffer zone in which it can still be believed such tricks will work if you avoid bright sunlight.

I think of Cordelia examining the growing pouches under her eyes, the skin, up close, loosened and crinkled like elbows. She sighs, pats in cream, which is the right kind. Cordelia would know the right kind. She takes stock of her hands, which are shrinking a little, warping a little, as mine are. Gnarling has set in, the withering of the mouth; the outlines of dewlaps are beginning to be visible, down toward the chin, in the dark glass of subway windows. Nobody else notices these things yet, unless they look closely; but Cordelia and I are in the habit of looking closely.

She drops the bath towel, which is green, a muted sea-green to match her eyes, looks over her shoulder, sees in the mirror the dog's-neck folds of skin above the waist, the buttocks drooping like wattles, and, turning, the dried fern of hair. I think of her in a sweatsuit, sea-green as well, working out in some gym or other, sweating like a pig. I know what she would say about this, about all of this. How we giggled, with repugnance and delight, when we found the wax her older sisters used on their legs, congealed in a little pot, stuck full of bristles. The grotesqueries of the body were always of interest to her.

I think of encountering her without warning. Perhaps in a worn coat and a knitted hat like a tea cosy, sitting on a curb, with two plastic bags filled with her only possessions, muttering to herself. *Cordelia! Don't you recognize me?* I say. And she does, but pretends not to. She gets up and shambles away on swollen feet, old socks poking through the holes in her rubber boots, glancing back over her shoulder.

There's some satisfaction in that, more in worse things. I watch from a window, or a balcony so I can see better, as some man chases Cordelia along the sidewalk below me, catches up with her, punches her in the ribs—I can't handle the face—throws her down. But I can't go any farther.

Better to switch to an oxygen tent. Cordelia is unconscious. I have been summoned, too late, to her hospital bedside. There are flowers, sickly smelling, wilting in a vase, tubes going into her arms and nose, the sound of terminal breathing. I hold her hand. Her face is puffy, white, like an unbaked biscuit, with yellowish circles under the closed eyes. Her eyelids don't flicker but there's a faint twitching of her fingers, or do I imagine it? I sit there wondering whether to pull the tubes out of her arms, the plug out of the wall. No brain activity, the doctors say. Am I crying? And who would have summoned me?

Even better: an iron lung. I've never seen an iron lung, but the newspapers had pictures of children in iron lungs, back when people still got polio. These pictures—the iron lung a cylinder, a gigantic sausage roll of metal, with a head sticking out one end of it, always a girl's head, the hair flowing across the pillow, the eyes large, nocturnal—fascinated me, more than stories about children who went out on thin ice and fell through and were drowned, or children who played on the railroad tracks and had their arms and legs cut off by trains. You could get polio without knowing how or where, end up in an iron lung without knowing why. Something you breathed in or ate, or picked up from the dirty money other people had touched. You never knew.

The iron lungs were used to frighten us, and as reasons why we couldn't do things we wanted to. No public swimming pools, no crowds in summer. *Do you want to spend the rest of your life in an iron lung?* they would say. A stupid question; though for me such a life, with its inertia and pity, had its secret attractions.

Cordelia in an iron lung, then, being breathed, as an accordion is played. A mechanical wheezing sound comes from around her. She is fully conscious, but unable to move or speak. I come into the room, moving, speaking. Our eyes meet.

Cordelia must be living somewhere. She could be within a mile of me, she could be right on the next block. But finally I have no idea what I would do if I bumped into her by accident, on the subway for instance, sitting across from me, or waiting on the platform reading the ads. We would stand side by side, looking at a large red mouth stretching itself around a chocolate bar, and I would turn to her and say: *Cordelia. It's me, it's Elaine.* Would she turn, give a theatrical shriek? Would she ignore me?

Or would I ignore her, given the chance? Or would I go up to her wordlessly, throw my arms around her? Or take her by the shoulders, and shake and shake.

I've been walking for hours it seems, down the hill to the downtown, where the streetcars no longer run. It's evening, one of those gray watercolor washes, like liquid dust, the city comes up with in fall. The weather at any rate is still familiar.

Now I've reached the place where we used to get off the streetcar, stepping into the curbside mounds of January slush, into the grating wind that cut up from the lake between the flat-roofed dowdy buildings that were for us the closest thing to urbanity. But this part of the city is no longer flat, dowdy, shabby-genteel. Tubular neon in cursive script decorates the restored brick façades, and there's a lot of brass trim, a lot of real estate, a lot of money. Up ahead there are huge oblong towers, all of glass, lit up, like enormous gravestones of cold light. Frozen assets.

I don't look much at the towers though, or the people passing me in their fashionable getups, imports, handcrafted leather, suede, whatever. Instead I look down at the sidewalk, like a tracker.

I can feel my throat tightening, a pain along the jawline. I've started to chew my fingers again. There's blood, a taste I remember. It tastes of orange Popsicles, penny gumballs, red licorice, gnawed hair, dirty ice.

SILVER PAPER

I'm lying on the floor, on a futon, covered by a duvet. *Futon, duvet:* this is how far we've come. I wonder if Stephen ever figured out what futons and duvets were. Most likely not. Most likely if you'd said *futon* to him, he'd have looked at you as if he was deaf or you were brain-damaged. He did not exist in the futon dimension.

When there were no futons and no duvets, the price of an ice cream cone was five cents. Now it's a dollar if you're lucky, and not as big either. That's the bottom-line difference between then and now: ninety-five cents.

This is the middle of my life. I think of it as a place, like the middle of a river, the middle of a bridge, halfway across, halfway over. I'm supposed to have accumulated things by now: possessions, responsibilities, achievements, experience and wisdom. I'm supposed to be a person of substance.

But since coming back here I don't feel weightier. I feel lighter, as if I'm shedding matter, losing molecules, calcium from my bones, cells from my blood; as if I'm shrinking, as if I'm filling with cold air, or gently falling snow.

With all this lightness I do not rise, I descend. Or rather I am dragged downward, into the layers of this place as into liquefied mud.

The fact is that I hate this city. I've hated it so long I can hardly remember feeling any other way about it.

Once it was fashionable to say how dull it was. First prize a week in Toronto, second prize two weeks in Toronto, Toronto the Good, Toronto the Blue, where you couldn't get wine on Sundays. Everyone who lived here said those things: provincial, self-satisfied, boring. If you said that, it showed you recognized these qualities but did not partake of them yourself.

Now you're supposed to say how much it's changed. *World-class city* is a phrase they use in magazines these days, a great deal too much. All those ethnic restaurants, and the theater and the boutiques. New York without the garbage and muggings, it's supposed to be. People from Toronto used to go to Buffalo for the weekends, the men to watch girlie shows and drink after-hours beer, the women to shop; they'd come back jumped-up and pissed and wearing several layers of clothes to smuggle them through Customs. Now the weekend traffic is the other way around.

I've never believed either version, the dull, the world-class. Toronto was never dull, for me. Dull isn't a word you'd use to describe such misery, and enchantment.

And I can't believe it's changed. Driving in from the airport yesterday in the taxi, past the flat neat factories and warehouses that were once flat neat farms, mile after mile of caution and utilitarianism, and then through the center of the city with the glitz and the European-style awnings and the paving stones, I could see it's still the same. Underneath the flourish and ostentation is the old city, street after street of thick red brick houses, with their front porch pillars like the off-white stems of toadstools and their watchful, calculating windows. Malicious, grudging, vindictive, implacable.

In my dreams of this city I am always lost.

Apart from all this, I do of course have a real life. I sometimes have trouble believing in it, because it doesn't seem like the kind of life I could ever get away with, or deserve. This goes along with another belief of mine: that everyone else my age is an adult, whereas I am merely in disguise.

I live in a house, with window curtains and a lawn, in British Columbia, which is as far away from Toronto as I could get without drowning. The unreality of the landscape there encourages me: the greeting-card mountains, of the sunset-and-sloppy-message variety, the cottagy houses that look as if they were built by the Seven Dwarfs in the thirties, the giant slugs, so much larger than a slug needs to be. Even the rain is overdone, I can't take it seriously. I suppose these things are as real, and as oppressive, to the people who grew up there as this place is to me. But on good days it still feels like a vacation, an evasion. On bad days I don't notice it, or much else.

I have a husband, not my first, whose name is Ben. He is not any sort of an artist, for which I am thankful. He runs a travel agency, specializing in Mexico. Among his other sterling qualities are cheap tickets to the Yucatán. The travel agency is why he hasn't come with me on this trip: the months before Christmas are a hectic time in the travel business.

I also have two daughters, by now grown up. Their names are Sarah and Anne, good sensible names. One of them is almost a doctor, the other an accountant. These are sensible choices. I am a believer in sensible choices, so different from many of my own. Also in sensible names for children, because look what happened to Cordelia.

Alongside my real life I have a career, which may not qualify as exactly real. I am a painter. I even put that on my passport, in a moment of bravado, since the other choice would have been *housewife*. It's an unlikely thing for me to have become; on some days it still makes me cringe. Respectable people do not become painters: only overblown, pretentious, theatrical people. The word *artist* embarrasses me; I prefer *painter*, because it's more like a valid job. An artist is a tawdry, lazy sort of thing to be, as most people in this country will tell you. If you say you are a painter, you will be looked at strangely. Unless you paint wildlife, or make a lot of money at it, of course. But I only make enough to generate envy, among other painters, not enough so I can tell everyone else to stuff it.

Most of the time though I exult, and think I have had a narrow escape.

My career is why I'm here, on this futon, under this duvet. I'm having a retrospective, my first. The name of the gallery is Sub-Versions, one of those puns that used to delight me before they became so fashionable. I ought to be pleased by this retrospective, but my feelings are mixed;

I don't like admitting I'm old enough and established enough to have such a thing, even at an alternative gallery run by a bunch of women. I find it improbable, and ominous: first the retrospective, then the morgue. But also I'm cheesed off because the Art Gallery of Ontario wouldn't do it. Their bias is toward dead, foreign men.

The duvet is in a studio that belongs to my first husband Jon. It interests me that he would have a duvet here, although his house is elsewhere. So far I've restrained myself from going through his medicine cabinet, in search of hairpins and female deodorants, as I once would have done. This is none of my business any longer, I can leave the hairpins to his iron-clad wife.

Staying here is possibly a silly thing to do, too retrospective. But we've always kept in touch because of Sarah, who is his daughter also, and after we got through the shouting and the broken glass we settled into being friends of a sort, by long distance, which is always easier than up close. When he heard about the retrospective, he offered. The price of a hotel in Toronto, he said, even a second-rate hotel, is becoming offensive. Sub-Versions would have put me up, but I didn't mention that. I don't like the neatness of hotels, the squeaky-clean bathtubs. I don't like hearing my own voice echo in there, especially at night. I prefer the shedding and disorder and personal dirt of people like myself, people like Jon. Transients and nomads.

Jon's studio is down on King Street, near the waterfront. King Street used to be one of those places you never went, a place of dingy warehouses and rumbling trucks and dubious alleyways. Now it's come up in the world. Artists have infested it; in fact the first wave of artists has almost come and gone, and brass lettering and heating pipes painted fire-engine red and firms of lawyers are taking over. Jon's studio, on the fifth and top floor of one of the warehouses, doesn't have long to live in its present form. Track lighting is spreading over the ceilings, the lower floors are being stripped of their old linoleum, smelling of Pine Sol with an obscure base note of ancient throwup and pee, and the wide boards underneath are being sandblasted. I know all this because I walk up the five floors; they haven't got around to an elevator yet.

Jon left me the key in an envelope under the mat, and a note saying *Blessings*, which is a measure of how much he's softened, or mellowed. *Blessings* was not his former style. He's temporarily in Los Angeles, doing a chain-saw murder, but he'll be back before my opening.

I last saw him at Sarah's college graduation four years ago. He flew out to the coast, luckily without the wife, who is not fond of me. Although we haven't met, I know about her lack of fondness. During the proceedings, the ritual mumbo-jumbo and the tea and cookies afterward, we acted like responsible, grown-up parents. We took both the girls out to dinner and behaved ourselves. We even dressed the way we knew Sarah wanted us to: I had on an outfit, matching shoes and all, and Jon wore a suit and an actual tie. I told him he looked like an undertaker.

But the next day we snuck out to lunch, alone, and got plastered. That word, *plastered*, on the brink of obsolescence, indicates to me what sort of an event that was. It was a retrospective. And I still think of it as sneaking out, though of course Ben knew all about it. Though he would never go to lunch with his own first wife.

"You've always said it was such a disaster," Ben said to me, puzzled.

"It was," I said. "It was horrible."

"Then why would you want to have lunch with him?"

"That's hard to explain," I said, though it may not be. What we share, Jon and I, may be a lot like a traffic accident, but we do share it. We are survivors, of each other. We have been shark to one another, but also lifeboat. That counts for something.

In the old days Jon did constructions. He made them out of bits of wood and leather he'd pick out of people's trash, or else he'd smash things—violins, glassware—and glue the pieces into the position of the smash; shatter patterns, he called them. At one time he wrapped pieces of colored tape around tree trunks and took photographs of them, at another he made a replica of a mold-covered loaf of bread that breathed in and out with the aid of a small electric motor. The mold was made from the hair clippings of himself and friends. I think there's even some of my hair on that loaf of bread; I caught him snitching some out of my hairbrush.

He does special effects for movies now, to support his artist habit. The studio is scattered with his half-finished doings. On the workbench where he keeps his paints, glues, knives, and pliers, there's a hand and arm, done in plastic resin, arteries worming from the cut end, straps to strap it on. There are hollow casts of legs and feet standing around on the floor like elephant-foot umbrella stands; in one of them there's an umbrella. Also there's part of a face, with the skin blackened and withered, made to fit over the actor's real face. A monster, warped by others, bent on revenge.

Jon has told me he isn't sure this hacked-up body-part stuff is the sort of thing he should be doing. It's too violent, it doesn't contribute to human goodness. He's coming to believe in human goodness in his old age, which is certainly a change; I've even found some herbal tea in the cupboard. He claims he'd rather make friendly animals for children's shows. But as he says, you have to eat, and there's just more demand for cut-off limbs.

I wish he were here, or Ben, or any man I know. I'm losing the appetite for strangers. Once I would have focused on the excitement, the hazard; now it's the mess, the bother. Getting your clothes off gracefully, always such an impossibility; thinking up what to say afterward, without setting the echoes going in your head. Worse, the encounter with another set of particularities: the toenails, the ear holes, the nosehairs. Perhaps at this age we return to the prudishness we had as children.

I get up off the duvet, feeling as if I haven't slept. I riffle through the herbal tea bags in the kitchenette, Lemon Mist, Morning Thunder, and bypass them in favor of some thick, jolting, poisonous coffee. I find myself standing in the middle of the main room, not knowing exactly how I got in here from the kitchenette. A little time jump, a little static on the screen, probably jet lag: up too late at night, drugged in the morning. Early Alzheimer's.

I sit at the window, drinking my coffee, biting my fingers, looking down the five stories. From this angle the pedestrians appear squashed from above, like deformed children. All around are flat-roofed, boxy warehouse buildings, and beyond them the flat railroad lands where the trains used to shunt back and forth, once the only entertainment available here on Sundays. Beyond that is flat Lake Ontario, a zero at the beginning and a zero at the end, slate-gray and brimming with venoms. Even the rain from it is carcinogenic.

I wash in Jon's tiny, greasy bathroom, resisting the medicine cabinet. The bathroom is smeared with fingerprints and painted dingy white, not the most flattering light. Jon wouldn't feel like an artist without a certain amount of dinge around. I squint into the mirror, preparing my face: with my contact lenses in I'm too close to the mirror, without them I'm too far away. I've taken to doing these mirror things with one lens in my mouth, glassy and thin like the tag end of a lemon drop. I could choke on it by mistake, an undignified way to die. I should get bifocals. But then I'd look like an old biddy.

I pull on my powder-blue sweatsuit, my disguise as a nonartist, and go down the four flights of stairs, trying to look brisk and purposeful. I could be a businesswoman out jogging, I could be a bank manager, on her day off. I head north, then east along Queen Street, which is another place we never used to go. It was rumored to be the haunt of grubby drunks, rubby-dubs we called them; they were said to drink rubbing alcohol and sleep in telephone booths and vomit on your shoes in the streetcar. But now it's art galleries and bookshops, boutiques filled with black clothing and weird footgear, the saw-toothed edge of trend.

I decide I'll go and have a look at the gallery, which I have never seen because all of this has been arranged by phone and mail. I don't intend to go in, make myself known, not yet. I just want to look at it from the outside. I'll walk past, glance casually, pretending to be a housewife, a tourist, someone window shopping. Galleries are frightening places, places of evaluation, of judgment. I have to work up to them.

But before I reach the gallery I come to a wall of plywood, concealing a demolition. On it is spray-painted, in defiance of squeaky-clean Toronto: *It's Bacon or Me, Babe.* And underneath: *What Is This Bacon and Where Can I Get Some?* Beside this there's a poster. Or not a poster, more like a flier: a violent shade of purple, with green accents and black lettering. RISLEY IN RETROSPECT, it says; just the last name, like a boy. The name is mine and so is the face, more or less. It's the photo I sent the gallery. Except that now I have a mustache.

Whoever drew this mustache knew what he was doing. Or she: nothing precludes that. It's a curled, flowing mustache, like a cavalier's, with a graceful goatee to match. It goes with my hair.

I suppose I should be worried about this mustache. Is it just doodling, or is it political commentary, an act of aggression? Is it more like *Kilroy Was Here* or more like *Fuck Off?* I can remember drawing such mustaches myself, and the spite that went into them, the desire to ridicule, to deflate, and the feeling of power. It was defacing, it was taking away someone's face. If I were younger I'd resent it.

As it is, I study the mustache and think: *That looks sort of good.* The mustache is like a costume. I examine it from several angles, as if I'm considering buying one for myself. It casts a different light. I think about men and their facial hair, and the opportunities for disguise and concealment they have always at their disposal. I think about mustache-covered men, and about how naked they must feel with the thing shaved off. How diminished. A lot of people would look better in a mustache.

Then, suddenly, I feel wonder. I have achieved, finally, a face that a mustache can be drawn on, a face that attracts mustaches. A public face, a face worth defacing. This is an accomplishment. I have made something of myself, something or other, after all.

I wonder if Cordelia will see this poster. I wonder if

she'll recognize me, despite the mustache. Maybe she'll come to the opening. She'll walk in through the door and I will turn, wearing black as a painter should, looking successful, holding a glass of only moderately bad wine. I won't spill a drop.

4.

*U*ntil we moved to Toronto I was happy.

Before that we didn't really live anywhere; or we lived so many places it was hard to remember them. We spent a lot of the time driving, in our low-slung, boat-sized Studebaker, over back roads or along two-lane highways up north, curving past lake after lake, hill after hill, with the white lines going down the middle of the road and the telephone poles along the sides, tall ones and shorter ones, the wires looking as if they were moving up and down.

I sit by myself in the back of the car, among the suitcases and the cardboard boxes of food and the coats, and the gassy, dry-cleaning smell of the car upholstery. My brother Stephen sits in the front seat, beside the partly open window. He smells of peppermint LifeSavers; underneath that is his ordinary smell, of cedarwood lead pencils and wet sand. Sometimes he throws up into paper bags, or beside the road if my father can stop the car in time. He gets carsick and I do not, which is why he has to sit in the front. It's his only weakness that I know of.

From my cramped vantage point in the back I have a good view of my family's ears. My father's, which stick out from under the brim of the old felt hat he wears to keep twigs and tree sap and caterpillars out of his hair, are large and soft-looking, with long lobes; they're like the ears of gnomes, or those of the flesh-colored, doglike minor characters in *Mickey Mouse* comic books. My mother wears her hair pinned back at the sides with bobby pins, so her ears are visible from the back. They're narrow, with fragile upper edges, like the handles of china cups, although she herself is not fragile. My brother's ears are round, like dried apricots, or like the ears of the green-tinged, oval-headed aliens from outer space he draws with his colored pencils. Around and over his round ears and down the back of his neck his hair, dark blond and straight, grows in thick wisps. He resists haircuts.

It's difficult for me to whisper into my brother's round ears when we're in the car. In any case he can't whisper back, because he has to look straight ahead at the horizon, or at the white lines of the road that washes toward us, wave after slowly undulating wave.

The roads are mostly empty, because it's the war, though once in a while there's a truck loaded with cut tree trunks or fresh lumber, trailing its perfume of sawdust. At lunchtime we stop by the roadside and spread a groundsheet among the white papery everlasting and the purple fireweed and eat the lunch our mother makes, bread and sardines or bread and cheese, or bread and molasses or bread and jam if we can't get anything else. Meat and cheese are scarce, they are rationed. That means you have a ration book with colored stamps in it.

Our father makes a small fire to boil water in a billy tin for tea. After lunch we disappear into the bushes, one by one, with pieces of toilet paper in our pockets. Sometimes there are other pieces of toilet paper there already, melting among the bracken and dead leaves, but mostly there are not. I squat, listening behind me for bears, aster leaves rough on the tops of my legs, then bury the toilet paper under sticks and bark and dried bracken. Our father says you should make it look as if you haven't been there at all.

Our father walks into the forest, carrying his ax, a pack-sack, and a large wooden box with a leather shoulder strap. He looks up, from tree to tree to tree, considering. Then he spreads a tarpaulin out on the ground, underneath the chosen tree, wrapping it around the trunk. He opens the wooden box, which is filled with racks of small bottles. He hits the tree trunk with the back of his ax. The tree shakes; leaves and twigs and caterpillars patter down, bouncing off his gray felt hat, hitting the tarpaulin. Stephen and I crouch, picking up the caterpillars, which are blue-striped, and velvety and cool, like the muzzles of dogs. We put them into the collecting bottles filled with pale alcohol. We watch them twist and sink.

My father looks at the harvest of caterpillars as if he's grown them himself. He examines the chewed leaves. "A beautiful infestation," he says. He's joyful, he's younger than I am now.

The alcohol smell is on my fingers, cold and remote, piercing, like a steel pin going in. It smells like white enamel basins. When I look up at the stars in the nighttime, cold and white and sharp, I think they must smell like that.

When it gets to the end of the day we stop again and put up our tent, heavy canvas with wooden poles. Our sleeping bags are khaki and thick and lumpy, and always feel a little damp. Underneath them we put groundsheets, and inflatable mattresses that make you feel dizzy while you blow them up and fill your nose and mouth with the taste of stale rain boots or spare tires piled in a garage. We eat around the fire, which turns brighter as the shadows grow out from the trees like darker branches. We crawl into the tent and take our clothes off inside our sleeping bags, the flashlight making a circle on the canvas, a light ring enclosing a darker one, like a target. The tent smells of tar and kapok and brown paper with cheese grease on it, and crushed grass. In the mornings the weeds outside are sprinkled with dew.

Sometimes we stay in motels, but only if it's too late at night to find a place to set up the tent. The motels are always far from anything, set against a dark wall of forest, their lights glimmering in the uniform obscuring night like those of ships, or oases. They have gas pumps outside, human-sized, with round discs on top, lit up like pale moons or haloes minus the head. On each disc is a shell or a star, an orange maple leaf, a white rose. The motels and the gas pumps are often empty or closed: gas is rationed, so people don't travel much unless they have to.

Or we stay in cabins belonging to other people or to the government, or we stay in abandoned logging camps, or we pitch two tents, one for sleeping and one for supplies. In the winters we stay in towns or cities up north, the Soo or North Bay or Sudbury, in apartments that are really the top floors of other people's houses, so that we have to be careful about the noise of our shoes on the wooden floors. We have furniture which comes from storage. It's always the same furniture but it always looks unfamiliar.

In these places there are flush toilets, white and alarming, where things vanish in an instant, with a roar. When we first get to cities my brother and I go to the bathroom a lot, and drop things in as well, such as pieces of macaroni, to see them disappear. There are air-raid sirens, and then we pull the curtains and turn off the lights, though our mother says the war will never come here. The war filters in over the radio, remote and crackly, the voices from London fading through the static. Our parents are dubious as they listen, their mouths tighten: it could be that we are losing.

My brother does not think so. He thinks our side is the good side, and therefore it will win. He collects cigarette cards with pictures of airplanes on them, and knows the names of all the planes.

My brother has a hammer and some wood, and his own jackknife. He whittles and hammers: he's making a gun. He nails two pieces of wood at right angles, with another nail for the trigger. He has several of these wooden guns, and daggers and swords also, with blood coloured onto the blades with red pencils. Some of the blood is orange, from when he ran out of red. He sings:

Coming in on a wing and a prayer,
Coming in on a wing and a prayer,
Though there's one motor gone
We will still carry on,
Coming in on a wing and a prayer.

He sings this cheerfully, but I think it's a sad song, because although I've seen the pictures of the airplanes on the cigarette cards I don't know how they fly. I think it's like birds, and a bird with one wing can't fly. This is what my father says in the winters, before dinner, lifting his glass when there are other men there at the table: "You can't fly on one wing." So in fact the prayer in the song is useless.

Stephen gives me a gun and a knife and we play war. This is his favorite game. While our parents are putting up the tent or making the fire or cooking, we sneak around behind the trees and bushes, aiming through the leaves. I am the infantry, which means I have to do what he says. He waves me forward, motions me back, tells me to keep my head down so the enemy won't blow it off.

"You're dead," he says.

"No I'm not."

"Yes you are. They got you. Lie down."

There is no arguing with him, since he can see the enemy and I can't. I have to lie down on the swampy ground, propped against a stump to avoid getting too wet, until it's time for me to be alive again.

Sometimes, instead of war, we hunt through the forest, turning over logs and rocks to see what's underneath. There are ants, grubs and beetles, frogs and toads, garter snakes, even salamanders if we're lucky. We don't do anything with the things we find. We know they will die if we put them into bottles and leave them by accident in the sun in the back window of the car, as we have done before. So we merely look at them, watching the ants hiding their pill-shaped eggs in panic, the snakes pouring themselves into darkness. Then we put the logs back where they were, unless we need some of these things for fishing.

Once in a while we fight. I don't win these fights: Stephen is bigger and more ruthless than I am, and I want to play with him more than he wants to play with me. We fight in whispers or well out of the way, because if we're caught we will both be punished. For this reason we don't tell on each other. We know from experience that the satisfactions of betrayal are scarcely worth it.

Because they're secret, these fights have an extra attraction. It's the attraction of dirty words we aren't supposed to say, words like *bum*; the attraction of conspiracy, of collusion. We step on each other's feet, pinch each other's arms, careful not to give away the pain, loyal even in outrage.

How long did we live this way, like nomads on the far edges of the war?

Today we've driven a long time, we're late setting up our tent. We're near the road, beside a raggedy anonymous lake. The trees around the shore are doubled in the water, the leaves of the poplars are yellowing towards fall. The sun sets in a long, chilly, lingering sunset, flamingo pink, then salmon, then the improbable vibrant red of Mercurochrome. The pink light rests on the surface, trembling, then fades and is gone. It's a clear night, moonless, filled with antiseptic stars. There is the Milky Way clear as can be, which predicts bad weather.

We pay no attention to any of this, because Stephen is teaching me to see in the dark, as commandos do. You never know when you might need to do this, he says. You can't use a flashlight; you have to stay still, in the darkness, waiting until your eyes become accustomed to no light. Then the shapes of things begin to emerge, grayish and glimmering and insubstantial, as if they're condensing from the air. Stephen tells me to move my feet slowly, balancing on one foot at a time, careful not to step on twigs. He tells me to breathe quietly. "If they hear you they'll get you," he whispers.

He crouches beside me, outlined against the lake, a blacker patch of water. I catch the glint of an eye, then he's gone. This is a trick of his.

I know he's sneaking up on the fire, on my parents, who are flickering, shadowy, their faces indistinct. I'm alone with my heartbeat and my too-loud breathing. But he's right: now I can see in the dark.

Such are my pictures of the dead.

5.

I have my eighth birthday in a motel. My present is a Brownie box camera, black and oblong, with a handle on top and a round hole at the back to look through.

The first picture taken with it is of me. I'm leaning against the doorframe of the motel cabin. The door behind me is white and closed, with the metal number on it showing: 9. I'm wearing pants, baggy at the knees, and a jacket too short in the sleeves. Under the jacket, I know though you can't see it, is a hand-me-down brown and yellow striped jersey of my brother's. Many of my clothes were once his. My skin is ultrawhite from overexposure of the film, my head is tilted to one side, my mittenless wrists dangle. I look like old photos of immigrants. I look as if I've been put there in front of the door and told to stand still.

What was I like, what did I want? It's hard to remember. Did I want a camera for my birthday? Probably not, although I was glad to have it.

I want some more cards from the Nabisco Shredded Wheat boxes, the gray cards with pictures on them that you color, cut out, and fold to make the houses in a town. Also I want some pipe cleaners. We have a book called *Rainy Day Hobbies* that shows how to make a walkie-talkie out of two cans and a piece of string, or how to make a boat that will go forward if you drop lubricating oil into a hole in it; also how to make a doll's chest of drawers out of miniature matchboxes, and how to make various animals—a dog, a sheep, a camel—out of pipe cleaners. The boat and the chest of drawers don't appeal to me, only the pipe cleaners. I've never seen a pipe cleaner.

I want some silver paper out of cigarette packages. I have several pieces already, but I want more. My parents don't smoke cigarettes, so I have to collect this paper where I can find it, on the edges of gas stations, in the weedy grass near motels. I am in the habit of scavenging along the ground this way. When I find some I clean it off and flatten it out and store it between the pages of my school reader. I don't know what I'll do with it when I have enough, but it will be something amazing.

I want a balloon. Balloons are coming back, now that the war is over. When I was sick with the mumps, once in the winter, my mother found one at the bottom of her steamer trunk. She must have tucked it away there before the war, suspecting perhaps that there would not be any more for a while. She blew it up for me. It was blue, translucent, round, like a private moon. The rubber was old and rotting and the balloon burst almost at once, and I was heartbroken. But I want another balloon, one that will not break.

I want some friends, friends who will be girls. Girl friends. I know that these exist, having read about them in books, but I've never had any girl friends because I've never been in one place long enough.

Much of the time it's raw and overcast, the low metallic sky of late autumn; or else it rains and we have to stay inside the motel. The motel is the kind we're used to: a row of cottages, flimsily built, strung together with Christmas tree lights, yellow or blue or green. These are called "house-keeping cottages," which means they have some kind of a stove in them, a pot or two and a tea kettle, and a table covered with oilcloth. The floor of our housekeeping cottage is linoleum, with a faded pattern of floral squares. The towels are skimpy and thin, the sheets have worn places in the middles, rubbed there by other people's bodies. There's a framed print of the woods in winter and another of ducks in flight. Some motels have outhouses, but this one has a real though smelly flush toilet, and a bathtub.

We've been living in this motel for weeks, which is unusual: we never stay in motels for more than a night at a time. We eat cans of Habitant pea soup, heated up on the two-burner stove in a dented pot, and slices of bread spread with molasses, and hunks of cheese. There's more cheese, now that the war is over. We wear our outdoor clothes indoors, and socks at night, because these cottages with their one-layer walls are supposed to be for summer tourists. The hot water is never more than lukewarm, and our mother heats water in the tea kettle and pours it into the tub for our baths. "Just to get the crust off," she says.

In the mornings we wrap blankets around our shoulders while we eat our breakfast. Sometimes we can see our own breath, even inside the cottage. All of this is irregular, and slightly festive. It isn't just that we aren't going to school. We've never gone to school for more than three or four months at a time anyway. I was in school the last time eight months ago and have only dim and temporary ideas of what it was like.

In the mornings we do our schoolwork, in our workbooks. Our mother tells us which pages to do. Then we read our school readers. Mine is about two children who live in a white house with ruffled curtains, a front lawn, and a picket fence. The father goes to work, the mother wears a dress and an apron, and the children play ball on the lawn with their dog and cat. Nothing in these stories is anything like my life. There are no tents, no highways, no peeing in the bushes, no lakes, no motels. There is no war. The children are always clean, and the little girl, whose name is Jane, wears pretty dresses and patent-leather shoes with straps.

These books have an exotic appeal for me. When Stephen and I draw with our colored pencils, he draws wars, ordinary wars and wars in space. His red and yellow and orange are worn to stubs, from the explosions, and his gold and silver are used up too, on the shining metal carapaces of the tanks and spaceships and on the helmets and the complicated guns. But I draw girls. I draw them in old-fashioned clothing, with long skirts, pinafores and puffed sleeves, or in dresses like Jane's, with big hairbows on their heads. This is the elegant, delicate picture I have in my mind, about other little girls. I don't think about what I might say to them if I actually met some. I haven't got that far.

In the evenings we're supposed to do the dishes—"Rattle them up," our mother calls it. We squabble in whispers and monosyllables about whose turn it is to wash: drying with a clammy tea towel isn't as good as washing, which warms up your hands. We float the plates and glasses in the dish pan and dive-bomb them with the spoons and knives, whispering "Bombs away." We try to aim as close as possible without actually hitting them. They aren't our dishes. This gets on our mother's nerves. If it gets on her nerves enough, she will do the dishes herself, which is intended to be a rebuke.

At night we lie in the saggy pull-out bed, head to toe, which is supposed to make us go to sleep sooner, and kick each other silently under the covers; or else we try to see how far we can get our sock feet up each other's pajama legs. Once in a while the headlights of a passing car show through the window, moving first along one wall, then along the next wall, then fading away. There's an engine sound, then the sizzle of tires on the wet road. Then silence.

6
.

I don't know who took that picture of me. It must have been my brother, because my mother is inside the cabin, behind the white door, wearing gray slacks and a dark-blue plaid shirt, packing our food into cardboard boxes and our clothes into suitcases. She has a system for packing; she talks to herself while she's doing it, reminding herself of details, and she likes us out of the way.

Right after the picture it begins to snow, small dry flakes falling singly out of the hard northern November sky. There's a kind of hush and lassitude until that first snow, with the light waning and the last moose-maple leaves dangling from the branches like seaweed. We felt sleepy until it began to snow. Now we feel exhilarated.

We're running around outside the motel, wearing nothing but our worn-out summer shoes, with our bare hands outstretched to the falling snowflakes, our heads thrown back, our mouths open, eating snow. If it were thick on the ground we would roll in it, like dogs in dirt. It fills us with the same kind of rapture. But our mother looks out the window and sees us, and the snow, and makes us come inside and dry our feet off with the skimpy towels. We have no winter boots that fit. While we're inside, the snow turns to sleet.

Our father is pacing the floor, jingling his keys in his pocket. He always wants things to happen sooner than they do, and now he wants to leave right away, but my mother says he'll have to hold his horses. We go outside and help him scrape the crust of ice off the car windows and then we carry boxes, and finally we squeeze into the car ourselves and drive south. I know it's south because of the direction of the sunlight, which is coming weakly through the clouds now, touching the icy trees with glitter, glaring off the ice patches at the sides of the road, making it hard to see.

Our parents say we're going to our new house. This time the house will be really ours, not rented. It's in a city called Toronto. This name means nothing to me. I think about the house in my school reader, white, with a picket fence and a lawn, and window curtains. I want to see what my bedroom will be like.

By the time we arrive at the house it's late afternoon. At first I think there must be some mistake; but no, this is the house all right, because my father is already opening the door with a key. The house is hardly on a street at all, more like a field. It's square-shaped, a bungalow, built of yellow brick and surrounded by raw mud. On one side of it is an enormous hole in the ground, with large mud piles heaped around it. The road in front is muddy too, unpaved, potholed. There are some concrete blocks sunk in the mud for stepping-stones so we can get to the door.

Inside, things are even more daunting. There are doors and windows, true, and walls, and the furnace works. There's a picture window in the living room, though the view is of a large expanse of rippling mud. The toilet actually flushes, though it has a yellowish-brown ring around the inside of the bowl and several floating cigarette butts; and reddish, warmish water comes from the hot water tap when I turn it on. But the floors are not polished wood or even linoleum. They're made of wide, rough boards with cracks in between, gray with plaster dust and scattered with white speckles, like bird droppings. Only a few of the rooms have light fixtures; the others have wires dangling out of the middle of the ceiling. There are no counters in the kitchen, only the bare sink; there's no stove. Nothing is painted. Dust is on everything: the windows, the window ledges, the fixtures, the floor. There are a lot of dead flies around.

"We all have to pitch in," says our mother, which means that we are not to complain. We will have to do the best we can, she says. We will have to finish the house ourselves, because the man who was supposed to do it has gone bankrupt. Flown the coop, is how she puts it. Our father is not so cheerful. He paces around the house, peering and prodding at it, muttering to himself and making small whistling noises. "Son of a gun, son of a gun," is what he says.

From somewhere in the depths of the car our mother unearths a primus stove, which she sets up on the kitchen floor, since there is no table. She begins to heat up some pea soup. My brother goes outside; I know he's climbing up the mountain of dirt next door, or assessing the possibilities of the large hole in the ground, but I don't have the heart to join him.

I wash my hands in the reddish water in the bathroom. There's a crack in the sink, which at this moment seems a disaster, worse than any of the other flaws and absences. I look at my face in the dust-smeared mirror. There's no shade on the light, just a bare bulb overhead, which makes my face look pallid and ill, with circles under the eyes. I rub my eyes; I know it would be wrong to be seen crying. Despite its rawness, the house feels too hot, maybe because I'm still wearing my outside clothes. I feel trapped. I want to be back in the motel, back on the road, in my old rootless life of impermanence and safety.

The first nights we sleep on the floor, in our sleeping bags, on top of our air mattresses. Then some army surplus cots appear, canvas stretched across a metal frame smaller at the base than at the top, so that if you roll over in the night you tip off onto the floor and the cot falls on top of you. Night after night I fall out and wake up lying on the rough dusty floor wondering where I am, and my brother is not there to snicker at me or order me to shut up, because I'm in a room by myself. At first I found the thought of my own room exciting—an empty space to be arranged as I wanted, without regard to Stephen and his strewn clothes and wooden guns—but now I'm lonely. I've never been in a room by myself at night before.

Each day new things appear in the house while we're at school: a stove, a refrigerator, a card table and four chairs, so that we can eat in the ordinary way, sitting at a table, instead of cross-legged on a groundsheet spread in front of the fireplace. The fireplace actually works; this is one part

of the house that has been finished. In it we burn scrap
pieces of wood left over from the construction.

In his spare time our father hammers away at the interior
of the house. Floor coverings spread across the floors: narrow hardwood boards in the living room, asphalt tiles in
our bedrooms, advancing row by row. The house begins to
look more like a house. But this takes a lot longer than I
would like: we are a far cry from picket fences and white
curtains, here in our lagoon of postwar mud.

!

We're used to seeing our father in windbreakers, battered gray felt hats, flannel shirts with the cuffs tightly buttoned to keep the blackflies from crawling up his arms, heavy pants tucked into the tops of woolen work socks. Except for the felt hats, what our mother wore wasn't all that different.

Now, however, our father wears jackets and ties and white shirts, and a tweed overcoat and a scarf. He has galoshes that buckle on over his shoes instead of leather boots waterproofed with bacon grease. Our mother's legs have appeared, sheathed in nylons with seams up the backs. She draws on a lipstick mouth when she goes out. She has a coat with a gray fur collar, and a hat with a feather in it that makes her nose look too long. Every time she puts on this hat, she looks into the mirror and says, "I look like the Witch of Endor."

Our father has changed his job: this explains things. Instead of being a forest-insect field researcher, he is now a university professor. The smelly jars and collecting bottles that once were everywhere have diminished in number. Instead, scattered around the house, there are stacks of drawings made by his students with colored pencils. All of them are of insects. There are grasshoppers, spruce budworms, forest tent caterpillars, wood-boring beetles, each one the size of a page, their parts neatly labeled: mandibles, palps, antennae, thorax, abdomen. Some of them are in section, which means they're cut open so you can see what's inside them: tunnels, branches, bulbs and delicate filaments. I like this kind the best.

My father sits in an armchair in the evenings with a board across the arms of the chair and the drawings on the board, going through them with a red pencil. Sometimes he laughs to himself while doing this, or shakes his head, or makes ticking noises through his teeth. "Idiot," he says, or "blockhead." I stand behind his chair, watching the drawings, and he points out that this person has put the mouth at the wrong end, that person has made no provision for a heart, yet another one cannot tell a male from a female. This is not how I judge the drawings: I find them better or worse depending on the colors.

On Saturdays we get into the car with him and drive down to the place where he works. It is actually the Zoology Building, but we don't call it that. It is just the building.

The building is enormous. Whenever we're there it's almost empty, because it's Saturday; this makes it seem even larger. It's of dark-brown weathered brick, and gives the impression of having turrets, although it has none. Ivy grows on it, leafless now in winter, covering it with skeletal veining. Inside it there are long hallways with hardwood floors, stained and worn from generations of students in slushy winter boots, but still kept polished. There are staircases, also of wood, which creak when we climb them, and banisters we aren't supposed to slide down, and iron radiators that make banging noises and are either stone cold or blazing hot.

On the second floor there are corridors leading into other corridors, lined with shelves that contain jars full of dead lizards or pickled ox eyeballs. In one room there are glass cages with snakes in them, snakes bigger than any we've ever seen before. One is a tame boa constrictor, and if the man in charge of it is there he gets it out and winds it around his arm, so we can see how it crushes things to death in order to eat them. We're allowed to stroke it. Its skin is cool and dry. Other cages have rattlesnakes, and the man shows us how he milks the venom out of their fangs. For this he wears a leather glove. The fangs are curved and hollow, the venom dripping from them is yellow.

In the same room is a cement pool filled with thick-looking greenish water in which large turtles sit and blink or clamber ponderously up onto the rocks provided for them, hissing if we get too close. This room is hotter and steamier than the others because the snakes and turtles need it to be; it smells musky. In yet another room is a cage full of gigantic African cockroaches, white-colored and so poisonous that their keeper has to gas them to make them unconscious every time he opens the cage to feed them or get one out.

Down in the cellar there are shelves and shelves of white rats and black mice, special ones that aren't wild. They eat food pellets from hoppers in their cages and drink from bottles fitted with eyedroppers. They have chewed-newspaper nests full of pink hairless baby mice. They run over and under one another and sleep in heaps, and sniff one another with their quivering noses. The mouse feeder tells us that if you put a strange mouse into one of their cages, one with the wrong, alien scent, they will bite it to death.

The cellar smells strongly of mouse droppings, a smell which wafts upward through the whole building, getting fainter as you go up, mingling with the smell of the green Dustbane used to clean the floors, and with the other smells, the floor polish and furniture wax and formaldehyde and snakes.

We don't find any of the things in the building repulsive. The general arrangements, though not the details, are familiar to us, though we've never seen so many mice in one place before and are awed by their numbers and stench. We would like to get the turtles out of their pool and play with them, but since they're snapping turtles and bad-tempered and can take your fingers off, we know enough not to. My brother wants an ox eyeball out of one of the jars: it's the sort of thing other boys find impressive.

Some of the upstairs rooms are labs. The labs have vast ceilings and blackboards across the front. They contain rows and rows of large dark desks, more like tables than desks, with high stools to sit on. Each desk has two lamps with green glass shades, and two microscopes, old microscopes, with heavy thin tubing and brass fittings.

We've seen microscopes before, but not at such length; we can spend a lot of time with them before getting tired of them. Sometimes we're given slides to look at: butterfly wings, cross-sections of worms, planaria stained with pink and purple dyes so you can see the different parts. At other times we put our fingers under the lenses and examine our fingernails, the pale parts curved like hills against their dark pink sky, the skin around them grainy and creased like the edge of a desert. Or we pull hairs out of our heads to look at them, hard and shiny like the bristles that grow out of the chitonous skins of insects, with the hair roots at the end like tiny onion bulbs.

We like scabs. We pick them off—there isn't room for a whole arm or leg under the microscope—and turn the magnification up as high as it will go. The scabs look like rocks, bumpy, with a sheen like silica; or else like some kind of fungus. If we can get a scab off a finger we put the finger under and watch the place where the blood oozes out, bright red, in a round button, like a berry. Afterward we lick off the blood. We look at earwax, or snot, or dirt from our toes, checking first to see that there's no one around: we know without asking that such things would not be approved of. Our curiosity is supposed to have limits, though these have never been defined exactly.

This is what we do on Saturday mornings, while our father attends to things in his office and our mother goes grocery shopping. She says it keeps us out of her hair.

The building overlooks University Avenue, which has lawns and some copper-green statues of men on horses. Right across the road is the Ontario Parliament Building, which is also old and dingy. I think it must be another building like the building, filled with long creaky corridors and shelves of pickled lizards and ox eyes.

It's from the building that we watch our first Santa Claus Parade. We've never seen a parade before. You can listen to this parade on the radio, but if you want to actually see it you have to bundle up in your winter clothes and stand on the sidewalk, stomping your feet and rubbing your hands to keep warm. Some people climb up onto the horse statues to get a better view. We don't have to do this, as we can sit on the window ledge of one of the main labs in the building, separated from the weather by a pane of dusty glass, with blasts of heat from the iron radiator going up our legs.

From there we watch as people dressed like snowflakes, like elves, like rabbits, like sugar plum fairies, march past us, strangely truncated because we're looking down on them. There are bands of bagpipers in kilts, and things like big cakes, with people on them waving, that slide past on wheels. It's begun to drizzle. Everyone down there looks cold.

Santa Claus is at the end, smaller than expected. His voice and his loudspeaker jingle bells are muted by the dusty glass; he rocks back and forth behind his mechanical reindeer, looking soggy, blowing kisses to the crowd.

I know he isn't the real Santa Claus, just someone dressed up like him. Still, my idea of Santa Claus has altered, has acquired a new dimension. After this it becomes hard for me to think of him without thinking also of the snakes and the turtles and the pickled eyes, and the lizards floating in their yellow jars, and of the vast, echoing, spicy, ancient and forlorn but also comforting smell of old wood, furniture polish, formaldehyde and distant mice.

EMPIRE BLOOMERS

8

There are days when I can hardly make it out of bed. I find it an effort to speak. I measure progress in steps, the next one and the next one, as far as the bathroom. These steps are major accomplishments. I focus on taking the cap off the toothpaste, getting the brush up to my mouth. I have difficulty lifting my arm to do even that. I feel I am without worth, that nothing I can do is of any value, least of all to myself.

What do you have to say for yourself? Cordelia used to ask. *Nothing*, I would say. It was a word I came to connect with myself, as if I was nothing, as if there was nothing there at all.

Last night I felt the approach of nothing. Not too close but on its way, like a wingbeat, like the cooling of the wind, the slight initial tug of an undertow. I wanted to talk to Ben. I phoned the house but he was out, the machine was on. It was my own voice I heard, cheerful and in control. *Hi there. Ben and I can't come to the phone at the moment, but leave a message and we'll get back to you as soon as we can.* Then a beep.

A disembodied voice, an angel voice, wafting through the air. If I died this minute it would go on like that, placid and helpful, like an electronic afterlife. Hearing it made me want to cry.

"Big hugs," I said into the empty space. I closed my eyes, thought about the mountains on the coast. That's home, I told myself. That's where you really live. Among all that stagey scenery, too beautiful, like a cardboard movie backdrop. It's not real, it's not drab, not flat, not grubby enough. They're working on it though. Go a few miles here, a few miles there, out of sight of the picture windows, and you come to the land of stumps.

Vancouver is the suicide capital of the country. You keep going west until you run out. You come to the edge. Then you fall off.

I crawl out from under the duvet. I am a busy person, in theory. There are things to be done, although none of them are things I want to do. I check through the refrigerator in the kitchenette, dig out an egg, boil it, dump it into a teacup, mush it up. I don't even glance at the herbal teas, I go straight for the real, vile coffee. Jitter in a cup. It cheers me up to know I'll soon be so tense.

I pace among the severed arms and hollow feet, drinking blackness. I like this studio, I could work here. There's the right amount of makeshift and dinginess for me. Things that are falling apart encourage me: whatever else, I'm in better shape than they are.

Today we hang. An unfortunate term.

I tuck myself into my clothes, handling my arms and legs as if they're someone else's, someone not very big or not very well. It's the powder-blue sweatsuit again today; I didn't bring very much clothing with me. I don't like to check things, I like to jam it all under the airplane seat. At the back of my mind is the idea that if something goes wrong, up there in the air, I'll be able to grab my bag out from under the seat and jump out the window, gracefully, without leaving any of my possessions behind.

I head out into the open, walk quickly along the street, mouth slightly open, keeping time in my head. *Keep happy with the Happy Gang.* I used to jog but it's bad for the knees. Too much beta carotene turns you orange, too much calcium gives you kidney stones. Health kills.

The old emptiness of Toronto is gone. Now it's chock-full: Toronto's bloating itself to death, that much is clear. The traffic is astonishing, there's honking and barging, people drive right out to the middle of the intersections and sit there when the light changes. I'm glad I'm on foot. Every building I pass down here among the warehouses seems to cry *Renovate me! Renovate me!* The first time I saw the word *Reno* in the real estate section I thought it meant the gambling resort. Language is leaving me behind.

I hit the corner of King and Spadina, walk north. This used to be where you came to get wholesale clothing, and it still is; but the old Jewish delis are disappearing, replaced by Chinese emporia, wicker furniture, cutwork tablecloths, bamboo wind chimes. Some of the street signs are subtitled in Chinese, multiculturalism on the march, others have *Fashion District* underneath the names. Everything is a district now. There never used to be districts.

It comes over me that I need a new dress for the opening. I've brought one with me, of course; I've already pressed it with my travel iron, clearing a corner of Jon's workbench for the ironing board, folding a towel across it. This dress is black, because black is the best thing for such occasions: a simple, sober black dress, like those of the women who play cellos in symphony orchestras. It doesn't do to outdress the clients.

But the thought of this dress is depressing to me now. Black attracts lint, and I've forgotten my clothes brush. I remember the Scotch tape ads from the forties: mummify your hand in inside-out Scotch tape, defuzz your clothing. I think of myself standing there in the gallery, surrounded by one-of-a-kind boutique wear and real pearls, widow-colored and linty where the Scotch tape has missed. There are other colors, pink for instance: pink is supposed to weaken your enemies, make them go soft on you, which must be why it's used for baby girls. It's a wonder the military hasn't got onto this. Pale-pink helmets, with rosettes, a whole battalion, onto the beachhead, over the top in pink. Now is the time for me to make the switch, I could use a little pink right now.

I cruise the cut-rate windows. Each one is like a shrine, lit up from within, its goddess on display, hand on hip or leg thrust out, the faces beige and inaccessible. Party dresses have come back, bows and flamenco ruffles, straplessness and crinolines, puffed sleeves like cloth marshmallows:

everything I thought was left behind forever. And miniskirts too, as bad as ever, but I draw the line at those. I didn't like them the last time around either: too much underpants. I can't wear the ruffled things, I'd look like a cabbage, and not the strapless ones either, not with my collarbone high and dry, my hen's-foot elbows sticking out. What I need is something vertical, maybe a little draped.

A SALE sign lures me in. The name of this store is The Sleek Boutique, though it's not really a boutique: it's crammed full of ends-of-line, low on overhead. It's crowded, which pleases me. Salesladies intimidate me, I don't like to be caught shopping. I riffle furtively through the SALE rack, bypassing sequins, angora roses, gold thread, grubby white leather, looking for something. What I'd like is to be transformed, which becomes less possible. Disguise is easier when you're young.

I take three things to the fitting room: salmon with dollar-sized white polka dots, electric blue with satin inserts, and, to be on the safe side, something in black that will do if all else fails. The salmon is what I'd really like, but can I handle the dots? I slip it on, zipper and hook it, turn this way and that, in front of the mirror which is as usual badly lit. If I ran a store like this I'd paint all the cubicles pink and put some money into the mirrors: whatever else women want to see, it's not themselves; not in their worst light anyway.

I crane my neck, trying to get the rear view. Maybe with different shoes, or different earrings? The price tag dangles, pointing to my rump. There are the polka dots, rolling across a broad expanse. It's amazing how much bigger you always look from the back. Maybe because there are fewer distracting features to break up the wide monotony of hill and plane.

As I turn back, I see my purse, lying on the floor where I put it, and after all these years I should know better. It's open. The cubicle wall only comes down to a foot above the floor, and back through the gap a noiseless arm is retreating, the hand clutching my wallet. The fingernails are painted Day-Glo green.

I bring my shoeless foot down hard on the wrist. There's

a shriek, some loud plural giggling: youth on the fast track, schoolgirls on the prowl. My wallet is dropped, the hand shoots back like a tentacle.

I jerk open the door. *Damn you, Cordelia!* I think.

But Cordelia is long gone.

9
.

*T*he school we are sent to is some distance away, past a cemetery, across a ravine, along a wide curving street lined with older houses. The name of it is Queen Mary Public School. In the mornings we walk across the freezing mud in our new winter overshoes, carrying our lunches in paper bags, and down through the remains of an orchard to the nearest paved road, where we wait for the school bus to come lurching toward us, up the hill and over the potholes. I wear my new snowsuit, my skirt wrapped around my legs and stuffed down into the bulgy legs of the snowpants, which whisk together as I walk. You can't wear pants to school, you have to wear skirts. I'm not used to this, or to sitting still at a desk.

We eat our lunches in the chilly dimly lit cellar of the schoolhouse, where we sit in supervised rows on long scarred wooden benches under a festoon of heating pipes. Most of the children go home for lunch, it's only the school bus ones that have to stay. We're issued small bottles of milk which we drink through straws stuck in through a hole in the cardboard bottle tops. These are my first drinking straws, and they amaze me.

The school building itself is old and tall, made of liver-colored brick, with high ceilings, long ominous wood-floored hallways, and radiators that are either on full blast or not at all, so that we're either shivering with cold or too hot. The windows are high and thin and many-paned, and decorated with cutouts made of construction paper; right now there are snowflakes, for winter. There's a front door which is never used by children. At the back are two grandiose entranceways with carvings around them and ornate insets above the doors, inscribed in curvy, solemn lettering:

GIRLS and BOYS. When the teacher in the yard rings her brass handbell we have to line up in twos by classrooms, girls in one line, boys in another, and file into our separate doors. The girls hold hands; the boys don't. If you go in the wrong door you get the strap, or so everyone says.

I am very curious about the BOYS door. How is going in through a door different if you're a boy? What's in there that merits the strap, just for seeing it? My brother says there's nothing special about the stairs inside, they're plain ordinary stairs. The boys don't have a separate classroom, they're in with us. They go in the BOYS door and end up in the same place we do. I can see the point of the boys' washroom, because they pee differently, and also the boys' yard, because of all the kicking and punching that goes on among them. But the door baffles me. I would like to have a look inside.

Just as there are separate doors for boys and girls, there are also separate parts of the schoolyard. At the front, outside the teachers' entrance, is a dirt field covered with cinders, the boys' playing field. At the side of the school facing away from the street is a hill, with wooden steps going up it and eroded runnels worn down the side, and a few stunted evergreens on top. By custom this is reserved for the girls, and the older ones stand around up there in groups of three or four, their heads bent inward, whispering, although boys sometimes make charges up the hill, yelling and waving their arms. The cement-paved area outside the BOYS and GIRLS is common territory, since the boys have to cross it in order to go in their door.

Lining up is the only time I see my brother at school. At home we've rigged up a walkie-talkie with two tin cans and a piece of string, which runs between our two bedroom windows and doesn't work very well. We push messages under each other's doors, written in the cryptic language of the aliens, which is filled with x's and z's and must be decoded. We nudge and kick each other under the table, keeping our faces straight above the tablecloth; sometimes we tie our shoelaces together, for signaling. These are my main communications with my brother now, these raspy tin can words, sentences without vowels, the Morse of feet.

But in the daytime I lose sight of him as soon as we go out the door. He's up ahead, throwing snowballs; and on the bus he's at the back, in a noisy whirlpool of older boys. After school, after he's gone through the fights that are required of any new boy at any school, he's off helping to wage war on the boys from the Catholic school nearby. It's called Our Lady of Perpetual Help, but the boys from our school have renamed it Our Lady of Perpetual Hell. It's said that the boys from this Catholic school are very tough and that they conceal rocks inside their snowballs.

I know better than to speak to my brother during these times, or to call his or any boy's attention to me. Boys get teased for having younger sisters, or sisters of any kind, or mothers; it's like having new clothes. When he gets anything new my brother dirties it as soon as possible, to avoid having it noticed; and if he has to go anywhere with me and my mother, he walks ahead of us or crosses to the other side of the street. If he's teased about me, he will have to fight some more. For me to contact him, or even to call him by name, would be disloyal. I understand these things, and do my best.

So I am left to the girls, real girls at last, in the flesh. But I'm not used to girls, or familiar with their customs. I feel awkward around them, I don't know what to say. I know the unspoken rules of boys, but with girls I sense that I am always on the verge of some unforeseen, calamitous blunder.

A girl called Carol Campbell makes friends with me. In a way she has to, because she's the only school bus girl in my grade. The children who come on the school bus, who eat their lunches in the cellar instead of going home, are considered a little foreign, and are in danger of finding themselves without a partner when the bell rings and it's time to line up. So Carol sits beside me on the school bus, holds my hand in line, whispers to me, eats her lunch beside me on the wooden bench in the cellar.

Carol lives in one of the older houses on the other side of the abandoned orchard, closer to the school, a yellow brick house with two stories and green-painted shutters framing the windows. She's a stubby girl with a frequent laugh. She tells me her hair is honey-blond, that her haircut is called a pageboy, that she has to go to the hairdresser's every two months to get it done. I haven't known there are such things as pageboys and hairdressers. My mother doesn't go to the hairdresser's. She wears her hair long, pinned up at the sides, like the women in wartime posters, and my own hair has never been cut.

Carol and her younger sister have matching outfits for Sundays: fitted brown tweed coats with velvet collars, round brown velvet hats with an elastic under the chin to hold them on. They have brown gloves and little brown purses. She tells me all this. They are Anglicans. Carol asks me what church I go to, and I say I don't know. In fact we never go to church.

After school Carol and I walk home, not the way the school bus goes in the morning but a different way, along back streets and across a decaying wooden footbridge over the ravine. We've been told not to do this alone, and not to go down into the ravine by ourselves. There might be men down there, is what Carol says. These are not ordinary men but the other kind, the shadowy, nameless kind who do things to you. She smiles and whispers when saying *men*, as if they are a special, thrilling joke. We cross the bridge lightly, avoiding the places where the boards have rotted through, on the lookout for men.

Carol invites me to her house after school, where she shows me her cupboard with all her clothes hanging in it. She has a lot of dresses and skirts; she even has a dressing gown, with fuzzy slippers to match. I have never seen so many girls' clothes in one place.

She lets me look at her living room from the doorway, although we aren't allowed to go into it. She herself can't go in except to practice the piano. The living room has a sofa and two chairs and matching drapes, all of a flowered rose and beige material Carol says is chintz. She pronounces this word with awe, as if it's the name of something sacred, and I repeat it silently to myself: *chintz*. It sounds like the name of a kind of crayfish, or of one of the aliens on my brother's distant planet.

Carol tells me that her piano teacher hits her fingers with

a ruler if she gets a note wrong, and that her mother spanks her with the back of a hairbrush or else a slipper. When she's really in for it she has to wait until her father comes home and whacks her with his belt, right on the bare bum. All of these things are secrets. She says her mother sings on a radio program, under a different name, and we do overhear her mother practicing scales in the living room, in a loud quavery voice. She says her father takes some of his teeth out at night and puts them into a glass of water beside his bed. She shows me the glass, although the teeth aren't in it. There seems to be nothing she won't tell.

She tells me which boys at school are in love with her, making me promise not to tell. She asks me which ones are in love with me. I've never thought about this before, but I can see that some sort of an answer is expected. I say I'm not sure.

Carol comes to my house and takes it all in—the unpainted walls, the wires dangling from the ceilings, the unfinished floors, the army cots—with incredulous glee. "This is where you *sleep?*" she says. "This is where you *eat?* These are your *clothes?*" Most of my clothes, which are not many in number, are pants and jersey tops. I have two dresses, one for summer and one for winter, and a tunic and a wool skirt, for school. I begin to suspect that more may be required.

Carol tells everyone at school that our family sleeps on the floor. She gives the impression that we do this on purpose, because we're from outside the city; that it's a belief of ours. She's disappointed when our real beds arrive from storage, four-legged and with mattresses, like everybody else's. She puts it around that I don't know what church I go to, and that we eat off a card table. She doesn't repeat these items with scorn, but as exotic specialties. I am, after all, her lining-up partner, and she wants me to be marveled at. More accurate: she wants herself to be marveled at, for revealing such wonders. It's as if she's reporting on the antics of some primitive tribe: true, but incredible.

10

On Saturday we take Carol Campbell to the building. When we walk into it she says, wrinkling up her nose, "Is this where your father *works?*" We show her the snakes and the turtles; she makes a noise that sounds like "Ew," and says she wouldn't want to touch them. I'm surprised by this; I've been discouraged from having such feelings for so long that I no longer have them. Neither does Stephen. There's not much we won't touch, given the chance.

I think Carol Campbell is a sissy. At the same time I find myself being a little proud of her delicacy. My brother looks at her in an odd way: with contempt, true, and if I myself said such a thing he would make fun of me. But there's an undertone, like an invisible nod, as if something he wants to suspect has come true after all.

By rights he should ignore her after this, but he tries her out on the jars of lizards and ox eyeballs. "Ew," she says. "What if they put one down your *back?*" My brother says how would she like some for dinner? He makes chewing and slurping noises.

"Ew," says Carol, screwing up her face and wriggling all over. I can't pretend to be shocked and disgusted too: my brother wouldn't be convinced. Neither can I join in the game of making up revolting foods, such as toadburgers and leech chewing gum, although if we were alone or with other boys I would do it without a second thought. So I say nothing.

After we get back from the building I go to Carol's house again. She asks me if I want to see her mother's new twin set. I don't know what this is, but it sounds intriguing, so I say yes. She takes me stealthily into her mother's bedroom, saying that she'll really get it if we're caught, and shows me the twin set, folded on a shelf. The twin set is just two sweaters, both the same color, one with buttons down the front, the other without. I've already seen Mrs. Campbell wearing a different twin set, a beige one, her breasts pronging out, the buttoned sweater draped over her shoulders like a cape. So this is all twin sets are. I'm disappointed, because I was expecting something to do with twins.

Carol's mother and father don't sleep in one big bed, the way mine do. Instead they sleep in two little beds, exactly alike, with matching pink chenille bedspreads and matching night tables. These beds are called twin beds, which makes more sense to me than the twin set. Still, it's strange to think of Mr. and Mrs. Campbell lying in them at night, with different heads—his with a mustache, hers without—but nevertheless twinlike, identical, under the sheets and blankets. It's the matching bedspreads, the night tables, the lamps, the bureaus, the doubleness of everything in their room, that gives me this impression. My own parents' room is less symmetrical, and also less neat.

Carol says her mother wears rubber gloves while washing the dishes. She shows me the rubber gloves and a spray thing attached to the water tap. She turns on the tap and sprays the inside of the sink, and part of the floor by accident, until Mrs. Campbell comes in, wearing her beige twin set and frowning, and says hadn't we better go upstairs to play. Possibly she isn't frowning. She has a mouth that turns slightly down even when she's smiling, so it's hard to tell whether she's pleased or not. Her hair is the same color as Carol's, but done in a cold wave all over her head. It's Carol who points out that this is a cold wave. A cold wave has nothing to do with water. It's like doll hair, very tidy and arranged, as if sewn into place.

Carol is more and more gratified the more bewildered I am. "You didn't know what a *cold wave* is?" she says, delighted. She's eager to explain things to me, name them, display them. She shows me around her house as if it's a museum, as if she personally has collected everything in it. Standing in the downstairs hall, where there is a coat tree—

"You've never seen a *coat tree?*"—she says I am her best friend.

Carol has another best friend, who is sometimes her best friend and sometimes not. Her name is Grace Smeath. Carol points her out to me, on the bus, the same way she's pointed out the twin set and the coat tree: as an object to be admired.

Grace Smeath is a year older and in the next grade up. At school she plays with the other girls in her class. But after school and on Saturdays she plays with Carol. There are no girls in her class on our side of the ravine.

Grace lives in a two-story shoebox-shaped red brick house with a front porch that has two thick round white pillars holding it up. She's taller than Carol, with dark thick coarse hair done into two braids. Her skin is extremely pale, like a body under a bathing suit, but covered with freckles. She wears glasses. Usually she wears a gray skirt with two straps over the shoulders, and a red sweater pebbled with little balls of wool. Her clothes smell faintly of the Smeaths' house, a mixture of scouring powder and cooked turnips and slightly rancid laundry, and the earth under porches. I think she is beautiful.

On Saturdays I no longer go to the building. Instead I play with Carol and Grace. Because it's winter, we play mostly inside. Playing with girls is different and at first I feel strange as I do it, self-conscious, as if I'm only doing an imitation of a girl. But I soon get more used to it.

The things we play are mostly Grace's ideas, because if we try to play anything she doesn't like she says she has a headache and goes home, or else tells us to go home. She never raises her voice, gets angry, or cries; she is quietly reproachful, as if her headache is our fault. Because we want to play with her more than she wants to play with us, she gets her way in everything.

We color in Grace's movie star coloring books, which show the movie stars in different outfits, doing different things: walking their dogs, going sailing in sailor suits, swirling around in evening dresses at parties. Grace's favorite movie star is Esther Williams. I have no favorite movie

star—I've never been to a movie—but I say mine is Veronica Lake, because I like the name. The Veronica Lake
book is paper doll cutouts, with Veronica Lake in her bath-
ing-suit and dozens of outfits you can stick onto her with
tabs that fold around her neck. Grace won't let us cut out
these outfits, although we can put them on and take them
off once she's done it, but we're allowed to work away at
her coloring books as long as we stay inside the lines. She
likes to get these books all colored in. She tells us what
colors to use, on which parts. I know what my brother
would do—green skin for Esther, with beetle antennae, and
hairy legs for Veronica, eight of them—but I refrain from
doing it. Anyway I like the clothes.

We play school. Grace has a couple of chairs and a
wooden table in her cellar, and a small blackboard and chalk.
These are set up underneath the indoors clothesline where
the Smeath underwear is hung up to dry when it rains or
snows. The cellar isn't a finished cellar: the floor is cement,
the pillars holding up the house are brick, the water pipes
and wires are showing, and the air smells of coal dust because the coal bin is right beside the blackboard.

Grace is always the teacher, Carol and I the students.
We have to do spelling tests and sums in arithmetic; it's like
real school, but worse, because we never get to draw pictures. We can't pretend to be bad, because Grace doesn't
like disorder.

Or we sit on the floor in Grace's room with piles of old
Eaton's Catalogues. I've seen lots of *Eaton's Catalogues* before:
up north they're hung in outhouses for use as toilet paper.
Eaton's Catalogues remind me of the stench of such outhouses,
the buzzing of the flies down the hole underneath, the box
of lime and the wooden paddle for dumping the lime down,
onto the piles of old and recent droppings, of all shapes and
colors of brown. But here we treat these catalogues with
reverence. We cut the small colored figures out of them and
paste them into scrapbooks. Then we cut out other things—
cookware, furniture—and paste them around the figures.
The figures themselves are always women. We call them
"my lady." "My lady's going to have this refrigerator," we

say. "My lady's getting this rug." "This is my lady's umbrella."

Grace and Carol look at each other's scrapbook pages and say, "Oh, yours is so good. Mine's no good. Mine's *awful*." They say this every time we play the scrapbook game. Their voices are wheedling and false; I can tell they don't mean it, each one thinks her own lady on her own page is good. But it's the thing you have to say, so I begin to say it too.

I find this game tiring—it's the weight, the accumulation of all these objects, these possessions that would have to be taken care of, packed, stuffed into cars, unpacked. I know a lot about moving house. But Carol and Grace have never moved anywhere. Their ladies live in a single house each and have always lived there. They can add more and more, stuff the pages of their scrapbooks with dining room suites, beds, stacks of towels, one set of dishes after another, and think nothing of it.

I begin to want things I've never wanted before: braids, a dressing gown, a purse of my own. Something is unfolding, being revealed to me. I see that there's a whole world of girls and their doings that has been unknown to me, and that I can be part of it without making any effort at all. I don't have to keep up with anyone, run as fast, aim as well, make loud explosive noises, decode messages, die on cue. I don't have to think about whether I've done these things well, as well as a boy. All I have to do is sit on the floor and cut frying pans out of the *Eaton's Catalogue* with embroidery scissors, and say I've done it badly. Partly this is a relief.

For Christmas, Carol gives me some Friendship's Garden bath salts and Grace gives me a coloring book of Virginia Mayo. I open their presents before anyone else's.

I also get a photo album, to go with my camera. The pages and covers are black, tied together with something that looks like a big black shoelace; there's a package of black triangles with glue on them to stick the photos in with. So far I have taken only one roll of film with my camera. I think about what each picture will look like as I press the button. I don't want to waste any. When the pictures come back from being developed, the negatives come too. I hold them up to the light: everything that's white in the real picture is black in the negative. Snow for instance is black, and people's eyeballs and teeth.

I stick my photos into the album with the black triangles. Some of the pictures are of my brother, making threatening gestures with snowballs. Some are of Carol, some of Grace. There's only that one picture of me, standing in front of the motel door with 9 on it, long ago, a month ago. Already that child seems much younger, poorer, farther away, a shrunken, ignorant version of myself.

Another thing I get for Christmas is a red plastic purse, oval in shape, with a gold-colored clasp and a handle at the top end. It's soft and pliable inside the house, but hardens outside in the cold, so that things rattle in it. I keep my allowance in it, five cents a week.

By this time we have a living room floor, hardwood, waxed by my mother down on her knees, polished with a long-handled weighted brush that she pushes back and forth, making a sound like waves. The living room has been painted, the fixtures installed, the baseboards added on.

There are even curtains; drapes, they're called. The public, visible parts of the house have been finished first.

Our bedrooms remain in a rawer state. The windows there do not yet have drapes. Lying in bed at night, I can look out of my window at the snow falling, illuminated by the light from my brother's bedroom window beside mine.

It's the darkest time of the year. Even in the daytime it seems dark; and at night, when the lights are on, this darkness pervades everything, like a fog. Outside there are only a few streetlights, and they're far apart and not very bright. The lamps in people's houses cast a yellowish light, not cold and greenish but a buttery dim yellow with a tinge of brown. The colors of things in houses have darkness mixed into them: maroon, mushroom beige, a muted green, a dusty rose. These colors look a little dirty, like the squares in a paint box when you forget to rinse the brush.

We have a maroon chesterfield which has come out of storage, with an oriental-style maroon and purple rug in front of it. We have a tri-light floor lamp. The air in the evening lamplight is coagulated, like a custard thickening; heavier sediments of light collect in the corners of the living room. The drapes are kept closed at night, folds and folds of cloth drawn against the winter, hoarding the dim heavy light, keeping it in.

In this light I spread the evening paper out on the polished hardwood floor and rest on my knees and elbows, reading the comics. In the comics there are people with round holes for eyes, others who can hypnotize you instantly, others with secret identities, others who can stretch their faces into any shape at all. Around me is the scent of newsprint and floor wax, the bureau drawer smell of my itchy stockings mingled with that of grimy knees, the scratchy hot smell of wool plaid and the cat box aroma of cotton underpants. Behind me the radio plays square dance music from the Maritimes, Don Messer and His Islanders, in preparation for the six o'clock news. The radio is of dark varnished wood with a single green eye that moves along the dial as you turn the knob. Between the stations this eye makes eerie noises from outer space. Radio waves, says Stephen.

. . .

Often, now, Grace Smeath asks me over to her house after school without asking Carol. She tells Carol there's a reason why she isn't invited: it's because of her mother. Her mother is tired, so Grace can only have one best friend over that day.

Grace's mother has a bad heart. Grace doesn't treat this as a secret, as Carol would. She says it unemotionally, politely, as if requesting you to wipe your feet on the mat; but also smugly, as if she has something, some privilege or moral superiority that the two of us don't share. It's the attitude she takes toward the rubber plant that stands on the landing halfway up her stairs. This is the only plant in Grace's house, and we aren't allowed to touch it. It's very old and has to be wiped off leaf by leaf with milk. Mrs. Smeath's bad heart is like that. It's because of this heart that we have to tiptoe, walk quietly, stifle our laughter, do what Grace says. Bad hearts have their uses; even I can see that.

Every afternoon Mrs. Smeath has to take a rest. She does this, not in her bedroom, but on the chesterfield in the living room, stretched out with her shoes off and a knitted afghan covering her. That is how she is always to be found when we go there to play after school. We come in through the side door, up the steps to the kitchen, trying to be as quiet as possible, and into the dining room as far as the double French doors, where we peer in through the glass panes, trying to see whether her eyes are open or closed. She's never asleep. But there's always the possibility—put into our heads by Grace, in that same factual way—that on any given day she may be dead.

Mrs. Smeath is not like Mrs. Campbell. For instance, she has no twin sets, and views them with contempt. I know this because once, when Carol was bragging about her mother's twin sets, Mrs. Smeath said, "Is that so," not as a question but as a way of making Carol shut up. She doesn't wear lipstick or face powder, even when she goes out. She has big bones, square teeth with little gaps between them so that you can see each tooth distinctly, skin that looks rubbed raw as if scrubbed with a potato brush. Her face is

rounded and bland, with that white skin of Grace's, though
without the freckles. She wears glasses like Grace too, but
hers have steel rims instead of brown ones. Her hair is parted
down the middle and graying at the temples, braided and
wound over her head into a flat hair crown crisscrossed with
hairpins.

She wears print housedresses, not only in the mornings
but most of the time. Over the dresses she wears bibbed
aprons that sag at the bosom and make it look as if she
doesn't have two breasts but only one, a single breast that
goes all the way across her front and continues down until
it joins her waist. She wears lisle stockings with seams,
which make her legs look stuffed and sewn up the backs.
She wears brown Oxfords. Sometimes, instead of the stockings, she has thin cotton socks, above which her legs rise
white and sparsely haired, like a woman's mustache. She
has a mustache too, though not very much of one, just a
sprinkling of hairs around the corners of the mouth. She
smiles a lot, with her lips closed over her large teeth; but,
like Grace, she does not laugh.

She has big hands, knuckly and red from the wash.
There's a lot of wash, because Grace has two younger sisters
who get her skirts and blouses and also her underpants
passed down to them. I'm used to getting my brother's
jerseys, but not his underpants. It's these underpants, thin
and gray with use, that hang dripping on the line over our
heads as we sit in Grace's cellar pretending to be school-
children.

Before Valentine's Day we have to cut out hearts of red
construction paper at school and decorate them with pieces
of paper doily to stick on the tall thin windows. While I
am cutting mine I think about Mrs. Smeath's bad heart.
What exactly is wrong with it? I picture it hidden, underneath her woolen afghan and the billow of her apron bib,
pumping in the thick fleshy darkness of the inside of her
body: something taboo, intimate. It would be red, but with
a reddish-black patch on it, like rot in an apple or a bruise.
It hurts when I think about it. A little sharp wince of pain
goes through me, as it did when I watched my brother cut

his finger once on a piece of glass. But the bad heart is also compelling. It's a curiosity, a deformity. A horrible treasure.

Day after day I press my nose against the glass of the French doors, trying to see if Mrs. Smeath is still alive. This is how I will see her forever: lying unmoving, like something in a museum, with her head on the antimacassar pinned to the arm of the chesterfield, a bed pillow under her neck, the rubber plant on the landing visible behind her, turning her head to look at us, her scrubbed face, without her glasses, white and strangely luminous in the dim space, like a phosphorescent mushroom. She is ten years younger than I am now. Why do I hate her so much? Why do I care, in any way, what went on in her head?

12

*T*he snow erodes, leaving the potholes in the roads near our house filled with muddy water. Thin bubbles of ice form across these puddles overnight; we shatter them with the heels of our boots. Icicles crash down from the eaves of roofs, and we pick them up and lick them like Popsicles. We wear our mittens dangling. On the lawns, as we walk home from school, we can see damp pieces of paper under the hedges, old dog turds, crocuses poking up through the grainy, soot-colored snow. The gutters run with brownish water; the wooden bridge over the ravine is slippery and soft and has regained its smell of rot.

Our house looks like something left over from the war: all around it spreads rubble, devastation. My parents stand in their backyard, hands on their hips, looking out over the expanse of raw mud, planning their garden. Already clumps of couch grass are beginning to thrust up. Couch grass can grow in anything, my father says. He also says that the contractor, the same one that flew the coop, took the dense clay from where our cellar went in and spread it around the house, over what should have been the topsoil. "An idiot as well as a crook," says my father.

My brother watches the water level in the giant hole next door, waiting for the hole to dry up so he can use it for a bunker. He would like to roof it over, with sticks and old planks, but he knows this isn't possible because the hole is too big and also he wouldn't be allowed. Instead he plans to dig a tunnel down there, into the side of the hole, and to get up and down to it by a rope ladder. He has no rope ladder, but he says he'll make one, if he can get some rope.

He and the other boys run around in the mud; large extra feet of clay stick to the soles of their boots, leaving tracks like monsters. They crouch behind the trees in the old orchard, sniping at one another, shouting:

"You're dead!"

"I am not!"

"You're dead!"

At other times they crowd into my brother's room, lying on their stomachs on his bed or on the floor, reading his huge piles of comic books. I sometimes do this too, wallowing among the pages of colored paper, surrounded by the fuggy scent of boys. Boys don't smell the same as girls. They have a pungent, leathery, underneath smell, like old rope, like damp dogs. We keep the door closed because my mother doesn't approve of comics. The reading of comics is done in reverential silence, with now and then a few monosyllables of trade.

Comic books are what my brother is collecting now. He's always collected something. Once it was milk bottle tops, from dozens of dairies; he carried sheafs of them around in his pockets, held together with rubber bands, and stood them up against walls and threw other milk bottle tops at them to win more. Then it was pop bottle tops, then cigarette cards, then sightings of license plates from different provinces and states. There is no way of winning comic books. Instead you trade them, one good one for three or four of lesser value.

At school we make Easter eggs out of construction paper, pink and purple and blue, and stick them onto the windows. After that it's tulips, and soon there are real tulips. It seems to be a rule that the paper things always appear before the real ones.

Grace produces a long skipping rope, and she and Carol teach me how to turn it. As we turn, we chant, in monotonous minor-key voices:

Salome was a dancer, she did the hoochie kootch;
And when she did the hoochie kootch, she didn't wear very
*   mooch.*

Grace puts one hand on her head, the other on her hip, and wiggles her bum. She does this with perfect decorum;

she's wearing her pleated skirt with the straps over the shoulders. I know Salome is supposed to be more like the movie stars in our paper doll books. I think of gauzy skirts, high heels with stars on the toes, hats covered with fruit and feathers, lifted eyebrows, pencil-thin; gaiety and excess. But Grace in her pleats and woolen straps can wipe out all that.

Our other game is ball. We play it against the side wall of Carol's house. We throw our rubber balls up against the wall and catch them as they come down, clapping and twirling in time to the chant:

> *Ordinary, moving, laughing, talking, one hand, the other hand, one foot, the other foot, clap front, clap back, back and front, front and back, tweedle, twydle, curtsy, salute, and roundabout.*

For *roundabout* you throw the ball and twirl all the way around before catching it. This is the hardest thing, harder even than the left hand.

The sun lasts longer and longer and goes down golden-red. The willow trees drop yellow catkins over the bridge; the maple keys fall twirling to the sidewalks and we split the sticky seed part and pinch the keys onto our noses. The air is warm, humid, like invisible mist. We wear cotton dresses to school, and cardigans, which we take off walking home. The old trees in the orchard are in flower, white and pink; we climb up into them, breathing in their hand lotion smells, or we sit in the grass making chains of dandelions. We unbraid Grace's hair, which falls down her back in coarse brown ripples, and wind the chains around her head like a crown. "You're a princess," says Carol, stroking the hair. I take a picture of Grace and stick it into my photo album. There she sits, smiling primly, festooned with blossoms.

The field across from Carol's house is sprouting new houses, and in the evenings groups of children, boys and girls alike, clamber about inside them, in the fresh wood smell of shavings, walking through walls that don't yet exist, climbing ladders where there will soon be stairs. This is forbidden.

Carol won't climb to the higher floors because she's afraid. Grace won't climb either, but not because of fear: she doesn't want anybody, any boy, to see her underpants. No girl can wear slacks to school, but Grace never wears them at any time. So the two of them stay on the ground floor while I climb, up and along the beams with no ceiling covering them, up again to the attic. I sit on the top floor where there is no floor, among the rafters in this house of air, basking in the red-gold sunset, looking down. I don't think about falling. I am not yet afraid of heights.

One day someone appears in the schoolyard with a bag of marbles, and the next day everyone has them. The boys desert the boys' playground and throng into the common playground in front of the BOYS and GIRLS doors; they need to come to this side of the playground, because marbles have to be played on a smooth surface and the boys' yard is all cinders.

For marbles you're either the person setting up the target or the person shooting. To shoot you kneel down, sight, and roll your marble at the target marble like a bowling ball. If you hit it you keep it, and your own marble too. If you miss, you lose your marble. If you're setting up, you sit on the cement with your legs spread open and put a marble on a crack in front of you. It can be an ordinary marble, but these don't get many shooters, unless you offer two for one. Usually the targets are more valuable: cat's eyes, clear glass with a bloom of colored petals in the center, red or yellow or green or blue; puries, flawless like colored water or sapphires or rubies; waterbabies, with undersea filaments of color suspended in them; metal bowlies; aggies, like marbles only bigger. These exotics are passed from winner to winner. It's cheating to buy them; they have to be won.

Those with target marbles call out the names of their wares: *purie, purie, bowlie, bowlie,* the two-syllable words drawn out into a singsong, the voice descending, the way you call dogs, or children when they're lost. These cries are mournful, although they aren't meant to be. I sit that way myself, the cold marbles rolling in between my legs, gathering in my outspread skirt, calling out *cat's eye, cat's eye*, in a regretful tone, feeling nothing but avarice and a pleasurable terror.

The cat's eyes are my favorites. If I win a new one I wait until I'm by myself, then take it out and examine it, turning it over and over in the light. The cat's eyes really are like eyes, but not the eyes of cats. They're the eyes of something that isn't known but exists anyway; like the green eye of the radio; like the eyes of aliens from a distant planet. My favorite one is blue. I put it into my red plastic purse to keep it safe. I risk my other cat's eyes to be shot at, but not this one.

I don't collect many marbles because I'm not a very good shot. My brother is deadly. He takes five common marbles to school with him in a blue Crown Royal Whisky bag and comes back with the bag and his pockets bulging. He keeps his winnings in screw-top Crown preserving jars, donated by my mother, which he lines up on his desk. He never talks about his skill though. He just lines up the jars.

One Saturday afternoon he puts all his best marbles—his puries, his waterbabies and cat's eyes, his gems and wonders—into a single jar. He takes it down into the ravine somewhere, in under the wooden bridge, and buries it. Then he makes an elaborate treasure map of where it's buried, puts it in another jar, and buries that one too. He tells me he's done these things but he doesn't say why, or where the jars are buried.

13

*T*he raw house and its lawn of mud and the mountain of earth beside it recede behind us; I watch them out of the back window of the car, from where I sit jammed in among the boxes of food, the sleeping bags and raincoats. I'm wearing a blue-striped jersey of my brother's, a worn pair of corduroy pants. Grace and Carol stand under the apple trees, in their skirts, waving, disappearing. They still have to go to school; I don't. I envy them. Already the tarry, rubbery travel smell is wrapping itself around me, but I don't welcome it. I'm being wrenched away from my new life, the life of girls.

I settle back into the familiar perspective, the backs of heads, the ears, and past them the white line of the highway. We drive up through the meadowy farmlands, with their silos and elms and their smell of cut hay. The broad-leafed trees become smaller, there are more pines, the air cools, the sky turns an icier blue: we're heading away from spring. We hit the first ridges of granite, the first lakes; there's snow in the shadows. I sit forward, leaning my arms on the back of the front seat. I feel like a dog, ears pricked and sniffing.

The north smells different from the city: clearer, thinner. You can see farther. A sawmill, a hill of sawdust, the teepee shape of a sawdust burner; the smokestacks of the copper smelters, the rocks around them bare of trees, burnt-looking, the heaps of blackened slag: I've forgotten about these things all winter, but here they are again, and when I see them I remember them, I know them, I greet them as if they are home.

Men stand on corners, outside general stores, outside small banks, outside beer parlors with gray asphalt shingles on the walls. They have their hands in their windbreaker pockets. Some have dark, Indian-looking faces, others are merely tanned. They walk differently from men in the south, slower, more considering; they say less and their words are farther apart. My father jingles his keys and the change in his pockets while he talks with them. They talk about water levels, the dryness of the forest, how the fish are biting. "Chewing the fat," he calls it. He comes back to the car with a brown paper bag of groceries and packs it in behind my feet.

My brother and I stand at the end of a ramshackle dock beside a long blue craggy lake. It's evening, with a melon-colored sunset, loons calling in the distance, the drawn-out rising note that sounds like wolves. We're fishing. There are mosquitoes, but I'm used to them, I hardly bother to slap them. The fishing goes on without commentary: a cast, the plop of the lure, the sound of reeling in. We watch the lure to see if anything is following it. If there's a fish, we'll do our best to net it, step on it to hold it down, whack it over the head, stick a knife in back of its eyes. I do the stepping, my brother does the whacking, the sticking. Despite his silence he is poised, alert, the corners of his mouth tensed. I wonder if my eyes are gleaming like his, like some animal's, in the pink dusk.

We're living in an abandoned logging camp. We sleep on our air mattresses, in our sleeping bags, in the wooden bunks where the loggers used to sleep. Already the logging camp has a feeling of great age about it, although it's only been empty for two years. Some of the loggers have left inscriptions, their names, their initials, intertwined hearts, short dirty words and crude pictures of women, carved or penciled in the wood of the two-by-fours of the walls. I find an old tin of maple syrup, the lid rusted shut, but when Stephen and I get it open the syrup is moldy. I think of this syrup tin as an ancient artifact, like something dug up out of a tomb.

We prowl around among the trees, looking for bones, for hummocks in the earth that could mark diggings, the outlines of buildings, turning over logs and rocks to see what's underneath them. We would like to discover a lost civilization. We find a beetle, many small yellow and white roots, a toad. Nothing human.

Our father has shed his city clothing, turned back into himself. He has on his old jacket again, his baggy pants, his squashed felt hat with the fishing flies stuck into it. He tromps through the woods in his heavy lace-up bacongreased workboots, with his ax in its leather sheath, us in his wake. There's an outbreak of forest tent caterpillars, the biggest in years: this is what fills him with glee, makes his eyes of a gnome shine in his head like blue-gray buttons. The caterpillars are everywhere in the woods, striped and bristly. They dangle from the branches on threads of silk, forming a hanging curtain you have to brush out of the way; they river along the ground like a rug come to life, they cross roads, turning to greasy mush under the tires of the logging trucks. The trees around are denuded, as if they've been burnt; webbing sheathes their trunks.

"Remember this," our father says. "This is a classic infestation. You won't see an infestation like this again for a long time." It's the way I've heard people talk about forest fires, or the war: respect and wonderment mixed in with the sense of catastrophe.

My brother stands still and lets the caterpillars wash up over his feet, down on the other side of him, like a wave. "When you were a baby I caught you trying to eat those," says our mother. "You had a whole handful, you were squashing them around. You were just about to pop them into your mouth when I caught you."

"In some respects they're like one animal," our father says. He sits at the table made of planks left over from the loggers, eating fried Spam and potatoes. All during this meal he talks about the caterpillars: their numbers, their ingenuity, the various methods of defeating them. It's wrong to spray them with DDT and other insecticides, he says. That merely poisons the birds which are their natural enemies, whereas they themselves, being insects and therefore resourceful, more resourceful than humans in fact, will merely develop a resistance to the sprays, so all you get is dead birds and more caterpillars later on. He's working on something else: a growth hormone that will throw their systems out of whack and make them pupate before they're supposed to. Premature aging. But in the end, if he were a betting man, he says, he'd put his money on the insects. The insects are older than people, they have more experience at surviving, and there are a lot more of them than there are of us. Anyway, we'll probably blow ourselves sky-high before the end of the century, given the atom bomb and the way things are going. The future belongs to the insects.

"Cockroaches," my father says. "That's all that'll be left, once they get through with it." He says this jovially, skewering a potato.

I sit eating my fried Spam, drinking my milk mixed from powder. What I relish the most are the lumps that float on the top. I'm thinking about Carol and Grace, my two best friends. At the same time I can't remember exactly what they look like. Did I really sit on the floor of Grace's bedroom, on her braided bedside rug, cutting out pictures of frying pans and washing machines from the *Eaton's Catalogue* and pasting them into a scrapbook? Already it seems implausible, and yet I know I did it.

Out behind the logging camp is a huge cutover where they've taken off the trees. Only the roots and stumps remain. There's a lot of sand out there. The blueberry bushes have come up, as they do after a fire: first the fireweed, then the blueberries. We pick the berries into tin cups. Our mother pays us a cent a cup. She makes blueberry puddings, blueberry sauce, canned blueberries, boiling the jars in a large canning kettle over the outdoor fire.

The sun beats down, the heat comes wavering up off the sand. I wear a cotton kerchief on my head, folded into a triangle and tied behind my ears, the front of it damp with sweat. Around us is the drone of flies. I try to listen through it, behind it, for the sound of bears. I'm not sure what they would sound like, but I know that bears like blueberries, and they're unpredictable. They may run away. Or they may come after you. If they come you should lie down and pretend to be dead. This is what my brother says. Then they might go away, he says; or they might scoop out your innards. I've seen fish guts, I can picture this. My brother finds a bear turd, blue and speckled and human-looking, and pokes a stick into it to see how fresh it is.

In the afternoons, when it's too hot to pick berries, we swim in the lake, in the same water the fish come out of. I'm not supposed to go over my depth. The water is gelid, murky; down there, past where the sand drops away and it's deep, there are old rocks covered with slime, sunken logs, crayfish, leeches, huge pike with undershot jaws. Stephen tells me fish can smell. He says they'll smell us, and keep out of the way.

We sit on the shore, on rocks that poke up from the narrow beach, and toss bits of bread into the water, seeing what we can entice: minnows, a few perch. We search for flat stones and skip them, or we practice burping at will, or we put our mouths against the insides of our arms and blow to make farting noises, or we fill our mouths with water and see how far we can spit. In these contests I am not the winner, I am more like an audience; though my brother does not brag, and would probably do the same things, by himself, if I weren't there.

Sometimes he writes in pee, on the thin edge of sand or on the surface of the water. He does this methodically, as if it's important to do it well, the pee arching delicately out from the front of his swim trunks, from his hand and its extra finger, the writing angular, like his real writing, and ending always with a period. He doesn't write his name, or dirty words, as other boys do, as I know from snowbanks. Instead he writes: MARS. Or, if he's feeling up to it, something longer: JUPITER. By the end of the summer he has done the whole solar system, three times over, in pee.

It's the middle of September; the leaves are already turning, dark red, bright yellow. At night when I walk to the outhouse, in the dark with no flashlight because I can see better that way, the stars are sharp and crystalline and my breath goes before me. I see my parents, in through the window, sitting beside the kerosene lamp, and they are like a faraway picture with a frame of blackness. It's disquieting to look at them, in through the window, and know that they don't know I can see them. It's as if I don't exist; or as if they don't.

When we come back down from the north it's like coming down from a mountain. We descend through layers of clarity, of coolness and uncluttered light, down past the last granite outcrop, the last small raggedy-edged lake, into the thicker air, the dampness and warm heaviness, the cricket noises and weedy meadow smells of the south.

We reach our house in the afternoon. It looks strange, different, as if enchanted. Thistles and goldenrod have grown up around it, like a thorny hedge, out of the mud. The huge hole and the mountain of earth next door have vanished, and in their place is a new house. How has this happened? I wasn't expecting such changes.

Grace and Carol are standing among the apple trees, just where I left them. But they don't look the same. They don't look at all like the pictures of them I've carried around in my head for the past four months, shifting pictures in which only a few features stand out. For one thing they're bigger; and they have on different clothes.

They don't come running over, but stop what they're doing and stare, as if we're new people, as if I've never lived here. A third girl is with them. I look at her, empty of premonition. I've never seen her before.

14

Grace waves. After a moment Carol waves too. The third girl doesn't wave. They stand among the asters and gold-enrod, waiting as I go toward them. The apple trees are covered with scabby apples, red ones and yellow ones; some of the apples have fallen off and are rotting on the ground. There's a sweet, cidery smell, and the buzz of drunken yellowjackets. The apples mush under my feet.

Grace and Carol are browner, less pasty; their features are farther apart, their hair lighter. The third girl is the tallest. Unlike Grace and Carol, who are in summer skirts, she wears corduroys and a pullover. Both Carol and Grace are stubby-shaped, but this girl is thin without being fragile: lanky, sinewy. She has dark-blond hair cut in a long page-boy, with bangs falling half into her greenish eyes. Her face is long, her mouth slightly lopsided; something about the top lip is a little skewed, as if it's been cut open and sewn up crooked.

But her mouth evens out when she smiles. She has a smile like a grown-up's, as if she's learned it and is doing it out of politeness. She holds out her hand. "Hi, I'm Cordelia. And you must be . . ."

I stare at her. If she were an adult, I would take the hand, shake it, I would know what to say. But children do not shake hands like this.

"Elaine," Grace says.

I feel shy with Cordelia. I've been riding in the back of the car for two days, sleeping in a tent; I'm conscious of my grubbiness, my unbrushed hair. Cordelia is looking past me to where my parents are unloading the car. Her eyes are measuring, amused. I can see, without turning around, my father's old felt hat, his boots, the stubble on his face, my brother's uncut hair and seedy sweater and baggy knees, my mother's gray slacks, her manlike plaid shirt, her face blank of makeup.

"There's dog poop on your shoe," Cordelia says.

I look down. "It's only a rotten apple."

"It's the same color though, isn't it?" Cordelia says. "Not the hard kind, the soft squooshy kind, like peanut butter." This time her voice is confiding, as if she's talking about something intimate that only she and I know about and agree on. She creates a circle of two, takes me in.

Cordelia lives farther east than I do, in a region of houses even newer than ours, with the same surrounding mud. But her house is not a bungalow, it has two stories. It has a dining room separated by a curtain which you can pull back to make the living room and the dining room into one big room, and a bathroom on the ground floor with no bathtub in it which is called the powder room.

The colors in Cordelia's house are not dark, like those in other houses. They're light grays and light greens and whites. The sofa, for instance, is apple-green. There's nothing flowered or maroon or velvet. There's a picture, framed in light gray, of Cordelia's two older sisters, done in pastels when they were younger, both wearing smocked dresses, their hair feathery, their eyes like mist. There are real flowers, several different kinds at once, in chunky, flowing vases of Swedish glass. It's Cordelia who tells us the glass is Swedish. Swedish glass is the best kind, she says.

Cordelia's mother arranges the flowers herself, wearing gardening gloves. My own mother doesn't arrange flowers. Sometimes she sticks a few into a pot and puts them on the dinner table, but these are flowers she picks herself, during her exercise walks, in her slacks, along the road or in the ravine. Really they are weeds. She would never think of spending money on flowers. It occurs to me for the first time that we are not rich.

Cordelia's mother has a cleaning lady. She is the only

one of our mothers who has one. The cleaning lady is not called the cleaning lady, however. She is called the woman. On the days when the woman comes, we have to stay out of her way.

"The woman before this one," Cordelia tells us, in a hushed, scandalized voice, "was caught stealing potatoes. She put her bag down and they rolled out, all over the floor. It was so embarrassing." She means for them, not for the woman. "Of course we had to let her go."

Cordelia's family does not eat boiled eggs mushed up in a bowl but out of egg cups. Each egg cup has an initial on it, one for each person in the family. There are napkin rings too, also with initials. I have never heard of an egg cup before and I can tell Grace hasn't either, by the way she keeps silent about it. Carol says uncertainly that she has them at home.

"After you eat the egg," Cordelia tells us, "you have to put a hole in the bottom of the shell."

"Why?" we say.

"So the witches can't put out to sea." She says this lightly but scornfully, as if only a fool would need to ask. But there's the possibility she's joking, or teasing. Her two older sisters have this habit also. It's hard to tell when they mean to be taken seriously. They have an extravagant, mocking way of talking, which seems like an imitation of something, only it's unclear what they're imitating.

"I almost *died*," they say. Or, "I look like the wrath of God." Sometimes they say, "I look like an absolute hag," and sometimes, "I look like Haggis McBaggis." This is an ugly old woman they seem to have made up. But they don't really believe they almost died, or that they look ugly. Both of them are beautiful: one dark and intense, the other blond and kind-eyed and soulful. Cordelia is not beautiful in the same way.

Cordelia's two older sisters are Perdita and Miranda, but nobody calls them that. They're called Perdie and Mirrie. Perdie is the dark one; she takes ballet, and Mirrie plays the viola. The viola is kept in the coat closet, and Cordelia takes it out and shows it to us, lying there mysterious and important in its velvet-lined case. Perdie and Mirrie make drawling, gentle fun of each other and of themselves for doing these things, but Cordelia says they are gifted. This sounds like vaccinated, something that's done to you and leaves a mark. I ask Cordelia if she is gifted, but she puts her tongue in the corner of her mouth and turns away, as if she's concentrating on something else.

Cordelia ought to be Cordie, but she's not. She insists, always, on being called by her full name: Cordelia. All three of these names are peculiar; none of the girls at school have names like that. Cordelia says they're out of Shakespeare. She seems proud of this, as though it's something we should all recognize. "It was Mummie's idea," she says.

All three of them call their mother Mummie, and speak of her with affection and indulgence, as if she's a bright but willful child who has to be humored. She's tiny, fragile, absent-minded; she wears glasses on a silver chain around her neck and takes painting classes. Some of her paintings hang in the upstairs hall, greenish paintings of flowers, of lawns, of bottles and vases.

The girls have spun a web of conspiracy around Mummie. They agree not to tell her certain things. "Mummie isn't supposed to know that," they remind one another. But they don't like to disappoint her. Perdie and Mirrie try to do what they like as much as they can, but without disappointing Mummie. Cordelia is less agile at this: less able to do what she likes, more disappointing. This is what Mummie says when she's angry: "I am disappointed in you." If she gets very disappointed, Cordelia's father will be called into it, and that is serious. None of the girls jokes or drawls when mentioning him. He is large, craggy, charming, but we have heard him shouting, upstairs.

We sit in the kitchen, avoiding the dust mop of the woman, waiting for Cordelia to come down to play. She has been disappointing again, she has to finish tidying her room. Perdie strolls in, her camel's-hair coat thrown loosely, gracefully over one shoulder, her schoolbooks balanced on one hip. "Do you know what Cordelia says she wants to be when she grows up?" she says, in her husky, mock-serious, confiding voice. "A horse!" And we can't tell at all whether or not it's true.

. . .

Cordelia has a whole cupboard filled with dress-up costumes: old dresses of Mummie's, old shawls, old sheets you can cut up and drape around yourself. These used to be Perdie's and Mirrie's, but they've outgrown them. Cordelia wants us to act out plays, with her dining room and its curtain for the stage. She has an idea that we'll put these plays on and charge money for them. She turns out the lights, holds a flashlight under her chin, laughs in an eerie manner: this is how such things are done. Cordelia has been to plays, and even the ballet, once: *Giselle*, she says, offhand, as if we know. But somehow these plays never take shape the way she wants them to. Carol giggles and can't remember what she's supposed to say. Grace doesn't like being told what to do, and says she has a headache. Made-up stories don't interest her unless they contain a lot of real things: toasters, ironing boards, the wardrobes of movie stars. Cordelia's melodramas are beyond her.

"Now you kill yourself," says Cordelia.

"Why?" says Grace.

"Because you've been deserted," says Cordelia.

"I don't want to," says Grace. Carol, who is playing the maid, starts to giggle.

So we merely dress up and then trail down the stairs and out across the newly sodded front lawn, our shawls dragging behind us, uncertain what's supposed to happen next. Nobody wants to take boys' parts because there are no good clothes for them, though from time to time Cordelia draws a mustache on herself with Perdie's eyebrow pencil and wraps herself up in an old velvet curtain, in a last-ditch attempt at plot.

We walk home from school together, four now instead of three. There's a little shop on a side street halfway home where we stop and spend our allowances on penny gumballs, red licorice whips, orange Popsicles, sharing everything out equally. There are horse chestnuts in the gutters, wet-looking and glossy; we fill the pockets of our cardigans with them, uncertain what to use them for. The boys of our school and the Catholic boys from Our Lady of Perpetual Help throw them at one another, but we would not do that. They could put out your eye.

The dirt path going down to the wooden footbridge is dry, dusty; the leaves of the trees which hang over it are dull green and worn-out from the summer. Along the edge of the path is a thicket of weeds: goldenrod, ragweed, asters, burdocks, deadly nightshade, its berries red as valentine candies. Cordelia says that if you want to poison someone this would be a good way. The nightshade smells of earth, damp, loamy, pungent, and of cat piss. Cats prowl around in there, we see them every day, crouching, squatting, scratching up the dirt, staring out at us with their yellow eyes as if we're something they're hunting.

There are empty liquor bottles tossed into this thicket, and pieces of Kleenex. One day we find a safe. Cordelia knows it's called a safe, Perdie told her that once, when she was little and mistook one for a balloon. She knows it's a thing men use, the kind of men we're supposed to watch out for, though she doesn't know why it's called that. We pick it up on the end of a stick and examine it: whitish, limp, rubbery, like something inside a fish. Carol says "Ew." We carry it furtively back up the hill and shove it through a grating in the pavement; it floats down there on the surface of the dark water, pallid and drowned-looking. Even finding such a thing is dirty; even concealing it.

The wooden bridge is more askew, rottener than I remember. There are more places where the boards have fallen away. As a rule we walk down the middle, but today Cordelia goes right to the railing and leans on it, looking over. One by one and gingerly we follow. The stream below is shallow at this time of year; we can see the junk people have dumped into it, the worn-out tires, the broken bottles and rusty pieces of metal.

Cordelia says that because the stream flows right out of the cemetery it's made of dissolved dead people. She says that if you drink it or step into it or even get too close to it, the dead people will come out of the stream, all covered with mist, and take you with them. She says the only reason this hasn't happened to us is that we're on the bridge and the bridge is wooden. Bridges are safe, over dead people streams like this one.

Carol gets frightened, or acts frightened. Grace says Cordelia is being silly.

"Try it and see," says Cordelia. "Go on down there. I dare you." But we don't.

I know this is a game. My mother goes down there for her walks, my brother goes there with other older boys. They slosh through the culverts in their rubber boots and swing from trees and from the lower beams of the bridge. The reason the ravine is forbidden to us is not the dead people but the men. All the same, I wonder what the dead people would look like. I believe in them and I don't believe in them, both at the same time.

We pick blue and white weed flowers and some of the nightshade berries, and arrange them on burdock leaves by the side of the path, a horse chestnut on each. They are pretend meals, but it isn't clear who they're for. When we're finished we walk up the hill, leaving these arrangements behind us, half wreath, half lunch. Cordelia says we have to wash our hands really well because of the deadly nightshade berries; we have to wash off the poisonous juice. She says one drop could turn you into a zombie.

The next day when we come home from school these flower meals of ours are gone. Probably boys have destroyed them, it's the sort of thing boys destroy; or else the lurking men. But Cordelia makes her eyes wide, lowers her voice, looks over her shoulder.

"It's the dead people," she says. "Who else could it be?"

15

*W*hen the handbell rings we line up in front of GIRLS, two by two, holding hands: Carol and me, then Grace and Cordelia back behind us because they're a grade ahead. My brother is over there in front of BOYS. During recess he disappears into the cinder playground, where last week he had his lip kicked open during a soccer game and had to have stitches. I've seen the stitches, up close, black thread surrounded by swollen purple. I admire them. I know about the status conferred by wounds.

Now that I've changed back from pants to skirts, I have to remember the moves. You can't sit with your legs spread apart, or jump too high or hang upside-down, without ridicule. I've had to relearn the importance of underwear, which has a liturgy of its own:

> *I see England, I see France,*
> *I can see your underpants.*

Or else:

> *Me no know, me no care,*
> *Me no wear no underwear.*

This is said by boys, while making faces like monkeys.

There's a lot of speculation about underwear, especially the underwear of the teachers; but only that of the female teachers. Male underwear is of no importance. There aren't very many male teachers anyway, and the few that do exist are elderly; there are no young men, because the war has eaten them. The teachers are mostly women over a certain age, women who aren't married. Married women don't have jobs; we know this from our own mothers. There's something strange and laughable about older, unmarried women.

At recess, Cordelia doles out underwear: lavender frills for Miss Pigeon, who's fat and saccharine; plaid for Miss Stuart, lace-edged to go with her hankies; red satin long johns for Miss Hatchett, who's over sixty and wears garnet brooches. We don't believe any of this underwear actually exists, but thinking about it is a nasty joy.

My own teacher is Miss Lumley. It's said that every morning before the bell rings, even in late spring when it's warm, she goes to the back of the classroom and takes off her bloomers, which are rumored to be of heavy navy-blue wool and to smell of mothballs and of other, less definable things. This isn't repeated as speculation or as part of the underwear invention, but as fact. Several girls claim they've seen Miss Lumley putting her bloomers on again when they've had to stay in after school, and several others say they've seen them hanging in the cloakroom. The aura of Miss Lumley's dark, mysterious, repulsive bloomers clings around her and colors the air in which she moves. It makes her more terrifying; but she is terrifying in any case.

My teacher of the year before was kindly but so unmemorable that Cordelia doesn't even mention her in the underwear game. She had a face like a dinner roll and blancmange-colored skin, and ruled by wheedling. Miss Lumley rules by fear. She's short, and oblong in shape, so that her iron-gray cardigan falls straight from shoulder to hip with no pause in between for a waist. She always wears this cardigan, and a succession of dark skirts, which can't possibly be the same one. She has steel-rimmed glasses, behind which her eyes are hard to see, and black shoes with Cuban heels, and a tiny lipless smile. She does not send children to the principal for the strap, but does it herself, in front of the class, holding the hand out flat, bringing the black rubber strap down in sharp quick efficient strokes, her face white and quivering, while we watch, wincing, our eyes filling with involuntary tears. Some girls snivel audibly while she does this, even though she isn't doing it to them, but this isn't wise: Miss Lumley hates sniveling, and is likely to say, "I'll give you something to cry about." We learn to sit up straight, eyes front, faces blank, both feet on the floor, listening to the whack of rubber on cringing flesh.

Mostly it's boys who get the strap. They are thought to need it more. Also they fidget, especially during sewing. We are supposed to sew pot holders, for our mothers. The boys cannot seem to do this right; their stitches are large and clumsy, and they stick one another with the needles. Miss Lumley stalks the aisles, whacking their knuckles with a ruler.

The schoolroom is high-ceilinged, yellowy-brown, with blackboards at the front and along one side and tall many-paned windows above the radiators on the other side. Over the door to the cloakroom, so that you feel you're being watched from behind, there's a large photograph of the King and Queen, the King with medals, the Queen in a white ballgown and diamond tiara. High wooden desks that sit two, with slanted tops and holes for inkwells, are arranged in rows. It's like all the other schoolrooms at Queen Mary, but it seems darker, possibly because there's less decoration. Our old teacher brought paper doilies to school in her many efforts at appeasement, and her windows were always crawling with paper vegetation. But although Miss Lumley observes the seasons in this way too, the plants we bring forth under her glittering steel-rimmed eyes are smaller, shriveled-looking, so that there are never enough of them to cover the bare spaces of wall and glass. Also, if your fall foliage leaf or your pumpkin is not symmetrical, Miss Lumley won't put it up. She has standards.

Things are more British than they were last year. We learn to draw the Union Jack, using a ruler and memorizing the various crosses, for St. George of England, St. Patrick of Ireland, St. Andrew of Scotland, St. David of Wales. Our own flag is red and has a Union Jack in one corner, although there's no saint for Canada. We learn to name all the pink parts of the map.

"The sun never sets on the British Empire," says Miss Lumley, tapping the roll-down map with her long wooden pointer. In countries that are not the British Empire, they cut out children's tongues, especially those of boys. Before the British Empire there were no railroads or postal services in India, and Africa was full of tribal warfare, with spears, and had no proper clothing. The Indians in Canada did not have the wheel or telephones, and ate the hearts of their enemies in the heathenish belief that it would give them courage. The British Empire changed all that. It brought in electric lights.

Every morning, after Miss Lumley blows a thin metallic note on her pitch pipe, we stand up to sing "God Save the King." We also sing,

> Rule Britannia, Britannia rules the waves;
> Britons never, never, never shall be slaves!

Because we're Britons, we will never be slaves. But we aren't real Britons, because we are also Canadians. This isn't quite as good, although it has its own song:

> In days of yore, from Britain's shore,
> Wolfe, the dauntless hero, came
> And planted firm Britannia's flag
> On Canada's fair domain.
> Here may it wave, our boast, our pride
> And join in love together
> The thistle, shamrock, rose entwine
> The Maple Leaf forever.

Miss Lumley's jaw quivers in a frightening way when we sing this. Wolfe's name sounds like something you'd call a dog, but he conquered the French. This is puzzling, because I've seen French people, there are lots of them up north, so he couldn't have conquered all of them. As for maple leaves, they're the hardest part to draw on our red flag. Nobody ever gets them right.

Miss Lumley brings newspaper clippings about the Royal Family and sticks them to the side blackboard. Some of them are old clippings, and show Princess Elizabeth and Princess Margaret Rose, in Girl Guide uniforms, making radio and other speeches during the Blitz. This is what we should be like, Miss Lumley implies: steadfast, loyal, courageous, heroic.

There are other newspaper pictures too, showing thin-looking children in scruffy clothes, standing in front of piles of rubble. These are to remind us that there are many starving war orphans in Europe, and we should remember that and eat our bread crusts and potato skins and everything else on our plates, because waste is a sin. Also we should not complain. We are not really entitled to complain, because we are lucky children: English children got their houses bombed and we did not. We bring our used clothing, from home, and Miss Lumley ties it up into brown paper packages and sends it to England. There isn't much I can bring, because my mother tears our worn-out clothing up for dusters, but I manage to salvage a pair of corduroy pants, once my brother's, then mine, now too small, and a Viyella shirt of my father's that got washed wrong by mistake and shrank. It gives me a strange feeling on my skin to think of someone else, someone in England, walking around in my clothes. My clothes seem a part of me, even the ones I've outgrown.

All these things—the flags, the pitch pipe songs, the British Empire and the princesses, the war orphans, even the strappings—are superimposed against the ominous navy-blue background of Miss Lumley's invisible bloomers. I can't draw the Union Jack or sing "God Save the King" without thinking about them. Do they really exist, or not? Will I ever be in the classroom when she puts them on or—unthinkable—takes them off?

I'm not afraid of snakes or worms but I am afraid of these bloomers. I know it will be the worse for me if I ever actually catch sight of them. They're sacrosanct, at the same time holy and deeply shameful. Whatever is wrong with them may be wrong with me also, because although Miss Lumley is not what anyone thinks of as a girl, she is also not a boy. When the brass handbell clangs and we line up outside our GIRLS door, whatever category we are in also includes her.

iV
DEADLY NIGHTSHADE

16

I walk along Queen Street, past used comic book stores, windows full of crystal eggs and seashells, a lot of sulky black clothing. I wish I were back in Vancouver, in front of the fireplace with Ben, looking out over the harbor, while the giant slugs munch away at the greenery in the back garden. Fireplaces, back gardens: I wasn't thinking about them when I used to come down here to visit Jon, over the wholesale luggage store. Around the corner was the Maple Leaf Tavern, where I drank draft beer in the dark, two stoplights away from the art school where I drew naked women and ate my heart out. The streetcars rattled the front windows. There are still streetcars.

"I don't want to go," I said to Ben.

"You don't have to," he said. "Call it off. Come down to Mexico."

"They've gone to all the trouble," I said. "Listen, you know how hard it is to get a retrospective anywhere, if you're female?"

"Why is it important?" he said. "You sell anyway."

"I have to go," I said. "It wouldn't be right." I was brought up to say please and thank you.

"Okay," he said. "You know what you're doing." He gave me a hug.

I wish it were true.

Here is Sub-Versions, between a restaurant supply store and a tattoo parlor. Both of these will go, in time: once places like Sub-Versions move in, the handwriting's on the wall.

I open the gallery door, walk in with that sinking feeling I always have in galleries. It's the carpets that do it to me, the hush, the sanctimoniousness of it all: galleries are too much like churches, there's too much reverence, you feel there should be some genuflecting going on. Also I don't like it that this is where paintings end up, on these neutral-toned walls with the track lighting, sterilized, rendered safe and acceptable. It's as if somebody's been around spraying the paintings with air freshener, to kill the smell. The smell of blood on the wall.

This gallery is not totally sterilized, there are touches of cutting edge: a heating pipe shows, one wall is black. I don't give a glance to what's still on the walls, I hate those neo-expressionist dirty greens and putrid oranges, post this, post that. Everything is post these days, as if we're all just a footnote to something earlier that was real enough to have a name of its own.

Several of my own paintings have been uncrated and are leaning against the wall. They've been tracked down, requested, gathered in from whoever owns them. Whoever owns them is not me; worse luck, I'd get a better price now. The owners' names will be on little white cards beside the paintings, along with mine, as if mere ownership is on a par with creation. Which they think it is.

If I cut off my ear, would the market value go up? Better still, stick my head in the oven, blow out my brains. What rich art collectors like to buy, among other things, is a little vicarious craziness.

Face out is a piece I painted twenty years ago: Mrs. Smeath, beautifully rendered in egg tempera, with her gray hairpin crown and her potato face and her spectacles, wearing nothing but her flowered one-breast bib apron. She's reclining on her maroon velvet sofa, rising to Heaven, which is full of rubber plants, while a moon shaped like a doily floats in the sky. *Rubber Plant: The Ascension*, it's called. The angels around her are 1940s Christmas stickers, laundered little girls in white, with rag-set curly hair. The word *Heaven* is stenciled at the top of the painting with a child's school stencil set. I thought that was a nifty thing to do, at the time.

I caught some shit for that piece, as I recall. But not because of the stencil.

I don't look at this painting for very long, or at any of them. If I do I'll start finding things wrong with them. I'll want to take an Exacto knife to them, torch them, clear the walls. Begin again.

A woman strides toward me from the back, in a modified blond porcupine haircut, a purple jumpsuit and green leather boots. I know immediately that I should not have worn this powder-blue jogging outfit. Powder-blue is light-weight. I should've worn nun black, Dracula black, like all proper female painters. I should have some clotted-neck vampire lipstick, instead of wimping out with Rose Perfection. But that really would make me look like Haggis McBaggis. At this age the complexion can't stand those grape-jelly reds, I'd look all white and wrinkly.

But I will tough out the jogging suit, I'll pretend I meant it. It could be iconoclasm, how do they know? A powder-blue jogging suit lacks pretensions. The good thing about being out of fashion is that you're never in fashion either, so you can never be last year's model. That's my excuse for my painting, too; or it was for years.

"Hi," says the woman. "You must be Elaine! You don't look much like your picture." What does that mean, I think: better or worse? "We've talked a lot on the phone. My name is Charna." Toronto didn't used to have names like Charna. My hand gets crunched, this woman's got about ten heavy silver rings strung onto her fingers like knuckle dusters. "We were just wondering about the order." There are two more women; each of them looks five times more artistic than I do. They have abstract art earrings, hair arrangements. I am feeling dowdy.

They've got take-out gourmet sprout and avocado sand-wiches and coffee with steamed milk, and we eat those and drink that while we discuss the arrangement of the pictures. I say I favor a chronological approach, but Charna has other ideas, she wants things to go together tonally and resonate and make statements that amplify one another. I get more nervous, this kind of talk makes me twitch. I'm putting some energy into silence, resisting the impulse to say I have a headache and want to go home. I should be grateful, these women are on my side, they planned this whole thing for me, they're doing me an honor, they like what I do. But still I feel outnumbered, as if they are a species of which I am not a member.

Jon comes back tomorrow, from Los Angeles and his chain-saw murder. I can hardly wait. We'll circumvent his wife, go out for lunch, both of us feeling sneaky. But it's merely a civilized thing to do, having lunch with an ex-husband in a comradely way: a good coda to all that smashed crockery and mayhem. We've known each other since the year zot; at my age, our age, that's becoming important. And from here he looks like relief.

Someone else comes in, another woman. "Andrea!" says Charna, stalking over to her. "You're late!" She gives Andrea a kiss on the cheek and walks her over to me, holding her arm. "Andrea wants to do a piece on you," she says. "For the opening."

"I wasn't told about this," I say. I've been ambushed.

"It came up at the last minute," says Charna. "Lucky for us! I'll put you two in the back room, okay? I'll bring you some coffee. Getting the word out, they call it," she adds, to me, with a wry smile. I allow myself to be herded down the corridor; I can still be bossed around by women like Charna.

"I thought you would be different," says Andrea as we settle.

"Different how?" I ask.

"Bigger," she says.

I smile at her. "I am bigger."

Andrea checks out my powder-blue jogging suit. She herself is wearing black, approved, glossy black, not early-sixties holdover as mine would be. She has red hair out of a spray can and no apologies, cut into a cap like an acorn. She's upsettingly young; to me she doesn't look more than a teen-ager, though I know she must be in her twenties. Probably she thinks I'm a weird middle-aged frump, sort of like her high school teacher. Probably she's out to get me. Probably she'll succeed.

We sit across from each other at Charna's desk and Andrea sets down her camera and fiddles with her tape recorder. Andrea writes for a newspaper. "This is for the Living section," she says. I know what that means, it used to be the Women's Pages. It's funny that they now call it Living, as if only women are alive and the other things, such as the Sports, are for the dead.

"Living, eh?" I say. "I'm the mother of two. I bake cookies." All true. Andrea gives me a dirty look and flicks on her machine.

"How do you handle fame?" she says.

"This isn't fame," I say. "Fame is Elizabeth Taylor's cleavage. This stuff is just a media pimple."

She grins at that. "Well, could you maybe say something about your generation of artists—your generation of woman artists—and their aspirations and goals?"

"Painters, you mean," I say. "What generation is that?"

"The seventies, I suppose," she says. "That's when the women's—that's when you started getting attention."

"The seventies isn't my generation," I say.

She smiles. "Well," she says, "what is?"

"The forties."

"The forties?" This is archaeology as far as she's concerned. "But you couldn't have been . . ."

"That was when I grew up," I say.

"Oh right," she says. "You mean it was *formative*. Can you talk about the ways, how it reflects in your work?"

"The colors," I say. "A lot of my colors are forties colors." I'm softening up. At least she doesn't say *like* and *you know* all the time. "The war. There are people who remember the war and people who don't. There's a cut-off point, there's a difference."

"You mean the Vietnam War?" she says.

"No," I say coldly. "The Second World War." She looks a bit scared, as if I've just resurrected from the dead, and incompletely at that. She didn't know I was *that* old. "So," she says. "What is the difference?"

"We have long attention spans," I say. "We eat everything on our plates. We save string. We make do."

She looks puzzled. That's all I want to say about the

forties. I'm beginning to sweat. I feel as if I'm at the dentist, mouth gracelessly open while some stranger with a light and mirror gazes down my throat at something I can't see.

Brightly and neatly she veers away from the war and back toward women, which was where she wanted to be in the first place. Is it harder for a woman, was I discriminated against, undervalued? What about having children? I give unhelpful replies: all painters feel undervalued. You can do it while they're at school. My husband's been terrific, he gives me a lot of support, some of which has been financial. I don't say which husband.

"So you don't feel it's sort of demeaning to be propped up by a man?" she says.

"Women prop up men all the time," I say. "What's wrong with a little reverse propping?"

What I have to say is not altogether what she wants to hear. She'd prefer stories of outrage, although she'd be unlikely to tell them about herself, she's too young. Still, people my age are supposed to have stories of outrage; at least insult, at least put-down. Male art teachers pinching your bum, calling you *baby*, asking you why there are no great female painters, that sort of thing. She would like me to be furious, and quaint.

"Did you have any female mentors?" she asks.

"Female what?"

"Like, teachers, or other woman painters you admired."

"Shouldn't that be mentresses?" I say nastily. "There weren't any. My teacher was a man."

"Who was that?" she says.

"Josef Hrbik. He was very kind to me," I add quickly. He'd fit the bill for her, but she won't hear that from me. "He taught me to draw naked women."

That startles her. "Well, what about, you know, feminism?" she says. "A lot of people call you a feminist painter."

"What indeed," I say. "I hate party lines, I hate ghettos. Anyway, I'm too old to have invented it and you're too young to understand it, so what's the point of discussing it at all?"

"So it's not a meaningful classification for you?" she says.

"I like it that women like my work. Why shouldn't I?"

"Do men like your work?" she asks slyly. She's been going through the back files, she's seen some of those witch-and-succubus pieces.

"Which men?" I say. "Not everyone likes my work. It's not because I'm a woman. If they don't like a man's work it's not because he's a man. They just don't like it." I am on dubious ground, and this enrages me. My voice is calm; the coffee seethes within me.

She frowns, diddles with the tape recorder. "Why do you paint all those women then?"

"What should I paint, men?" I say. "I'm a painter. Painters paint women. Rubens painted women, Renoir painted women, Picasso painted women. Everyone paints women. Is there something wrong with painting women?"

"But not like that," she says.

"Like what?" I say. "Anyway, why should my women be the same as everyone else's women?" I catch myself picking at my fingers, and stop. In a minute my teeth will be chattering like those of cornered mice. Her voice is getting farther and farther away, I can hardly hear her. But I see her, very clearly: the ribbing on the neck of her sweater, the fine hairs of her cheek, the shine of a button. What I hear is what she isn't saying. *Your clothes are stupid. Your art is crap. Sit up straight and don't answer back.*

"Why do you paint?" she says, and I can hear her again as clear as anything. I hear her exasperation, with me and my refusals.

"Why does anyone do anything?" I say.

17

*T*he light fades earlier; on the way home from school we walk through the smoke from burning leaves. It rains, and we have to play inside. We sit on the floor of Grace's room, being quiet because of Mrs. Smeath's bad heart, and cut out rolling pins and frying pans and paste them around our paper ladies.

But Cordelia makes short work of this game. She knows, instantly it seems, why Grace's house has so many *Eaton's Catalogues* in it. It's because the Smeaths get their clothes that way, the whole family—order them out of the *Eaton's Catalogue*. There in the Girls' Clothing section are the plaid dresses, the skirts with straps, the winter coats worn by Grace and her sisters, three colors of them, in lumpy, serviceable wool, with hoods: Kelly Green, Royal Blue, Maroon. Cordelia manages to convey that she herself would never wear a coat ordered from the *Eaton's Catalogue*. She doesn't say this out loud though. Like the rest of us, she wants to stay on the good side of Grace.

She bypasses the cookware, flips through the pages. She turns to the brassieres, to the elaborately laced and gusseted corsets—foundation garments, they're called—and draws mustaches on the models, whose flesh looks as if it's been painted over with a thin coat of beige plaster. She pencils hair in, under their arms, and on their chests between the breasts. She reads out the descriptions, snorting with stifled laughter: "'Delightfully trimmed in dainty lace, with extra support for the mature figure.' That means big bazooms. Look at this—*cup* sizes! Like teacups!"

Breasts fascinate Cordelia, and fill her with scorn. Both of her older sisters have them by now. Perdie and Mirrie sit in their room with its twin beds and sprigged-muslin flounces, filing their nails, laughing softly; or they heat brown wax in little pots in the kitchen and take it upstairs to spread on their legs. They look into their mirrors, making sad faces—"I look like Haggis McBaggis! It's the curse!" Their wastebaskets smell of decaying flowers.

They tell Cordelia there are some things she's too young to understand, and then they tell these things to her anyway. Cordelia, her voice lowered, her eyes big, passes on the truth: the curse is when blood comes out between your legs. We don't believe her. She produces evidence: a sanitary pad, filched from Perdie's wastebasket. On it is a brown crust, like dried gravy. "That's not blood," Grace says with disgust, and she's right, it's nothing like when you cut your finger. Cordelia is indignant. But she can prove nothing.

I haven't thought much about grown-up women's bodies before. But now these bodies are revealed in their true, upsetting light: alien and bizarre, hairy, squashy, monstrous. We hang around outside the room where Perdie and Mirrie are peeling the wax off their legs while they utter yelps of pain, trying to see through the keyhole, giggling: they embarrass us, although we don't know why. They know they're being laughed at and come to the door to shoo us away. "Cordelia, why don't you and your little friends bug off!" They smile a little ominously, as if they know already what is in store for us. "Just wait and see," they say.

This frightens us. Whatever has happened to them, bulging them, softening them, causing them to walk rather than run, as if there's some invisible leash around their necks, holding them in check—whatever it is, it may happen to us too. We look surreptitiously at the breasts of women on the street, of our teachers; though not of our mothers, that would be too close for comfort. We examine our legs and underarms for sprouting hairs, our chests for swellings. But nothing is happening: so far we are safe.

Cordelia turns to the back pages of the catalogue, where the pictures are in gray and black and there are crutches and trusses and prosthetic devices. "Breast pumps," she says. "See this? It's for pumping your titties up bigger, like a bicycle pump." And we don't know what to believe.

We can't ask our mothers. It's hard to imagine them without clothes, to think of them as having bodies at all, under their dresses. There's a great deal they don't say. Between us and them is a gulf, an abyss, that goes down and down. It's filled with wordlessness. They wrap up the garbage in several layers of newspaper and tie it with string, and even so it drips onto the freshly waxed floor. Their clotheslines are strung with underpants, nighties, socks, a display of soiled intimacy, which they have washed and rinsed, plunging their hands into the gray curdled water. They know about toilet brushes, about toilet seats, about germs. The world is dirty, no matter how much they clean, and we know they will not welcome our grubby little questions. So instead a long whisper runs among us, from child to child, gathering horror.

Cordelia says that men have carrots, between their legs. They aren't really carrots but something worse. They're covered with hair. Seeds come out the end and get into women's stomachs and grow into babies, whether you want it or not. Some men have their carrots pierced and rings set into them as if they are ears.

Cordelia's unclear about how the seeds get out or what they're like. She says they're invisible, but I think this can't be so. If there are seeds at all they must be more like bird seeds, or carrot seeds, long and fine. Also she can't say how the carrot gets in, to plant the seeds. Belly buttons are the obvious choice, but there would have to be a cut, a tear. The whole story is questionable, and the idea that we ourselves could have been produced by such an act is an outrage. I think of beds, where all of this is supposed to take place: the twin beds at Carol's house, always so tidy, the elegant canopy bed at Cordelia's, the dark mahogany-colored bed in Grace's house, heavily respectable with its crocheted spread and layers of woolen blankets. Such beds are a denial in themselves, a repudiation. I think of Carol's wry-mouthed mother, of Mrs. Smeath with her hairpinned crown of graying braids. They would purse their lips, draw themselves up in a dignified manner. They would not permit it.

Grace says, "God makes babies," in that final way of

hers, which means there is nothing more to be discussed. She smiles her buttoned-up disdainful smile, and we are reassured. Better God than us.

But there are doubts. I know, for instance, a lot of things. I know that *carrot* is not the right word. I've seen dragonflies and beetles, flying around, stuck together, one on the back of the other; I know it's called *mating*. I know about ovipositors, for laying eggs, on leaves, on caterpillars, on the surface of the water; they're right out on the page, clearly labeled, on the diagrams of insects my father corrects at home. I know about queen ants, and about the female praying mantises eating the males. None of this is much help. I think of Mr. and Mrs. Smeath, stark-naked, with Mr. Smeath stuck to the back of Mrs. Smeath. Such an image, even without the addition of flight, will not do.

I could ask my brother. But, although we've examined scabs and toe jam under the microscope, although we aren't worried by pickled ox eyes and gutted fish and whatever can be found under dead logs, putting this question to him would be indelicate, perhaps hurtful. I think of JUPITER scrolled on the sand in his angular script, by his extra, dextrous finger. In Cordelia's version it will end up covered with hair. Maybe he doesn't know.

Cordelia says boys put their tongues in your mouth when they kiss you. Not any boys we know, older ones. She says this the same way my brother says "slug juice" or "snot" when Carol's around, and Carol does the same thing, the same wrinkle of the nose, the same wriggle. Grace says that Cordelia is being disgusting.

I think about the spit you sometimes see, downtown, on the sidewalk; or cow's tongues in butcher's shops. Why would they want to do such a thing, put their tongues in other people's mouths? Just to be repulsive, of course. Just to see what you would do.

18

I go up the cellar stairs, which have black rubber stair treads nailed onto them. Mrs. Smeath is standing at the kitchen sink in her bib apron. She's finished her nap and now she's upright, getting supper. She's peeling potatoes; she often peels things. The peel falls from her large knuckly hands in a long pale spiral. The paring knife she uses is worn so thin its blade is barely more than a crescent moon sliver. The kitchen is steamy, and smells of marrow fat and stewing bones.

Mrs. Smeath turns and looks at me, a skinless potato in her left hand, the knife in her right. She smiles. "Grace says your family don't go to church," she says. "Maybe you'd like to come with us. To our church."

"Yes," says Grace, who has come up the stairs behind me. And the idea is pleasing. I'll have Grace all to myself on Sunday mornings, without Carol or Cordelia. Grace is still the desirable one, the one we all want.

When I tell my parents about this plan they become anxious. "Are you sure you really want to go?" my mother says. When she was young, she says, she had to go to church whether she liked it or not. Her father was very strict. She couldn't whistle on Sundays. "Are you really sure?"

My father says he doesn't believe in brainwashing children. When you're grown up, then you can make up your own mind about religion, which has been responsible for a lot of wars and massacres in his opinion, as well as bigotry and intolerance. "Every educated person should know the Bible," he says. "But she's only eight."

"Almost nine," I say.

"Well," says my father. "Don't believe everything you hear."

On Sunday I put on the clothes my mother and I have picked out, a dress of dark-blue and green wool plaid, white ribbed stockings that attach with garters onto my stiff white cotton waist. I have more dresses than I once had, but I don't go shopping with my mother to help pick them out, the way Carol does. My mother hates shopping, nor does she sew. My girls' clothes are secondhand, donated by a distant friend of my mother's who has a larger daughter. None of these dresses fits me very well; the hems droop, or the sleeves bunch up under my arms. I think this is the norm, for dresses. The white stockings are new though, and even itchier than the brown ones I wear to school.

I take my blue cat's eye marble out of my red plastic purse and leave it in my bureau drawer, and put the nickel my mother's given me for the collection plate into my purse instead. I walk along the rutted streets toward Grace's house, in my shoes; it isn't time for boots yet. Grace opens her front door when I ring. She must have been waiting for me. She has a dress on too and white stockings, and navy-blue bows at the ends of her braids. She looks me over. "She doesn't have a hat," she says.

Mrs. Smeath, standing in the hallway, considers me as if I'm an orphan left on her doorstep. She sends Grace upstairs to search for another hat, and Grace comes back down with an old one of dark-blue velvet with an elastic under the chin. It's too small for me but Mrs. Smeath says it will do for now. "We don't go into our church with our heads uncovered," she says. She emphasizes *our*, as if there are other, inferior, bareheaded churches.

Mrs. Smeath has a sister, who is going with us to church. Her name is Aunt Mildred. She's older and has been a missionary in China. She has the same knuckly red hands, the same metal-rimmed glasses, the same hair crown as Mrs. Smeath, only hers is all gray, and the hairs on her face are gray too and more numerous. Both of them have hats that look like packages of felt carelessly done up, with several ends sticking into the air. I've seen such hats in the *Eaton's Catalogues* of several years back, worn by models with sleeked-back hair and high cheekbones and dark-red, glossy mouths. On Mrs. Smeath and her sister they don't have the same effect.

When all of the Smeaths have their coats and hats on we climb into their car: Mrs. Smeath and Aunt Mildred in the front, me and Grace and her two little sisters in the back. Although I still worship Grace, this worship is not at all physical, and being squashed into the back seat of her car, so close to her, embarrasses me. Right in front of my face Mr. Smeath is driving. He is short and bald and hardly ever seen. It's the same with Carol's father, with Cordelia's: in the daily life of houses, fathers are largely invisible.

We drive through the nearly empty Sunday streets, following the streetcar tracks west. The air inside the car fills with the used breath of the Smeaths, a stale smell like dried saliva. The church is large and made of brick; on the top of it, instead of a cross, there's a thing that looks like an onion and goes around. I ask about this onion, which may mean something religious for all I know, but Grace says it's a ventilator.

Mr. Smeath parks the car and we get out of it and go inside. We sit in a row, on a long bench made of dark shiny wood, which Grace says is a pew. This is the first time I've ever been inside a church. There's a high ceiling, with lights shaped like morning glories hanging down on chains, and a plain gold cross up at the front with a vase of white flowers. Behind that there are three stained-glass windows. The biggest, middle one has Jesus in white, with his hands held out sideways and a white bird hovering over his head. Underneath it says in thick black Bible-type letters with dots in between the words: THE· KINGDOM·OF·GOD·IS·WITHIN· YOU. On the left side is Jesus sitting down, sideways in pinky-red, with two children leaning on his knees. It says: SUFFER·THE·LITTLE·CHILDREN. Both of the Jesuses have halos. On the other side is a woman in blue, with no halo and a white kerchief partly covering her face. She's carrying a basket and reaching down one hand. There's a man sitting down at her feet, with what looks like a bandage wound around his head. It says: THE·GREATEST· OF·THESE·IS·CHARITY. Around all these windows are borders, with vines twining around and bunches of grapes, and different flowers. The windows have light coming in behind them, which illuminates them. I can hardly take my eyes off them.

Then there's organ music and everyone stands up, and I become confused. I watch what Grace does, and stand up when she stands up, sit when she sits. During the songs she holds the hymnbook open and points, but I don't know any of the tunes. After a while it's time for us to go to Sunday school, and so we file out with the other children in a line and go down into the church basement.

At the entrance to the Sunday school place there's a blackboard, where someone has printed, in colored chalk: KILROY WAS HERE. Beside this is a drawing of a man's eyes and nose, looking over a fence.

Sunday school is in classes, like ordinary school. The teachers are younger though; ours is an older teenager with a light-blue hat and a veil. Our class is all girls. The teacher reads us a Bible story about Joseph and his coat of many colors. Then she listens as the girls recite things they're supposed to have memorized. I sit on my chair, dangling my legs. I haven't memorized anything. The teacher smiles at me and says she hopes I will come back every week.

After this all the different classes go into a large room with rows of gray wooden benches in it, like the benches we eat our lunches on at school. We sit on the benches, the lights are turned off, and colored slides are projected onto the bare wall at the far end of the room. The slides aren't photographs but paintings. They look old-fashioned. The first one shows a knight riding through the forest, gazing upward to where a shaft of light streams down through the trees. The skin of this knight is very white, his eyes are large like a girl's, and his hand is pressed to where his heart must be, under his armor, which looks like car fenders. Under his large, luminous face I can see the light switches and the top boards of the wainscoting, and the corner of the small piano, where it juts out.

The next picture has the same knight only smaller, and underneath him some words, which we sing to the heavy thumping of chords from the unseen piano:

I would be true, for there are those who trust me,
I would be pure, for there are those who care,
I would be strong, for there is much to suffer,
I would be brave, for there is much to dare.

Beside me, in the dark, I hear Grace's voice going up and up, thin and reedy, like a bird's. She knows all the words; she knew all the words to her memory passage from the Bible too. When we bend our heads to pray I feel suffused with goodness, I feel included, taken in. God loves me, whoever he is.

After Sunday school we go back into the regular church for the last part, and I put my nickel on the collection plate. Then there is something called the Doxology. Then we walk out of the church and stuff back into the Smeaths' car, and Grace says carefully, "Daddy, may we go and see the trains?" and the little girls, with a show of enthusiasm, say, "Yes, yes."

Mr. Smeath says, "Have you been good?" and the little girls say, "Yes, yes" again.

Mrs. Smeath makes an indeterminate sound. "Oh, all right," says Mr. Smeath to the little girls. He drives the car south through the empty streets, along the streetcar tracks, past a single streetcar like a gliding island, until finally we see the flat gray lake in the distance, and below us, over the edge of a sort of low cliff, a flat gray plain covered with train tracks. On this metal-covered plain several trains are shunting slowly back and forth. Because it is Sunday, and because this is evidently a routine after-church Sunday event for the Smeaths, I have the idea that the train tracks and the lethargic, ponderous trains have something to do with God. It is also clear to me that the person who really wants to see the trains is not Grace, or any of the little girls, but Mr. Smeath himself.

We sit there in the parked car watching the trains until Mrs. Smeath says that the dinner will be ruined. After that we drive back to Grace's house.

I am invited for Sunday dinner. It's the first time I've ever stayed for dinner at Grace's. Before dinner Grace takes me upstairs so we can wash our hands, and I learn a new thing about her house: you are only allowed four squares of toilet paper. The soap in the bathroom is black and rough. Grace says it's tar soap.

The dinner is baked ham and baked beans and baked potatoes and mashed squash. Mr. Smeath carves the ham, Mrs. Smeath adds the vegetables, the plates get passed around. Grace's little sisters look at me through their eyeglasses when I start to eat.

"We say grace in this house," says Aunt Mildred, smiling firmly, and I don't know what she's talking about. I look at Grace: why do they want to say her name? But they all bend their heads and put their hands together and Grace says, "For what we are about to receive may the Lord make us truly thankful, Amen," and Mr. Smeath says, "Good food, good drink, good God, let's eat," and winks at me. Mrs. Smeath says "Lloyd," and Mr. Smeath gives a small, conspiratorial laugh.

After dinner Grace and I sit in the living room, on the velvet chesterfield, the same one Mrs. Smeath takes her naps on. I've never sat on it before and feel I'm sitting on something reserved, like a throne or a coffin. We read our Sunday school paper, which has the story of Joseph in it and a modern story about a boy who steals from the collection plate but repents and collects wastepaper and old bottles for the church, to make reparations. The pictures are black-and-white pen-and-ink drawings, but on the front is a colored picture of Jesus, in pastel robes, surrounded by children, all of different colors, brown, yellow, white, clean and pretty, some holding his hand, others gazing up at him with large worshipful eyes. This Jesus does not have a halo.

Mr. Smeath dozes in the maroon easy chair, his round belly swelling up. From the kitchen comes the clatter of silverware. Mrs. Smeath and Aunt Mildred are doing the dishes.

I reach home in the late afternoon, with my red plastic purse and my Sunday school paper. "Did you like it?" says my mother, still with the same air of anxiety.

"Did you learn anything?" says my father.

"I have to memorize a psalm," I say importantly. The word *psalm* sounds like a secret password. I am a little resentful. There are things my parents have been keeping from me, things I need to know. The hats, for instance: how could my mother have forgotten about the hats? God is not an entirely new idea for me: they have him at school in the morning prayers, and even in "God Save the King." But it seems there is more to it, more things to be memorized, more songs to be sung, more nickels to be donated, before he can be truly appeased. I am worried about Heaven though. What age will I be when I get there? What if I'm old when I die? In Heaven I want to be the age I am.

I have a Bible, on loan from Grace, her second-best. I go to my room and begin to memorize: *The heavens declare the glory of God; and the firmament sheweth his handywork. Day unto day uttereth speech, and night unto night sheweth knowledge.*

I still don't have any bedroom curtains. I look out the window, look up: there are the heavens, there are the stars, where they usually are. They no longer look cold and white and remote, like alcohol and enamel trays. Now they look watchful.

19

*T*he girls stand in the schoolyard or up on top of the hill, in small clumps, whispering and whispering and doing spoolwork. It's now the fashion to have a spool with four nails pounded into one end, and a ball of wool. You loop the wool over each nail in turn, twice around, and use a fifth nail to hook the bottom loops over the top ones. Out of the other end of the spool dangles a round thick wool tail, which you're supposed to wind up like a flat snail shell and sew into a mat to put the teapot on. I have such a spool, and so do Grace and Carol, and even Cordelia, although her wool is a snarl.

These clumps of whispering girls with their spools and colored wool tails have to do with boys, with the separateness of boys. Each cluster of girls excludes some other girls, but all boys. The boys exclude us too, but their exclusion is active, they make a point of it. We don't need to.

Sometimes I still go into my brother's room and lie around on the floor reading comic books, but I never do this when any other girl is there. Alone I am tolerated, as part of a group of girls I would not be. This goes without saying.

Once I took boys for granted, I was used to them. But now I pay more attention, because boys are not the same. For example, they don't take baths as often as they're expected to. They smell of grubby flesh, of scalp, but also of leather, from the knee patches on their breeches, and wool, from the breeches themselves, which come down only to below the knee, and lace up there like football pants. On the bottom parts of their legs they wear thick wool socks, which are usually damp and falling down. On their heads, outdoors, they wear leather helmets that strap under the chin. Their clothing is khaki, or navy-blue or gray, or forest green, colors that don't show the dirt as much. All of this has a military feel to it. Boys pride themselves on their drab clothing, their drooping socks, their smeared and inky skin: dirt, for them, is almost as good as wounds. They work at acting like boys. They call each other by their last names, draw attention to any extra departures from cleanliness. "Hey, Robertson! Wipe off the snot!" "Who farted?" They punch one another on the arm, saying, "Got you!" "Got you back!" There always seem to be more of them in the room than there actually are.

My brother punches arms and makes remarks about smells like the rest of them, but he has a secret. He would never tell it to these other boys, because of the way they would laugh.

The secret is that he has a girlfriend. This girlfriend is so secret she doesn't even know about it herself. I'm the only one he's told, and I have been double-sworn not to tell anyone else. Even when we're alone I'm not allowed to refer to her by her name, only by her initials, which are B.W. My brother will sometimes murmur these initials when there are other people around, my parents for instance. When he says them he stares at me, waiting for me to nod or give some sign that I have heard and understood. He writes me notes in code, which he leaves where I'll find them, under my pillow, tucked into my top bureau drawer. When I translate these notes they turn out to be so unlike him, so lacking in invention, so moronic in fact, that I can hardly believe it: "Talked to B.W." "Saw HER today." He writes these notes in colored pencil, different colors, with exclamation marks. One night there's a freak early snowfall, and in the morning when I wake up and look out my bedroom window there are the supercharged initials, etched in pee on the white ground, already melting.

I can see that this girlfriend is causing him some anguish, as well as excitement, but I can't understand why. I know who she is. Her real name is Bertha Watson. She hangs around with the older girls, up on the hill under the stunted fir trees. She has straight brown hair with bangs and she's of ordinary size. There's no magic about her that I can see, or any abnormality. I'd like to know how she's done it, this trick with my brother that's turned him into a stupider, more nervous identical twin of himself.

Knowing this secret, being the only one chosen to know, makes me feel important in a way. But it's a negative importance, it's the importance of a blank sheet of paper. I can know because I don't count. I feel singled out, but also bereft. Also protective of him, because for the first time in my life I feel responsible for him. He is at risk, and I have power over him. It occurs to me that I could tell on him, lay him open to derision; I have that choice. He is at my mercy and I don't want it. I want him back the way he was, unchanged, invincible.

The girlfriend doesn't last long. After a while nothing more is heard of her. My brother makes fun of me again, or ignores me; he's back in charge. He gets a chemistry set and does experiments down in the basement. As an obsession I prefer the chemistry set to the girlfriend. There are things stewing, horrible stinks, little sulfurous explosions, amazing illusions. There's invisible writing that comes out when you hold the paper over a candle. You can make a hard-boiled egg rubbery so it will go into a milk bottle, although getting it out again is more difficult. *Turn Water to Blood*, the instructions say, *and Astound Your Friends*.

He still trades comic books, but effortlessly, absentmindedly. Because he cares less about them he makes better trades. The comic books pile up under his bed, stacks and stacks of them, but he seldom reads them any more when the other boys aren't around.

My brother exhausts the chemistry set. Now he has a star map, pinned to the wall of his room, and at night he turns out the lights and sits beside the darkened, open window, in the cold, with his maroon sweater pulled on over his pajamas, gazing skyward. He has a pair of my father's binoculars, which he's allowed to use as long as he keeps the strap around his neck so he won't drop them. What he really wants next is a telescope.

When he allows me to join him, and when he feels like

talking, he teaches me new names, charts the reference points: Orion, the Bear, the Dragon, the Swan. These are constellations. Every one of them is made up of a huge number of stars, hundreds of times bigger and hotter than our own sun. These stars are light-years away, he says. We aren't really seeing them at all, we're just seeing the light they sent out years, hundreds of years, thousands of years ago. The stars are like echoes. I sit there in my flannelette pajamas, shivering, the back of my neck hurting from the upward tilt, squinting into the cold and the infinitely receding darkness, into the black caldron where the fiery stars boil and boil. His stars are different from the ones in the Bible: they're wordless, they flame in an obliterating silence. I feel as if my body is dissolving and I am being drawn up and up, like thinning mist, into a vast emptying space.

"Arcturus," my brother says. It's a foreign word, one I don't know, but I know the tone of his voice: recognition, completion, something added to a set. I think of his jars of marbles in the spring, the way he dropped the marbles into the jar, one by one, counting. My brother is collecting again; he's collecting stars.

20

*B*lack cats and paper pumpkins gather on the school windows. On Halloween Grace wears an ordinary lady's dress, Carol a fairy outfit, Cordelia a clown suit. I wear a sheet, because that's what there is. We walk from door to door, our brown paper grocery bags filling with candy apples, popcorn balls, peanut brittle, chanting at each door: *Shell out! Shell out! The witches are out!* In the front windows, on the porches, the large orange heads of the pumpkins float, glowing, unbodied. The next day we take our pumpkins to the wooden bridge and throw them over the edge, watching them smash open on the ground below. Now it's November.

Cordelia is digging a hole, in her back garden where there's no sod. She has started several holes before, but they have been unsuccessful, they struck rock. This one is more promising. She digs with a pointed shovel; sometimes we help her. It isn't a small hole but a large, square hole; it gets deeper and deeper as the dirt piles up around it. She says we can use it for a clubhouse, we can put chairs down in the hole and sit on them. When it's deep enough she wants to cover it over with boards, for a roof. She's already collected the boards, scrap boards from the two new houses they're building near her house. She's very wrapped up in this hole, it's hard to get her to play anything else.

On the darkening streets the poppies blossom, for Remembrance Day. They're made of fuzzy cloth, red like valentine hearts, with a black spot and a pin through the center.

We wear them on our coats. We memorize a poem about them:

> *In Flanders fields the poppies blow,*
> *Between the crosses row on row*
> *That mark our place.*

At eleven o'clock we stand beside our desks in the dust motes of the weak November sunshine for the three minutes of silence, Miss Lumley grim at the front of the room, heads bowed, eyes closed, listening to the hush and the rustle of our own bodies and the booming of the guns in the distance. *We are the dead.* I keep my eyes closed, trying to feel pious and sorry for the dead soldiers, who died for us, whose faces I can't imagine. I have never known any dead people.

Cordelia and Grace and Carol take me to the deep hole in Cordelia's backyard. I'm wearing a black dress and a cloak, from the dress-up cupboard. I'm supposed to be Mary, Queen of Scots, headless already. They pick me up by the underarms and the feet and lower me into the hole. Then they arrange the boards over the top. The daylight air disappears, and there's the sound of dirt hitting the boards, shovelful after shovelful. Inside the hole it's dim and cold and damp and smells like toad burrows.

Up above, outside, I can hear their voices, and then I can't hear them. I lie there wondering when it will be time to come out. Nothing happens. When I was put into the hole I knew it was a game; now I know it is not one. I feel sadness, a sense of betrayal. Then I feel the darkness pressing down on me; then terror.

When I remember back to this time in the hole, I can't really remember what happened to me while I was in it. I can't remember what I really felt. Maybe nothing happened, maybe these emotions I remember are not the right emotions. I know the others came and got me out after a while, and the game or some other game continued. I have no image of myself in the hole; only a black square filled with nothing, a square like a door. Perhaps the square is empty; perhaps it's only a marker, a time marker that separates the time before it from the time after. The point at which I lost power. Was I crying when they took me out of the hole? It seems likely. On the other hand I doubt it. But I can't remember.

Shortly after this I became nine. I can remember my other birthdays, later and earlier ones, but not this one. There must have been a party, my first real one, because who would have come to the others? There must have been a cake, with candles and wishes and a quarter and a dime wrapped in wax paper hidden between the layers for someone to chip a tooth on, and presents. Cordelia would have been there, and Grace and Carol. These things must have occurred, but the only trace they've left on me has been a vague horror of birthday parties, not other people's, my own. I think of pastel icing, pink candles burning in the pale November afternoon light, and there is a sense of shame and failure.

I close my eyes, wait for pictures. I need to fill in the black square of time, go back to see what's in it. It's as if I vanish at that moment and reappear later, but different, not knowing why I have been changed. If I could even see the undersides of the boards above my head it might help. I close my eyes, wait for pictures.

At first there's nothing; just a receding darkness, like a tunnel. But after a while something begins to form: a thicket of dark-green leaves with purple blossoms, dark purple, a sad rich color, and clusters of red berries, translucent as water. The vines are intergrown, so tangled over the other plants they're like a hedge. A smell of loam and another, pungent scent rises from among the leaves, a smell of old things, dense and heavy, forgotten. There's no wind but the leaves are in motion, there's a ripple, as of unseen cats, or as if the leaves are moving by themselves.

Nightshade, I think. It's a dark word. There is no nightshade in November. The nightshade is a common weed. You pull it out of the garden and throw it away. The nightshade plant is related to the potato, which accounts for the similar shape of the flowers. Potatoes too can be poisonous, if left in the sun to turn green. This is the sort of thing it's my habit to know.

I can tell it's the wrong memory. But the flowers, the smell, the movement of the leaves persist, rich, mesmerizing, desolating, infused with grief.

WRINGER

21

I leave the gallery, walk east. I need to go shopping, get some decent food, organize. When I'm by myself I revert to the times when I would forget about eating, stay up all night working, go until I felt an odd sensation I'd identify after some thought as hunger. Then I'd go through the refrigerator like a vacuum cleaner, sucking in whatever there was. Leftovers.

This morning there were eggs, but there are no more eggs. There's no more bread, there's no more milk. Why were there eggs and bread and milk in the first place? It must have been a stash of Jon's, he must eat there sometimes. Or could it have been that he got this stuff for me? It's hardly believable.

I will buy oranges, yogurt without the jam. I will have a positive attitude, take care of myself, I'll feed myself enzymes, and friendly bacteria. These good thoughts carry me until I'm right downtown.

This is where Eaton's used to be, here on this corner, yellow and foursquare. Now there's a huge building in its place, what they call a shopping complex, as if shopping were a psychic disease. It's glassy and be-tiled, green as an iceberg.

Across the street from it is known territory: Simpsons department store. I know it has a food hall somewhere. In the plate-glass windows there are heaps of bath towels, overstuffed sofas and chairs, sheets in modern prints. I wonder where all this cloth ends up. People cart it away, stuff it into their houses: the nesting instinct. A less attractive concept if you've ever seen a nest up close. There must be a limit to how much cloth you can cram into any one house, but of course it's disposable. You used to buy for quality, things that would last. You kept your clothes until they were part of you, you checked the hemlines, the way the buttons were sewed on, you rubbed the cloth between your finger and thumb.

The next windows contain disgruntled mannequins, their pelvises thrust out, their shoulders flung this way and that, making them look like hunchbacked ax murderers. I guess this is the look now: surly aggression. On the sidewalks there are a lot of androgynes, in the flesh, the girls in the black leather jackets and tough boots of boys, crewcuts, ducktails, the boys with the sullen pouty look of those women on the front covers of fashion magazines, their hair gelled into quills. At a distance I can't tell the difference, though they themselves probably can. They make me feel outmoded.

What are they aiming for? Is each an imitation of the other? Or does it only seem that way to me because they're all so alarmingly young? Despite their cool poses they wear their cravings on the outside, like the suckers on a squid. They want it all.

But I suppose that's what Cordelia and I looked like then, to older people, crossing the street right here with our collars turned up and our eyebrows plucked into sceptical arches, swaggering in our rubber boots and straining for nonchalance, on our way down to Union Station where the trains came in, to put our quarters into the photo machine, four shots in monochrome, wallet-sized. Cordelia with a cigarette in the corner of her mouth, her eyelids half closed, trying for sultry. Ultrasharp.

I revolve through the revolving doors into Simpsons, where I become lost immediately. They've changed the whole thing over. It used to be sedate wood-rimmed glass counters, with gloves in standard models, appropriate wristwatches, accent scarves in floral prints. Serious-minded good taste. Now it's a cosmetic fairground: silver trim, gold pillars, marquee lights, brand-name letters the size of a human head. The air is saturated with the stink of perfumes at war. There are video screens on which flawless complexions turn, preen, sigh through their parted lips, are caressed. On other screens are close-ups of skin pores, before and after, details of regimes for everything, your hands, your neck, your thighs. Your elbows, especially your elbows: aging begins at the elbows and metastasizes.

This is religion. Voodoo and spells. I want to believe in it, the creams, the rejuvenating lotions, the transparent unguents in vials that slick on like roll-top glue. "Don't you know what that junk is made of?" Ben said once. "Ground-up cock's combs." But this doesn't deter me, I'd use anything if it worked—slug juice, toad spit, eye of newt, anything at all to mummify myself, stop the drip-drip of time, stay more or less the way I am.

But I own enough of this slop already to embalm all of the girls in my high school graduating class, who must need it by now as much as I do. I stop only long enough to allow myself to be sprayed by a girl giving away free squirts of some venomous new perfume. The femme fatale must be back, Veronica Lake slinks again. The stuff smells like grape Kool-Aid. I can't imagine it seducing anything but a fruit fly.

"You like this?" I say to the girl. They must get lonely, standing here all day in their high heels, spraying strangers.

"It's been very popular," she says evasively. Briefly I glimpse myself through her eyes: bloom off the rose, teetering on the brink of matronhood, hoping for the best. I am the market.

I ask her where the food hall is, and she tells me. It's down. I get on the escalator, but suddenly I'm going up. This is bad, confusing directions like that, or am I jumping time, did I go down already? I get off, and find myself wading through rack after rack of children's party dresses. They have the lace collars, the puffed sleeves, the sashes I remember; many of them are in plaid, the authentic somber blood-lit colors, dark greens with a stripe of red, dark blues, black. Black Watch. Have these people forgotten history, don't they know anything about the Scots, don't they know any better than to clothe small girls in the colors of despair, slaughter, treachery and murder? *My way of life*, new line,

Is fall'n into the sere and yellow leaf. Once we had to memorize things. Still, plaid was the fashion in my day too. The white socks, the Mary Janes, the always-inadequate birthday present swathed in tissue paper, and the little girls with their assessing eyes, their slippery deceitful smiles, tartaned up like Lady Macbeth.

In the endless time when Cordelia had such power over me, I peeled the skin off my feet. I did it at night, when I was supposed to be sleeping. My feet would be cool and slightly damp, smooth, like the skin of mushrooms. I would begin with the big toes. I would bend my foot up and bite a small opening in the thickest part of the skin, on the bottom, along the outside edge. Then, with my fingernails, which I never bit because why bite something that didn't hurt, I would pull the skin off in narrow strips. I would do the same to the other big toe, then to the ball of each foot, the heel of each. I would go down as far as the blood. Nobody but me ever looked at my feet, so nobody knew I was doing it. In the mornings I would pull my socks on, over my peeled feet. It was painful to walk, but not impossible. The pain gave me something definite to think about, something immediate. It was something to hold on to.

I chewed the ends of my hair, so that there was always one lock of hair that was pointed and wet. I gnawed the cuticles off from around my fingernails, leaving welts of exposed, oozing flesh which would harden into rinds and scale off. In the bathtub or in dishwater my fingers looked nibbled, as if by mice. I did these things constantly, without thinking about them. But the feet were more deliberate.

I remember thinking when the girls were born, first one and then the other, that I should have had sons and not daughters. I didn't feel up to daughters, I didn't know how they worked. I must have been afraid of hating them. With sons I would have known what to do: frog catching, fishing, war strategies, running around in the mud. I would have been able to teach them how to defend themselves, and what from. But the world of sons has changed; it's more likely to be the boys now with that baffled look, like a night dweller gone blind in sunlight. "Stand up for yourself like a man," I would have said. I would have been on shifty ground.

As for the girls, my girls at any rate, they seem to have been born with some kind of protective coating, some immunity I lacked. They look you in the eye, level and measuring, they sit at the kitchen table and the air around lights up with their lucidity. They are sane, or so I like to think. My saving graces.

They amaze me, they always have. When they were little I felt I had to protect them from certain things about myself, the fear, the messier parts of the marriages, the days of nothing. I didn't want to pass anything on to them, anything of mine they would be better off without. At those times I would lie on the floor in the dark, with the curtains drawn and the door closed. I would say, *Mummy has a headache. Mummy's working.* But they didn't seem to need that protection, they seemed to take everything in, look at it straight, accept everything. "Mummy's in there lying on the floor. She'll be fine tomorrow," I heard Sarah tell Anne when one was ten and the other was four. And so I was fine. Such faith, like the faith in sunrise or the phases of the moon, sustained me. It must be this sort of thing that keeps God going.

Who knows what they'll make of me later on, who knows what they've already made of me? I would like them to be the happy end of my story. But of course they are not the end of their own.

Someone comes up behind me, a sudden voice out of thin air. She startles me. "May I help you?" It's a saleslady, an older woman this time. Middle-aged. My age, I then think, discouraged. Mine and Cordelia's.

I'm standing among the plaid dresses, fingering a sleeve. God knows how long I've been doing it. Have I been talking out loud? My throat feels tight and my feet hurt. But whatever else may be in store for me, I do not intend to slide off my trolley tracks in the middle of Simpsons Girlswear.

"The food hall," I say.

She smiles gently. She is tired, and I am a disappointment to her, I don't want any plaid. "Oh, you need to be right downstairs," she says, "in the cellar." Kindly, she directs me.

22

The black door opens. I'm sitting in the mouse-dropping and formaldehyde smell of the building, on the window ledge, with the heat from the radiator going up my legs, watching out the window as the fairies and gnomes and snowballs below me slog through the drizzle to the tune of "Jingle Bells" played by a brass band. The fairies look foreshortened, damaged, streaked by the dust and rain on the window glass; my breath makes a foggy circle. My brother isn't here, he's too old for it. This is what he said. I have the whole window ledge to myself.

On the window ledge beside mine, Cordelia and Grace and Carol are sitting, jammed in together, whispering and giggling. I have to sit on a window ledge by myself because they aren't speaking to me. It's something I said wrong, but I don't know what it is because they won't tell me. Cordelia says it will be better for me to think back over everything I've said today and try to pick out the wrong thing. That way I will learn not to say such a thing again. When I've guessed the right answer, then they will speak to me again. All of this is for my own good, because they are my best friends and they want to help me improve. So this is what I'm thinking about as the pipe band goes past in sodden fur hats, and the drum majorettes with their bare wet legs and red smiles and dripping hair: what did I say wrong? I can't remember having said anything different from what I would ordinarily say.

My father walks into the room, wearing his white lab coat. He's working in another part of the building, but he's come to check on us. "Enjoying the parade, girls?" he says.

"Oh yes, thank you," Carol says, and giggles. Grace says, "Yes, thank you." I say nothing. Cordelia gets down off her windowsill and slides up onto mine, sitting close beside me.

"We're enjoying it extremely, thank you very much," she says in her voice for adults. My parents think she has beautiful manners. She puts an arm around me, gives me a little squeeze, a squeeze of complicity, of instruction. Everything will be all right as long as I sit still, say nothing, reveal nothing. I will be saved then, I will be acceptable once more. I smile, tremulous with relief, with gratitude.

But as soon as my father is out of the room Cordelia turns to face me. Her expression is sad rather than angry. She shakes her head. "How could you?" she says. "How could you be so impolite? You didn't even answer him. You know what this means, don't you? I'm afraid you'll have to be punished. What do you have to say for yourself?" And I have nothing to say.

I'm standing outside the closed door of Cordelia's room. Cordelia, Grace, and Carol are inside. They're having a meeting. The meeting is about me. I am just not measuring up, although they are giving me every chance. I will have to do better. But better at what?

Perdie and Mirrie come up the stairs, along the hall, in their armor of being older. I long to be as old as they are. They're the only people who have any real power over Cordelia, that I can see. I think of them as my allies; or I think they would be my allies if they only knew. Knew what? Even to myself I am mute.

"Hello, Elaine," they say. Now they say, "What's the little game today? Hide-and-seek?"

"I can't tell," I answer. They smile at me, condescending and kind, and head toward their room, to do their toenails and talk about older things.

I lean against the wall. From behind the door comes the indistinct murmur of voices, of laughter, exclusive and luxurious. Cordelia's Mummie drifts by, humming to herself. She's wearing her painting smock. There's a smudge of apple-green on her cheek. She smiles at me, the smile of an angel, benign but remote. "Hello, dear," she says. "You tell Cordelia there's a cookie for you girls, in the tin."

"You can come in now," says the voice of Cordelia from inside the room. I look at the closed door, at the doorknob, at my own hand moving up, as if it's no longer a part of me.

This is how it goes. It's the kind of thing girls of this age do to one another, or did then, but I'd had no practice in it. As my daughters approached this age, the age of nine, I watched them anxiously. I scrutinized their fingers for bites, their feet, the ends of their hair. I asked them leading questions: "Is everything all right, are your friends all right?" And they looked at me as if they had no idea what I was talking about, why I was so anxious. I thought they'd give themselves away somehow: nightmares, moping. But there was nothing I could see, which may only have meant they were good at deception, as good as I was. When their friends arrived at our house to play, I scanned their faces for signs of hypocrisy. Standing in the kitchen, I listened to their voices in the other room. I thought I would be able to tell. Or maybe it was worse. Maybe my daughters were doing this sort of thing themselves, to someone else. That would account for their blandness, the absence of bitten fingers, their level blue-eyed gaze.

Most mothers worry when their daughters reach adolescence, but I was the opposite. I relaxed, I sighed with relief. Little girls are cute and small only to adults. To one another they are not cute. They are life-sized.

It turns colder and colder. I lie with my knees up, as close to my body as I can get them. I'm peeling the skin off my feet; I can do it without looking, by touch. I worry about what I've said today, the expression on my face, how I walk, what I wear, because all of these things need improvement. I am not normal, I am not like other girls. Cordelia tells me so, but she will help me. Grace and Carol will help me too. It will take hard work and a long time.

In the mornings I get out of bed, put on my clothes, the stiff cotton waist with the garters, the ribbed stockings, the nubbled wool pullover, the plaid skirt. I remember these clothes as cold. Probably they were cold.

I put my shoes on, over my stockings and my peeled feet.

I go out to the kitchen, where my mother is cooking breakfast. There's a pot with porridge in it, Red River cereal or oatmeal or Cream of Wheat, and a glass coffee percolator. I rest my arms on the edge of the white stove and watch the porridge, simmering and thickening, the flaccid bubbles coming up out of it one at a time and releasing their small puffs of steam. The porridge is like boiling mud. I know that when it comes time to eat the porridge I will have trouble: my stomach will contract, my hands will get cold, it will be difficult to swallow. Something tight sits under my breastbone. But I will get the porridge down somehow, because it's required.

Or I watch the coffee percolator, which is better because I can see everything, the pinpoint bubbles gathering under the upside-down glass umbrella, then hesitating, then the column of water shooting upward through the stem, falling down over the coffee in its metal basket, the drops of coffee dripping down into the clear water, inking it brown.

Or I make toast, sitting at the table where the toaster is. Each of our spoons has a dark-yellow halibut liver oil capsule in it, shaped like a small football. There are the plates, gleaming whitely, and the glasses of juice. The toaster is on a silver heat pad. It has two doors, with a knob at the bottom of each, and a grid up the center that glows red-hot. When the toast is done on one side I turn the knobs and the doors open and the toast slides down and turns over, all by itself. I think about putting my finger in there, onto the red-hot grid.

All of these are ways of delaying time, slowing it down, so I won't have to go out through the kitchen door. But no matter what I do, and despite myself, I am pulling on my snowpants, wadding my skirt in between my legs, tugging thick woolen socks on over my shoes, stuffing my feet into boots. Coat, scarf, mittens, knitted hat, I am encased, I am kissed, the door opens, then closes behind me, frozen air shoots up my nose. I waddle through the orchard of leafless apple trees, the legs of my snowpants whisking against each other, down to the bus stop.

Grace is waiting there and Carol, and especially Cordelia. Once I'm outside the house there is no getting away from them. They are on the school bus, where Cordelia stands close beside me and whispers into my ear: "Stand up straight! People are looking!" Carol is in my classroom, and it's her job to report to Cordelia what I do and say all day. They're there at recess, and in the cellar at lunchtime. They comment on the kind of lunch I have, how I hold my sandwich, how I chew. On the way home from school I have to walk in front of them, or behind. In front is worse because they talk about how I'm walking, how I look from behind. "Don't hunch over," says Cordelia. "Don't move your arms like that."

They don't say any of the things they say to me in front of others, even other children: whatever is going on is going on in secret, among the four of us only. Secrecy is important, I know that: to violate it would be the greatest, the irreparable sin. If I tell I will be cast out forever.

But Cordelia doesn't do these things or have this power over me because she's my enemy. Far from it. I know about enemies. There are enemies in the schoolyard, they yell things at one another and if they're boys they fight. In the war there were enemies. Our boys and the boys from Our Lady of Perpetual Help are enemies. You throw snowballs at enemies and rejoice if they get hit. With enemies you can feel hatred, and anger. But Cordelia is my friend. She likes me, she wants to help me, they all do. They are my friends, my girl friends, my best friends. I have never had any before and I'm terrified of losing them. I want to please.

Hatred would have been easier. With hatred, I would have known what to do. Hatred is clear, metallic, one-handed, unwavering; unlike love.

23

None of this is unrelenting.

On some days Cordelia decides that it's Carol's turn to be improved. I am invited to join Grace and Cordelia as they walk ahead on the way home from school, with Carol trailing behind, and to think of things Carol has done wrong. "Carol is a smarty-pants," Cordelia says. At these times I don't pity Carol. She deserves what's happening to her, because of all the times she's done the same things to me. I rejoice that it's her turn instead of mine.

But these times don't last long. Carol cries too easily and noisily, she gets carried away with her own crying. She draws attention, she can't be depended on not to tell. There's a recklessness in her, she can be pushed just so far, she has a weak sense of honor, she's reliable only as an informer. If this is obvious to me, it must be even more obvious to Cordelia.

Other days appear normal. Cordelia seems to forget about improving anybody, and I think she may have given up on it. I'm expected to behave as if nothing has ever happened. But it's hard for me to do this, because I feel I'm always being watched. At any time I may step over some line I don't even know is there.

Last year I was hardly ever home, by myself, after school or on weekends. Now I want to be. I make excuses so I won't have to go out and play. I still call it playing.

"I have to help my mother," I say. This has a ring of truth to it. Girls do have to help their mothers, sometimes; Grace in particular has to help her mother. But it's less true than I would like it to be. My mother doesn't linger over housework, she'd rather be outside raking up leaves in the fall, shoveling snow in the winter, pulling weeds in the spring. When I help her I slow her down. But I dangle around the kitchen, saying, "Can I help?" until she gives me a duster and has me dust the scrolled legs of the dining table, or the edges of the bookcases; or I cut up dates, chop nuts, grease the muffin cups with a corner of waxed paper torn from the inside wrapper of the Crisco box; or I rinse the wash.

I like rinsing the wash. The laundry room is small and enclosed, secret, underground. On the shelves there are packages of odd, power-filled substances: laundry starch in white twisted shapes like bird droppings, bluing to make the whites look whiter, Sunlight soap in bars, Javex bleach with a skull and crossbones on it, reeking of sanitation and death.

The washing machine itself is tubular white enamel, a hulk on four spindly legs. It dances slowly across the floor, *chug-lug, chug-lug*, the clothes and the soapy water moving as if boiling sluggishly, like cloth porridge. I watch it, hands on the edge of the tub, chin on hands, my body dragging downward from this ledge, not thinking about anything. The water turns gray and I feel virtuous because of all the dirt that's coming out. It's as if I myself am doing this just by looking.

My job is to run the washed clothes through the wringer into the laundry sink full of clean water and then into the second laundry sink for the second rinse, and after that into the creaky laundry basket. After that my mother takes the clothes outside and hangs them onto the clothesline with wooden clothespins. Sometimes I do this too. In the cold the clothes freeze stiff, like plywood. One day a small neighborhood boy collects horse buns, from the milk wagon horse, and puts them along the bottom folds of the freshly washed double-hung white sheets. All sheets are white, all milk comes from horses.

The wringer is two rubber rollers, the color of pale flesh, that revolve around and around, the clothes squeezing in between them, water and suds squooshing out like juice. I roll up my sleeves, stand on tiptoe, rummage in the tub and haul up the sopping underpants and slips and pajamas, which feel like something you might touch just before you know it's a drowned person. I poke the corners of the clothes in between the wringers and they are grabbed and dragged through, the arms of the shirts ballooning with trapped air, suds dripping from the cuffs. I've been told to be very careful when doing this: women can get their hands caught in wringers, and other parts of their bodies, such as hair. I think about what would happen to my hand if it did get caught: the blood and flesh squeezing up my arm like a traveling bulge, the hand coming out the other side flat as a glove, white as paper. This would hurt a lot at first, I know that. But there's something compelling about it. A whole person could go through the wringer and come out flat, neat, completed, like a flower pressed in a book.

"You coming out to play?" says Cordelia, on our way home from school.

"I have to help my mother," I say.

"Again?" says Grace. "How come she does that so much? She never used to do it." Grace has begun talking about me in the third person, like one grown-up to another, when Cordelia is there.

I think of saying my mother is sick, but my mother is so obviously healthy I know I won't get away with this.

"She thinks she's too good for us," says Cordelia. Then, to me: "Do you think you're too good for us?"

"No," I say. Thinking you are too good is bad.

"We'll come and ask your mother if you can play," says Cordelia, switching back to her concerned, friendly voice. "She won't make you work all the time. It isn't fair."

And my mother smiles and says yes, as if she's pleased that I'm so much in demand, and I am pried away from the muffin cups and the washing machine wringer, expelled into the outside air.

On Sundays I go to the church with the onion on top of it, crammed into the Smeaths' car with all the Smeaths, Mr. Smeath, Mrs. Smeath, Aunt Mildred, Grace's younger sisters, whose nostrils in the winter season are forever plugged with yellowy-green snot. Mrs. Smeath seems pleased about this arrangement, but she is pleased with herself, for going out of her way, for displaying charity. She's not especially pleased with me. I can tell this by the line between her eyebrows when she looks at me, although she smiles with her closed lips, and by the way she keeps asking whether I wouldn't like to bring my brother next time, or my parents? I focus on her chest, on her single breast that goes all the way down to her waist, with her dark-red, black-spotted heart beating within it, gasping in out, in out, out of breath like a fish on shore, and shake my head, ashamed. My failure to produce these other members of my family tells against me.

I have memorized the names of all the books of the Bible, in order, and the Ten Commandments and the Lord's Prayer, and most of the Beatitudes. I've been getting ten out of ten on my Bible quizzes and my memory work, but I'm beginning to falter. In Sunday school we have to stand up and recite, out loud, in front of the others, and Grace watches me. She watches everything I do on Sundays, and reports on me, matter-of-factly, to Cordelia.

"She didn't stand up straight in Sunday school yesterday." Or: "She was a goody-goody." I believe each of these comments: my shoulders sag, my spine crumples, I exude the wrong kind of goodness; I see myself shambling crookedly, I make an effort to stand straighter, my body rigid with anxiety. And it's true that I got ten out of ten, again, and Grace only got nine. Is it wrong to be right? How right should I be, to be perfect? The next week I put five wrong answers, deliberately.

"She only got five out of ten on Bible," Grace says on Monday.

"She's getting stupider," Cordelia says. "You aren't really that stupid. You'll have to try harder than that!"

Today is White Gift Sunday. We have all brought cans of food from home for the poor, wrapped up in white tissue paper. Mine are Habitant pea soup and Spam. I suspect they are the wrong things, but they're what my mother had in the cupboard. The idea of white gifts bothers me: such hard gifts, made uniform, bleached of their identity and colors. They look dead. Inside those blank, sinister bundles of tissue paper piled up at the front of the church there could be anything.

Grace and I sit on the wooden benches in the church basement, watching the illuminated slides on the wall, singing the words to the songs, while the piano plods onward in the darkness.

> *Jesus bids us shine*
> *With a pure, clear light,*
> *Like a little candle*
> *Burning in the night:*
> *In this world is darkness;*
> *So let us shine,*
> *You in your small corner,*
> *And I in mine.*

I want to shine like a candle. I want to be good, to follow instructions, to do what Jesus bids. I want to believe you should love your neighbors as yourself and the Kingdom of God is within you. But all of this seems less and less possible.

In the darkness I can see a gleam of light, to the side. It's not a candle: it's light reflected back off Grace's glasses, from the light on the wall. She knows the words by heart, she doesn't have to look at the screen. She's watching me.

After church I go with the Smeaths through the vacant Sunday streets to watch the trains shunting monotonously back and forth along their tracks, on the gray plain beside the flat lake. Then I go back to their house for Sunday dinner. This happens every Sunday now, it's part of going to church; it would be very bad if I said no, to either thing.

I've learned the way things are done here. I climb the stairs past the rubber plant, not touching it, and go into the Smeaths' bathroom and count off four squares of toilet paper and wash my hands afterward with the gritty black Smeath soap. I no longer have to be admonished, I bow my head automatically when Grace says, "For what we are about to receive may the Lord make us truly thankful, Amen."

"Pork and beans the musical fruit, the more you eat, the more you toot," says Mr. Smeath, grinning round the table. Mrs. Smeath and Aunt Mildred do not think this is funny. The little girls regard him solemnly. They both have glasses and white freckled skin and Sunday bows on the ends of their brown wiry braids, like Grace.

"Lloyd," says Mrs. Smeath.

"Come on, it's harmless," Mr. Smeath says. He looks me in the eye. "Elaine thinks it's funny. Don't you, Elaine?"

I am trapped. What can I say? If I say no, it could be rudeness. If I say yes, I have sided with him, against Mrs. Smeath and Aunt Mildred and all three of the Smeath girls, including Grace. I feel myself turn hot, then cold. Mr. Smeath is grinning at me, a conspirator's grin.

"I don't know," I say. The real answer is no, because I don't in fact know what this joke means. But I can't abandon Mr. Smeath, not entirely. He is a squat, balding, flabby man, but still a man. He does not judge me.

Grace repeats this incident to Cordelia, next morning, in the school bus, her voice a near whisper. "She said she didn't know."

"What sort of an answer was that?" Cordelia asks me sharply. "Either you think it's funny or you don't. Why did you say 'I don't know'?"

I tell the truth. "I don't know what it means."

"You don't know what *what* means?"

"Musical fruit," I say. "The more you toot." I am now deeply embarrassed, because I don't know. Not knowing is the worst thing I could have done.

Cordelia gives a hoot of contemptuous laughter. "You don't know what *that* means?" she says. "What a stupe! It means *fart*. Beans make you fart. Everyone knows that."

I am doubly mortified, because I didn't know, and because Mr. Smeath said *fart* at the Sunday dinner table and enlisted me on his side, and I did not say no. It isn't the word itself that makes me ashamed. I'm used to it, my brother and his friends say it all the time, when there are no adults listening. It's the word at the Smeath dinner table, stronghold of righteousness.

But inwardly I do not recant. My loyalty to Mr. Smeath is similar to my loyalty to my brother: both are on the side of ox eyeballs, toe jam under the microscope, the outrageous, the subversive. Outrageous to whom, subversive of what? Of Grace and Mrs. Smeath, of tidy paper ladies pasted into scrapbooks. Cordelia ought to be on this side too. Sometimes she is, sometimes she isn't. It's hard to tell.

24

*I*n the mornings the milk is frozen, the cream risen in icy, granular columns out of the bottle necks. Miss Lumley bends over my desk, her invisible navy-blue bloomers casting their desolating aura around her. On either side of her nose the skin hangs down, like the jowls of a bulldog; there's a trace of dried spit in the corner of her mouth. "Your handwriting is deteriorating," she says. I look at my page in dismay. She's right: the letters are no longer round and beautiful, but spidery, frantic, and disfigured with blots of black rusty ink where I've pressed down too hard on the steel nib. "You must try harder." I curl my fingers under. I think she's looking at the ragged edges of skin. Everything she says, everything I do, is heard and seen by Carol and will be reported later.

Cordelia is in a play and we go to watch her. This is my first play and I ought to be excited. Instead I am filled with dread, because I know nothing of the etiquette of play-going and I'm sure I'll do something wrong. The play is at the Eaton's Auditorium; the stage has blue curtains with black velvet horizontal stripes on them. The curtains part to reveal *The Wind in the Willows*. All the actors are children. Cordelia is a weasel, but since she's in a weasel costume with a weasel head, it's impossible to tell her apart from all the other weasels. I sit in the plush theater seat, biting my fingers, craning my neck, looking for her. Knowing she's there but not knowing where is the worst thing. She could be anywhere.

The radio fills with sugary music: "I'm Dreaming of a White Christmas," "Rudolph the Red-Nosed Reindeer," which we have to sing in school, standing beside our desks with Miss Lumley tooting on her pitch pipe to give the note and keeping time with her wooden ruler, the same one she whacks the boys' hands with when they fidget. Rudolph bothers me, because there's something wrong with him; but at the same time he gives me hope, because he ended up beloved. My father says he is a nauseating commercial neologism. "A fool and his money are soon parted," he says.

We make red bells out of construction paper, folding the paper in half before cutting out the shape. We make snowmen the same way. It's Miss Lumley's recipe for symmetry: everything has to be folded, everything has two halves, a left and a right, identical.

I go through these festive tasks like a sleepwalker. I take no interest in bells or snowmen or for that matter in Santa Claus, in whom I've ceased to believe, since Cordelia has told me it's really just your parents. There's a class Christmas party, which consists of cookies brought from home and eaten silently at our desks, and different-colored jelly beans provided by Miss Lumley, five for each child. Miss Lumley knows what the conventions are and pays her own rigid tributes to them.

For Christmas I get a Barbara Ann Scott doll, which I've said I wanted. I had to say I wanted something and I did in a way want this doll. I haven't had any girl-shaped dolls before. Barbara Ann Scott is a famous figure skater, a very famous one. She has won prizes. I've studied the pictures of her in the newspaper.

The doll of her has little leatherette skates and a fur-trimmed costume, pink with white fur, and fringed eyes that open and close, but it looks nothing at all like the real Barbara Ann Scott. According to the pictures she's muscular, with big thighs, but the doll is a slender stick. Barbara is a woman, the doll is a girl. It has the worrying power of effigies, a lifeless life that fills me with creeping horror. I put it back into its cardboard box and tuck the tissue paper around it, over the face. I say I'm doing this to keep it safe, but in fact I don't want it watching me.

Over our chesterfield there's a badminton net, festooned across the wall. In the squares of this net my parents have hung their Christmas cards. No one else I know has a badminton net like this on their wall. Cordelia's Christmas tree is not like ot'.ers: it's covered in gauzy angel hair, and all the lights and decorations on it are blue. But she can get away with such differences, I can't. I know I'll be made to pay for the badminton net, sooner or later.

We sit around the table, eating our Christmas dinner. There's a student of my father's, a young man from India who's here to study insects and who's never seen snow before. We're having him to Christmas dinner because he's foreign, he's far from home, he will be lonely, and they don't even have Christmas in his country. This has been explained to us in advance by our mother. He's polite and ill at ease and he giggles frequently, looking with what I sense is terror at the array of food spread out before him, the mashed potatoes, the gravy, the lurid green and red Jell-O salad, the enormous turkey: my mother has said that the food is different there. I know he's miserable, underneath his smiles and politeness. I'm developing a knack for this, I can sniff out hidden misery in others now with hardly any effort at all.

My father sits at the head of the table, beaming like the Jolly Green Giant. He lifts his glass, his gnome's eyes twinkling. "Mr. Banerji, sir," he says. He always calls his students Mr. and Miss. "You can't fly on one wing."

Mr. Banerji giggles and says, "Very true, sir," in his voice that sounds like the BBC News. He lifts his own glass and sips. What is in the glass is wine. My brother and I have cranberry juice in our wineglasses. Last year or the year before we might have tied our shoelaces together, under the table, so we could signal each other with secret jerks and tugs, but we're both beyond this now for different reasons.

My father ladles out the stuffing, deals the slices of dark and light; my mother adds the mashed potatoes and cranberry sauce and asks Mr. Banerji, enunciating carefully, whether they have turkeys in his country. He says he doesn't believe so. I sit across the table from him, my feet dangling, staring at him, enthralled. His spindly wrists extend from his over-large cuffs, his hands are long and thin, ragged around the nails, like mine. I think he is very beautiful, with his brown skin and brilliant white teeth and his dark appalled eyes. There's a child these colors in the ring of children on the front of the Sunday school missionary paper, yellow children, brown children, all in different costumes, dancing around Jesus. Mr. Banerji doesn't have a costume, only a jacket and tie like other men. Nevertheless I can hardly believe he's a man, he seems so unlike one. He's a creature more like myself: alien and apprehensive. He's afraid of us. He has no idea what we will do next, what impossibilities we will expect of him, what we will make him eat. No wonder he bites his fingers.

"A little off the sternum, sir?" my father asks him, and Mr. Banerji brightens at the word.

"Ah, the sternum," he says, and I know they have entered together the shared world of biology, which offers refuge from the real, awkward world of manners and silences we're sitting in at the moment. As he slices away with the carving knife my father indicates to all of us, but especially to Mr. Banerji, the areas where the flight muscles attach, using the carving fork as a pointer. Of course, he says, the domestic turkey has lost the ability to fly.

"*Meleagris gallopavo*," he says, and Mr. Banerji leans forward; the Latin perks him up. "A pea-brained animal, or bird-brained you might say, bred for its ability to put on weight, especially on the drumsticks"—he points these out—"certainly not for intelligence. It was originally domesticated by the Mayans." He tells a story of a turkey farm where the turkeys all died because they were too stupid to go into their shed during a thunderstorm. Instead they stood around outside, looking up at the sky with their beaks wide open and the rain ran down their throats and drowned them. He says this is a story told by farmers and probably not true, although the stupidity of the bird is legendary. He says that the wild turkey, once abundant in the deciduous forests in these regions, is far more intelligent and can elude even practiced hunters. Also it can fly.

I sit picking at my Christmas dinner, as Mr. Banerji is picking at his. Both of us have messed the mashed potatoes around on our plates without actually eating much. Wild things are smarter than tame ones, that much is clear. Wild things are elusive and wily and look out for themselves. I divide the people I know into tame and wild. My mother, wild. My father and brother, also wild; Mr. Banerji, wild also, but in a more skittish way. Carol, tame. Grace, tame as well, though with sneaky vestiges of wild. Cordelia, wild, pure and simple.

"There are no limits to human greed," says my father.

"Indeed, sir?" says Mr. Banerji, as my father goes on to say that he's heard some son of a gun is working on an experiment to breed a turkey with four drumsticks, instead of two drumsticks and two wings, because there's more meat on a drumstick.

"How would such a creature walk, sir?" asks Mr. Banerji, and my father, approving, says, "Well may you ask." He tells Mr. Banerji that some darn fool scientists are working on a square tomato, which will supposedly pack more easily into crates than the round variety.

"All the flavor will be sacrificed, of course," he says. "They care nothing for flavor. They bred a naked chicken, thinking they'd get more eggs by utilizing the energy saved from feather production, but the thing shivered so much they had to double-heat the coop, so it cost more in the end."

"Fooling with Nature, sir," says Mr. Banerji. I know already that this is the right response. Investigating Nature is one thing and so is defending yourself against it, within limits, but fooling with it is quite another.

Mr. Banerji says he hears there is now a naked cat available, he's read about it in a magazine, though he himself does not see the point of it at all. This is the most he has said so far.

My brother asks if there are any poisonous snakes in India, and Mr. Banerji, now much more at ease, begins to enumerate them. My mother smiles, because this is going better than she thought it would. Poisonous snakes are fine with her, even at the dinner table, as long as they make people happy.

My father has eaten everything on his plate and is digging for more stuffing in the cavity of the turkey, which resembles a trussed, headless baby. It has thrown off its disguise as a meal and has revealed itself to me for what it is, a large dead bird. I'm eating a wing. It's the wing of a tame turkey, the stupidest bird in the world, so stupid it can't even fly any more. I am eating lost flight.

25

*A*fter Christmas I'm offered a job. The job is wheeling Brian Finestein around the block in his baby carriage after school, for an hour or a little longer if it isn't too cold, one day a week. For this I get twenty-five cents, which is a lot of money.

The Finesteins live in the house beside ours, the big house that was built suddenly where the mud mountain used to be. Mrs. Finestein is short for a woman, plump, with dark curly hair and lovely white teeth. These show often, as she laughs a lot, wrinkling up her nose like a puppy as she does it, shaking her head, which makes her gold earrings twinkle. I'm not sure, but I think these earrings actually go through little holes in her ears, unlike any earrings I have ever seen.

I ring the doorbell and Mrs. Finestein opens the door. "My little lifesaver," she says. I wait in the vestibule, my winter boots dripping onto the spread newspapers. Mrs. Finestein, wearing a flowered pink housecoat and slippers with high heels and real fur, bustles upstairs to get Brian. The vestibule smells of Brian's ammonia-soaked diapers, which are in a pail waiting to be collected by the diaper company. I'm intrigued by the idea that someone else can come and take away your laundry. Mrs. Finestein always has a bowl of oranges out, on a table up a few steps from the vestibule; no one else leaves oranges lying around like that when it isn't Christmas. There's a gold-colored candlestick like a tree behind the bowl. These things—the sickly sweet baby shit smell of the festering diapers, the bowl of oranges and the gold tree—blend in my mind into an image of ultrasophistication.

Mrs. Finestein clops down the stairs carrying Brian, who is zipped into a blue bunny suit with ears. She gives him a big kiss on his cheek, joggles him up and down, tucks him into the carriage, snaps up the waterproof carriage cover. "There, Bry-Bry," she says. "Now Mummy can hear herself think." She laughs, wrinkles her nose, shakes her gold earrings. Her skin is rounded out, milky-smelling. She's not like any mother I've ever seen.

I wheel Brian out into the cold air and we start off around the block, over the crunchy snow which is spread with cinders from people's furnaces and dotted here and there with frozen horse buns. I can't figure out how Brian would ever be able to interfere with Mrs. Finestein's thinking, because he never cries. Also he never laughs. He never makes any noises at all, nor does he go to sleep. He just lies there in his carriage, gazing solemnly up at me with his round blue eyes as his button of a nose gets redder and redder. I make no attempt to entertain him. But I like him: he's silent, but also uncritical.

When I think it's time I wheel him back, and Mrs. Finestein says, "Don't tell me it's five o'clock already!" I ask her to give me nickels instead of a quarter, because it looks like more. She laughs a lot at that, but she does it. I keep all my money in an old tin tea caddy with a picture of the desert on it, palm trees and camels. I like taking it out and spreading it over my bed. Instead of counting it, I arrange it by the year that's stamped on each piece of money: 1935, 1942, 1945. Every coin has a King's head on it, cut off neatly at the neck, but the Kings are different. The ones from before I was born have beards, but the ones now don't, because it's King George, the one at the back of the classroom. It gives me an odd comfort to sort this money into piles of cut-off heads.

Brian and I wheel around the block, around the block again. It's hard for me to tell when it's an hour because I don't have a watch. Cordelia and Grace come around the corner up ahead, with Carol trailing. They see me, walk over.

"What rhymes with Elaine?" Cordelia asks me. She doesn't wait for an answer. "Elaine is a pain."

Carol peers into the baby carriage. "Look at the bunny ears," she says. "What's his name?" Her voice is wistful. I see Brian in a new light. It isn't everyone who's allowed to wheel a baby.

"Brian," I say. "Brian Finestein."

"Finestein is a Jewish name," says Grace.

I don't know what Jewish is. I've seen the word *Jew*, the Bible is full of that word, but I didn't know there were any live, real ones, especially next door to me.

"Jews are kikes," says Carol, glancing at Cordelia for approval.

"Don't be vulgar," says Cordelia, in her adult voice. "Kike is not a word we use."

I ask my mother what *Jewish* is. She says it's a different kind of religion. Mr. Banerji is a different kind of religion as well, though not Jewish. There are many different kinds. As for the Jews, Hitler killed a great many of them, during the war.

"Why?" I say.

"He was demented," says my father. "A megalomaniac." Neither of these words is much help.

"A bad person," says my mother.

I wheel Brian over the cindery snow, easing him around the potholes. He goggles up at me, his nose red, his tiny mouth unsmiling. Brian has a new dimension: he is a Jew. There is something extra and a little heroic about him; not even the blue ears of his bunny suit can detract from that. *Jewish* goes with the diapers, the oranges in the bowl, Mrs. Finestein's gold earrings and her possibly real ear holes, but also with ancient, important matters. You wouldn't expect to see a Jew every day.

Cordelia and Grace and Carol are beside me. "How's the little baby today?" asks Cordelia.

"He's fine," I say guardedly.

"I didn't mean him, I meant you," says Cordelia.

"Can I have a turn?" asks Carol.

"I can't," I tell her. If she does it wrong, if she upsets Brian Finestein into a snowbank, it will be my fault.

"Who wants an old Jew baby anyway," she says.

"The Jews killed Christ," says Grace primly. "It's in the Bible."

But Jews don't interest Cordelia much. She has other things on her mind. "If a man who catches fish is a fisher, what's a man who catches bugs?" she says.

"I don't know," I say.

"You are so stupid," says Cordelia. "That's what your father is, right? Go on. Figure it out. It's really simple."

"A bugger," I say.

"Is that what you think of your own father?" Cordelia says. "He's an *entomologist*, stupid. You should be ashamed. You should have your mouth washed out with soap."

I know that *bugger* is a dirty word, but I don't know why. Nevertheless I have betrayed, I have been betrayed. "I have to go," I say. Wheeling Brian back to Mrs. Finestein's, I cry silently, while Brian watches me, expressionless. "Goodbye, Brian," I whisper to him.

I tell Mrs. Finestein that I can't do the job any more because I have too much schoolwork. I can't tell her the real reason: that in some obscure way Brian is not safe with me. I have images of Brian headfirst in a snowbank; Brian hurtling in his carriage down the icy hill by the side of the bridge, straight toward the creek full of dead people; Brian tossed into the air, his bunny ears flung upwards in terror. I have only a limited ability to say no.

"Honey, that's all right," she says, looking into my raw, watery eyes. She puts her arm around me and gives me a hug and an extra nickel. No one has ever called me *honey* before this.

I go home, knowing I have failed her, and also myself. *Bugger*, I think to myself. I say it over and over until it disappears into its own syllables. *Erbug, erbug.* It's a word with no meaning, like *kike*, but it reeks of ill will, it has power. What have I done to my father?

I take all of Mrs. Finestein's King's-head nickels and spend them at the store on the way home from school. I buy licorice whips, jelly beans, many-layered blackballs with the seed in the middle, packages of fizzy sherbet you suck up through a straw. I dole them out equally, these offerings, these atonements, into the waiting hands of my friends. In the moment just before giving, I am loved.

26

*I*t's Saturday. Nothing has happened all morning. Icicles form on the eaves trough above the south window, dripping in the sunlight with a steady sound like a leak. My mother is baking in the kitchen, my father and brother are elsewhere. I eat my lunch alone, watching the icicles.

The lunch is crackers and orange cheese and a glass of milk, and a bowl of alphabet soup. My mother thinks of alphabet soup as a cheerful treat for children. The alphabet soup has letters floating in it, white letters: capital A's and O's and S's and R's, the occasional X or Z. When I was younger I would fish the letters out and spell things with them on the edge of the plate, or eat my name, letter by letter. Now I just eat the soup, taking no particular interest. The soup is orangey-red and has a flavor, but the letters themselves taste like nothing.

The telephone rings. It's Grace. "You want to come out and play?" she says, in her neutral voice that is at the same time blank and unsoft, like glazed paper. I know Cordelia is standing beside her. If I say no, I will be accused of something. If I say yes, I will have to do it. I say yes.

"We'll come and get you," Grace says.

My stomach feels dull and heavy, as if it's full of earth. I put on my snowsuit and boots, my knitted hat and mittens. I tell my mother I'm going out to play. "Don't get chilled," she says.

The sun on the snow is blinding. There's a crust of ice over the drifts, where the top layer of snow has melted and then refrozen. My boots make clean-edged footprints through the crust. There's no one around. I walk through the white glare, toward Grace's house. The air is wavery, filled with light, overfilled; I can hear the pressure of it against my eyes. I feel translucent, like a hand held over a flashlight or the pictures of jellyfish I've seen in magazines, floating in the sea like watery flesh balloons.

At the end of the street I can see the three of them, very dark, walking toward me. Their coats look almost black. Even their faces when they come closer look too dark, as if they're in shadow.

Cordelia says, "We said we would come and get you. We didn't say you could come here."

I say nothing.

Grace says, "She should answer when we talk to her."

Cordelia says, "What's the matter, are you deaf?"

Their voices sound far away. I turn aside and throw up onto a snowbank. I didn't mean to do it and didn't know I was going to. I feel sick to my stomach every morning, I'm used to that, but this is the real thing, alphabet soup mixed with shards of chewed-up cheese, amazingly red and orange against the white of the snow, with here and there a ruined letter.

Cordelia doesn't say anything. Grace says, "You better go home." Carol, behind them, sounds as if she's going to cry. She says, "It's on her face." I walk back toward my house, smelling the vomit on the front of my snowsuit, tasting it in my nose and throat. It feels like bits of carrot.

I lie in bed with the scrub pail beside me, floating lightly on waves of fever. I throw up several times, until nothing but a little green juice comes out. My mother says, "I suppose we'll all get it," and she's right. During the night I can hear hurrying footsteps and retching and the toilet flushing. I feel safe, small, wrapped in my illness as if in cotton wool.

I begin to be sick more often. Sometimes my mother looks into my mouth with a flashlight and feels my forehead and takes my temperature and sends me to school, but sometimes I'm allowed to stay home. On these days I feel relief, as if I've been running for a long time and have reached a place where I can rest, not forever but for a while. Having a fever is pleasant, vacant. I enjoy the coolness of things, the flat ginger ale I'm given to drink, the delicacy of taste, afterward.

I lie in bed, propped up on pillows, a glass of water on a chair beside me, listening to the faraway sounds coming from my mother: the eggbeater, the vacuum cleaner, music from the radio, the lakeshore sound of the floor polisher. Winter sunlight slants in through the window, between the half-drawn curtains. I now have curtains. I look at the ceiling light fixture, opaque yellowish glass with the shadows of two or three dead flies caught inside it showing through as if through cloudy jelly. Or I look at the doorknob.

Sometimes I cut things out of magazines and paste them into a scrapbook with LePage's mucilage, from the bottle that looks like a chess bishop. I cut out pictures of women, from *Good Housekeeping*, *The Ladies' Home Journal*, *Chatelaine*. If I don't like their faces I cut off the heads and glue other heads on. These women have dresses with puffed sleeves and full skirts, and white aprons that tie very tightly around their waists. They put germ killers onto germs, in toilet bowls; they polish windows, or clean their spotty complexions with bars of soap, or shampoo their oily hair; they get rid of their unwanted odors, rub hand lotion onto their rough wrinkly hands, hug rolls of toilet paper against their cheeks.

Other pictures show women doing things they aren't supposed to do. Some of them gossip too much, some are sloppy, others bossy. Some of them knit too much. "Walking, riding, standing, sitting, Where she goes, there goes her knitting," says one. The picture shows a woman knitting on a streetcar, with the ends of her knitting needles poking into the people beside her and her ball of wool unrolling down the aisle. Some of the women have a Watchbird beside them, a red and black bird like a child's drawing, with big eyes and stick feet. "This is a Watchbird watching a Busybody," it says. "This is a Watchbird watching YOU."

I see that there will be no end to imperfection, or to doing things the wrong way. Even if you grow up, no matter how hard you scrub, whatever you do, there will always be some other stain or spot on your face or stupid act, somebody frowning. But it pleases me somehow to cut out all these imperfect women, with their forehead wrinkles that show how worried they are, and fix them into my scrapbook.

At noon there's the Happy Gang, on the radio, knocking at the door.

> *Knock knock knock.*
> *Who's there?*
> *It's the Happy Gang!*
> *Well, come ON IN!*
> > *Keep happy in the Happy Gang way,*
> > *Keep healthy, hope you're feeling okay,*
> > *Cause if you're happy, and healthy,*
> > *The heck with being wealthy,*
> > *So be happy with the Happy Gang!*

The Happy Gang fills me with anxiety. What happens to you if you aren't happy and healthy? They don't say. They themselves are always happy, or say they are; but I can't believe anyone can be always happy. So they must be lying some of the time. But when? How much of their fake-sounding laughter is really fake?

A little later there's the Dominion Observatory Official Time Signal: first a series of outer space beeps, then silence, then a long dash. The long dash means one o'clock. Time is passing; in the silence before the long dash the future is taking shape. I turn my head into the pillow. I don't want to hear it.

27

*T*he winter melts, leaving a grubby scum of cinders, wet paper, soggy old leaves. A huge pile of topsoil appears in our backyard, then a pile of rolled-up squares of grass. My parents, in muddy boots and earth-stained pants, lay them over our mud like bathroom tiles. They pull out couch grass and dandelions, plant green onions and a row of lettuce. Cats appear from nowhere, scratching and squatting in the soft, newly planted earth, and my father throws clumps of dug-up dandelions at them. "Dad-ratted cats," he says.

The buds turn yellow, the skipping ropes come out. We stand in Grace's driveway, beside her dark pink crab apple tree. I turn the rope, Carol turns the other end, Grace and Cordelia skip. We look like girls playing.

We chant:

> *Not last night but the night before*
> *Twenty-four robbers come to my back door*
> *And this is what they said . . . to . . . me!*
> *Lady turn around, turn around, turn around,*
> *Lady touch the ground, touch the ground, touch the ground;*
> *Lady show your shoe, show your shoe, show your shoe,*
> *Lady, lady, twenty-four skiddoo!*

Grace, skipping in the middle, turns around, touches the driveway, kicks up one foot sedately, smiling her little smile. She rarely trips.

This chant is menacing to me. It hints at an obscure dirtiness. Something is not understood: the robbers and their strange commands, the lady and her gyrations, the tricks she's compelled to perform, like a trained dog. And what does "twenty-four skiddoo" mean, at the end of it? Is she scooted out the door of her house while the robbers remain inside, free to take anything they like, break anything, do whatever they want? Or is it the end of her altogether? I see her dangling from the crab apple tree, the skipping rope noosed around her neck. I am not sorry for her.

The sun shines, the marbles return, from wherever they've been all winter. The voices of the children rise in the schoolyard: *purie, purie, bowlie, bowlie, two for one.* They sound to me like ghosts, or like animals caught in a trap: thin wails of exhausted pain.

We cross the wooden bridge on the way home from school. I am walking behind the others. Through the broken boards I can see the ground below. I remember my brother burying his jar full of puries, of waterbabies and cat's eyes, a long time ago, down there somewhere under the bridge. The jar is still there in the earth, shining in the dark, in secret. I think about myself going down there alone despite the sinister unseen men, digging up the treasure, having all that mystery in my hands. I could never find the jar, because I don't have the map. But I like to think about things the others know nothing about.

I retrieve my blue cat's eye from where it's been lying all winter in the corner of my bureau drawer. I examine it, holding it up so the sunlight burns through it. The eye part of it, inside its crystal sphere, is so blue, so pure. It's like something frozen in the ice. I take it to school with me, in my pocket, but I don't set it up to be shot at. I hold on to it, rolling it between my fingers.

"What's that in your pocket?" says Cordelia.

"Nothing," I say. "It's only a marble."

It's marble season; everyone has marbles in their pockets. Cordelia lets it pass. She doesn't know what power this cat's eye has, to protect me. Sometimes when I have it with me I can see the way it sees. I can see people moving like bright animated dolls, their mouths opening and closing but no real words coming out. I can look at their shapes and sizes, their colors, without feeling anything else about them. I am alive in my eyes only.

· · ·

We stay in the city later than we've ever stayed before. We stay until school ends for the summer and the daylight lasts past bedtime and wet heat descends over the streets like a steaming blanket. I drink grape Freshie, which does not taste like grapes but like something you might use to kill insects, and wonder when we're going to leave for the north. I tell myself it will be never, so I won't be disappointed. But despite my cat's eye I know I can't stand to be here in this place much longer. I will burst inward. I've read in the *National Geographic* about deepsea diving and why you have to wear a thick metal suit or the invisible pressure of the heavy undersea water will crush you like mud in a fist, until you implode. This is the word: *implode*. It has a dull final sound to it, like a lead door closing.

I sit in the car, packed into the back seat like a parcel. Grace and Cordelia and Carol are standing among the apple trees, watching. I hunch down, avoiding them. I don't want to pretend, to undergo goodbyes. As the car moves away they wave.

We drive north. Toronto is behind us, a smear of brownish air on the horizon, like smoke from a distant burning. Only now do I turn and look.

The leaves get smaller and yellower, folding back toward the bud, and the air crisps. I see a raven by the side of the road, picking at a porcupine that's been run over by a car, its quills like a huge burr, its guts pink and scrambled like eggs. I see the northern granite rock rising straight up out of the ground with the road cut through it. I see a raggedy lake with dead trees stuck into the marsh around the edges. A sawdust burner, a fire tower.

Three Indians stand beside the road. They aren't selling anything, no baskets and it's too early for blueberries. They just stand there as if they've been doing it for a long time. They're familiar to me but only as scenery. Do they see me as I stare at them out of the car window? Probably not. I'm a blur to them, one more face in a car that doesn't stop. I have no claim on them, or on any of this.

.   .   .

I sit in the back seat of the car that smells of gasoline and cheese, waiting for my parents, who are buying groceries. The car is beside a wooden general store, saggy and weathered gray, stuck together by the signs nailed all over the outside of it: BLACK CAT CIGARETTES, PLAYERS, COCA-COLA. This isn't even a village, just a wide place in the highway, beside a bridge beside a river. Once I would have wanted to know the river's name. Stephen stands on the bridge, dropping pieces of wood upstream, timing how long it takes them to come out the other side, calculating the rate of flow. The blackflies are out. Some of them are in the car, crawling up the window, jumping, crawling up again. I watch them do this: I can see their hunched backs, their abdomens like little black-red bulbs. I squash them against the glass, leaving red smears of my own blood.

I've begun to feel not gladness, but relief. My throat is no longer tight, I've stopped clenching my teeth, the skin on my feet has begun to grow back, my fingers have healed partially. I can walk without seeing how I look from the back, talk without hearing the way I sound. I go for long periods without saying anything at all. I can be free of words now, I can lapse back into wordlessness, I can sink back into the rhythms of transience as if into bed.

This summer we're in a rented cabin on the north shore of Lake Superior. There are a few other cottages around, most of them empty; there are no other children. The lake is huge and cold and blue and treacherous. It can sink freighters, drown people. In a wind the waves roll in with the crash of oceans. Swimming in it doesn't frighten me at all. I wade into the freezing water, watching my feet and then my legs go down into it, long and white and thinner than on land.

There's a wide beach, and to one end of it a colony of boulders. I spend time among them. They're rounded, like seals, only hard; they heat up in the sun, and stay warm in the evening when the air cools. I take pictures of them with my Brownie camera. I give them the names of cows.

Above the beach, on the dunes, there are beach plants,

fuzzy mulleins and vetch with its purple flowers and tiny bitter peapods, and grasses that will cut your legs; and behind that the forest, oak and moose maple and birch and poplar, with balsam and spruce among them. There's poison ivy sometimes. It's a secretive, watchful forest, though hard to get lost in, so close to the shore.

Walking in the forest I find a dead raven. It's bigger than they look alive. I poke it with a stick, turning it over, and see the maggots. It smells like rot, like rust, and, more strangely, like some sort of food I've eaten once but can't remember. It's black, but not like a color; more like a hole. Its beak is dingy, horn-colored, like old toenails. Its eyes are shriveled up.

I've seen dead animals before, dead frogs, dead rabbits, but this raven is deader. It looks at me with its shriveled-up eye. I could poke this stick right through it. No matter what I do to it, it won't feel a thing. No one can get at it.

It's hard to fish from the shore of this lake. There's nowhere to stand, no dock. We aren't allowed out in a boat by ourselves because of the currents; anyway we have no boat. Stephen is doing other things. He makes a collection of the boat funnels from the lake freighters, checking them through binoculars. He sets up chess problems and works them out, or splits kindling, or goes for long walks by himself with a butterfly book. He isn't interested in catching the butterflies and mounting them on a board with pins; he just wants to see them, identify them, count them. He writes them down in a list at the back of the book.

I like looking at the pictures of butterflies in his book. My favorite is the luna moth, huge and pale green, with crescents on the wings. My brother finds one of these, and shows it to me. "Don't touch it," he says. "Or the dust will come off its wings, and then it can't fly."

But I don't play chess with him. I don't start my own list of boat funnels or butterflies. I'm ceasing to be interested in games I can't win.

Along the edges of the forest, where there's open sunlight, there are chokecherry trees. The red chokecherries ripen and turn translucent. They're so sour they dry up the inside of your mouth. I pick them into a lard pail, then sort out the dead twigs and leaves, and my mother makes jelly from them, boiling them up, straining the pits out through a cloth jelly bag, adding sugar. She pours the jelly into hot jars, capping them with paraffin wax. I count the beautiful red jars. I helped make them. They look poisonous.

As if I've been given permission I begin to dream. My dreams are brightly colored and without sound.

I dream that the dead raven is alive, only it looks the same, it still looks dead. It hops around and flaps its decaying wings and I wake up, my heart beating fast.

I dream I'm putting on my winter clothes, in Toronto, but my dress doesn't fit. I pull it on over my head and struggle to get my arms into the sleeves. I'm walking along the street and parts of my body are sticking out through the dress, parts of my bare skin. I am ashamed.

I dream that my blue cat's eye is shining in the sky like the sun, or like the pictures of planets in our book on the solar system. But instead of being warm, it's cold. It starts to move nearer, but it doesn't get any bigger. It's falling down out of the sky, straight toward my head, brilliant and glassy. It hits me, passes right into me, but without hurting, except that it's cold. The cold wakes me up. My blankets are on the floor.

I dream that the wooden bridge over the ravine is falling apart. I'm standing on it, the boards crack and separate, the bridge sways. I walk along what's left, clinging to the railing, but I can't get onto the hill where the other people are standing because the bridge isn't attached to anything. My mother is on the hill, but she's talking to the other people.

I dream I'm picking the chokecherries off the chokecherry tree and putting them into the lard pail. Only they aren't chokecherries. They're deadly nightshade berries, translucent, brilliantly red. They're filled with blood, like the bodies of blackflies. As I touch them they burst, and the blood runs over my hands.

None of my dreams is about Cordelia.

Our father plays touch tag with us on the beach in the evenings, running lumberingly like a bear, laughing at the same time, *wuff wuff wuff*. Pennies and dimes fall out of his pockets into the sand. The lake boats go slowly by in the distance, their smoke trailing behind them, the sun sets to the left, pink and tranquil. I look in the mirror over the washbasin: my face is brown and rounder. My mother smiles at me, in the little kitchen with the woodstove, and hugs me with one arm. She thinks I am happy. Some nights we have marshmallows, for a treat.

V
i
──────
● CAT'S EYE

28

*S*impsons Basement used to be bargain clothes and wrenches. Now it's resplendent. There are pyramids of imported chocolates, an ice cream counter, aisles and aisles of fancy cookies and canned gourmet food, ticking away like little clocks toward the obsolescence dates stamped on their packages. There's even an espresso counter. It's all very world-class down here, where I used to buy cheap nighties in high school with my tiny clothes allowance, on sale at that and a size too large. I'm overwhelmed by all the chocolates. Just looking at them reminds me of Christmas and the sticky feeling after eating too many, the surfeit and glut.

I sit at the espresso counter and have a cappuccino, to deal with the inertia that's come over me at the sight of so much sugar-coated self-indulgence. The espresso counter is either fake or real dark-green marble; it has a cute canopy over it, someone's idea of Italy, and little swivel stools. The view from here is the shoe repair counter, which is not very world-class but is reassuring to me. People still get their shoes repaired, despite all this chocolate, they don't just toss them out at the first hint of wear.

I think about the shoes of my childhood, the brown Oxfords scuffed at the toes, half-soled, new-heeled, the falling-apart grubby white running shoes, the brown sandals with two buckles that you wore with socks. Most shoes were brown. They went with the pot roast done in the pressure cooker along with the limp carrots and the flaccid potatoes and the onions with their slippery layers. The pressure cooker had a whistle-shaped thing on the top. If you forgot to pay attention to it the lid would blow off like a bomb, and the carrots and potatoes would be hurled to the ceiling, where they'd stick like mush. This happened to my mother once. Luckily she was not in the kitchen at the time and was not scalded. When she saw what had happened she did not swear. She laughed, and said, "Wouldn't that take the gold-plated gingerbread."

My mother did most of the cooking but it was not her favorite thing. She was not fond of housework generally. In the steamer trunk in the cellar, along with a cut-velvet evening gown from the twenties and a pair of riding jodhpurs, there were several things made of real silver, ornate salt and pepper shakers, sugar tongs in the shape of chickens' feet, rose bowls lavish with silver flowers. They were down there, wrapped in tissue paper and turning black, because otherwise they would have to be polished. Our knives and forks and spoons had to be polished, with an old toothbrush for the decorations. The scrolled legs under the dining table were dust catchers, and so were the kinds of objects—doodads, my mother called them—other people kept on their mantelpieces. But she liked making cakes, though this may only be something I prefer to think.

What would I have done if I had been my mother? She must have realized what was happening to me, or that something was. Even toward the beginning she must have noted my silences, my bitten fingers, the dark scabs on my lips where I'd pulled off patches of the skin. If it were happening now, to a child of my own, I would know what to do. But then? There were fewer choices, and a great deal less was said.

I once did a series about my mother. It was six images, six panels, like a double triptych or a comic book, arranged in two groups, three on top, three underneath. The first was my mother in colored pencil, in her city house kitchen and her late-forties dress. Even she had a bib apron, blue flowers with navy piping, even she wore it, from time to time. The second image was the same figure in collage, made from the illustrations from old *Ladies' Home Journals* and *Chatelaines*, not the photos but the artwork, with those rancid greens and faded blues and dirty-looking pinks. The third was the same figure, white on white, the raised parts pipe cleaners contoured side by side and glued onto a white cloth-covered backing. Reading across from left to right it looked as if my mother was slowly dissolving, from real life into a Babylonian bas-relief shadow.

The bottom set of images went the other way: first the pipe-cleaners, then the same image in collage, then the final one in full-colored realistic detail. But this time my mother was in her slacks and boots and her man's jacket, making chokecherry jam over the outdoor fire. You could read it as a materialization, out of the white pipe cleaner mist into the solid light of day.

I called the whole series *Pressure Cooker*. Because of when it was done and what was going on in those years, some people thought it was about the Earth Goddess, which I found hilarious in view of my mother's dislike of housework. Other people thought it was about female slavery, others that it was a stereotyping of women in negative and trivial domestic roles. But it was only my mother cooking, in the ways and places she used to cook, in the late forties.

I made this right after she died. I suppose I wanted to bring her back to life. I suppose I wanted her timeless, though there is no such thing on earth. These pictures of her, like everything else, are drenched in time.

I finish my cappuccino, pay for it, leave a tip for the imitation Italian waiter who served it to me. I know I won't buy any food in the food hall, I'm too intimidated by it. Ordinarily, or in some other city, I would not be: I am a grown-up and used to shopping. But how could I find, down here, anything I want right now? I'll stop in at some corner store on the way back, some place where they sell milk till midnight and slightly stale sliced white bread. Such stores are run, now, by people the color of Mr. Banerji, or by Chinese people. They aren't necessarily any friendlier than the pasty-white people who used to run such stores, but the general content of their disapproval is more easily guessed; though not the details.

.   .   .

I head back up the escalator, into the perfumed fug of the ground floor. The air is bad here, there's too much musk, the overpowering scent of money. I make it into the open air and walk west, past the murderous mannequins in the windows, past the bivalvular City Hall.

Ahead of me there's a body lying on the sidewalk. People walk around it, look down, look away, keep going. I see their faces coming toward me bearing that careful rearrangement of the features that's meant to say, *This is none of my business.*

When I get up even, I see that this person is a woman. She's lying on her back, staring straight at me. "Lady," she says. "Lady. Lady."

That word has been through a lot. Noble lady, Dark Lady, she's a real lady, old-lady lace, Listen lady, Hey lady watch where you're going, Ladies' Room, run through with lipstick and replaced with Women. But still the final word of appeal. If you want something very badly you do not say *Woman, Woman,* you say *Lady, Lady.* As she is saying now.

I think, *What if she's had a heart attack?* I look: there's blood on her forehead, not much, but a cut. She must have hit her head falling. And no one's stopping, and she's lying there on her back, a bulky fifty-odd woman in a poor-person green coat, gabardine, and lamentable shoes all cracked, her arms outflung. The tanned-looking skin around her brown eyes is red and puffy, her long black and gray hair is splayed across the sidewalk.

"Lady," she says, or something, it's a mumble, but she's got me now.

I look over my shoulder to see if anyone else will do this, but there are no takers. I kneel, say to her, "Are you all right?" What a stupid question, she so obviously isn't. Vomit and alcohol are around here somewhere. I have visions of myself taking her for coffee, and then where? I won't be able to get rid of her, she'll follow me back to the studio, throw up in the bathtub, sleep on the futon. They get me every time, they can spot me coming, pick me out of the crowd no matter how hard I frown. Sidewalk rap artists, Moonies, guitar-playing young men who ask me for subway tokens. In the clutch of the helpless I am helpless.

"She's only drunk," a man says in passing. What does he mean, *only*? It's hell enough.

"Here," I say, "I'll help you up." Wimp, I tell myself. She'll ask you for money and you'll give it to her, and she'll spend it on cheap sweet wine. But I have her on her feet now, she's slumped against me. If I can lug her over to the nearest wall I can prop her up, dust her off a little, think how to get away.

"There," I say. But she won't lean against the wall, she's leaning against me instead. Her breath smells like a bad accident. She's crying now, the shameless abandoned weeping of a child; her fingers clutch my sleeve.

"Don't leave me," she says. "Oh God. Don't leave me all alone." Her eyes are closed, her voice is pure neediness, pure woe. It hits the weakest, most sorrowing part of me; but I am only a surrogate, for who knows what lack, what loss. There's nothing I can do.

"Here," I say. I fumble in my purse, find a ten, crumple it into her hand, paying her off. I'm a sucker, I'm a bleeding heart. There's a cut in my heart, it bleeds money.

"Bless you," she says. Her head rolls from side to side, back against the wall. "God bless you lady, Our Lady bless you." It's a slurred blessing, but who's to say I don't need it? She must be a Catholic. I could find a church, slide her in through the door like a packet. She's theirs, let them deal with her.

"I have to go now," I say. "You'll be all right." Lying through my teeth. She opens her eyes wide, trying to focus. Her face goes quiet.

"I know about you," she says. "You're Our Lady and you don't love me."

Full-blown booze madness, and absolutely the wrong person. I draw my hand back from her as if she's a live socket. "No," I say. She's right, I don't love her. Her eyes are not brown but green. Cordelia's.

I walk away from her, guilt on my hands, absolving myself: I'm a good person. She could have been dying. Nobody else stopped.

I'm a fool, to confuse this with goodness. I am not good. I know too much to be good. I know myself.

I know myself to be vengeful, greedy, secretive and sly.

29

*W*e come back in September. In the north the nights are cold and the leaves are beginning to turn, but the city is still hot, still damp. It's astonishingly noisy and stinks of gasoline and the tar of melting roads. The air inside our house is stale and flat, air that's been locked up in the heat all summer. The water's rusty at first, coming out of the taps. I take a bath in the reddish lukewarm water. Already my body is stiffening, emptying itself of feeling. The future is closing on me like a door.

Cordelia has been waiting for me. I know this as soon as I see her standing at the school bus stop. Before the summer she would alternate between kindness and malice, with periods of indifference; but now she's harsher, more relentless. It's as if she's driven by the urge to see how far she can go. She's backing me toward an edge, like the edge of a cliff: one step back, another step, and I'll be over and falling.

Carol and I are in Grade Five now. We have a new teacher, Miss Stuart. She's Scottish and has an accent. "Now ger-ruls," she says. She has a little bunch of dried heather stuck into a jelly jar on her desk, and a miniature of Bonnie Prince Charlie who was ruined by the English and whose last name is the same as her own, and a bottle of hand lotion in her desk drawer. She cooks this hand lotion herself.

In the afternoons she makes herself a cup of tea, which does not smell entirely like tea but of something else she puts into it, out of a small silver bottle. She has bluish-white hair, beautifully waved, and wears rustling, silky mauve dresses with a lace-edged handkerchief tucked into the sleeve. She often has a nurse's white gauze mask over her nose and mouth because she's allergic to chalk dust. This doesn't stop her from throwing the blackboard brushes at boys who aren't paying attention. Although she throws underhand and not hard, she never misses. After she hits the boy he has to bring the blackboard brush back to her. The boys don't seem to resent this habit of hers; they take it as a mark of distinction to get hit.

Everyone loves Miss Stuart. Carol says we are lucky to be in her class. I would love her too, if I had the energy. But I am too numb, too enthralled.

I keep my cat's eye in my pocket, where I can hold on to it. It rests in my hand, valuable as a jewel, looking out through bone and cloth with its impartial gaze. With the help of its power I retreat back into my eyes. Up ahead of me are Cordelia, Grace, and Carol. I look at their shapes as they walk, the way shadow moves from one leg to another, the blocks of color, a red square of cardigan, a blue triangle of skirt. They're like puppets up ahead, small and clear. I could see them or not, at will.

I reach the path to the bridge, start down, past the nightshade vines with their red berries, past the undulating leaves, the lurking cats. The three of them are already on the bridge but they've stopped, they're waiting for me. I look at the ovals of their faces, the outline of hair around each one. Their faces are like moldy eggs. My feet move down the hill.

I think about becoming invisible. I think about eating the deadly nightshade berries from the bushes beside the path. I think about drinking the Javex out of the skull and crossbones bottle in the laundry room, about jumping off the bridge, smashing down there like a pumpkin, half of an eye, half of a grin. I would come apart like that, I would be dead, like the dead people.

I don't want to do these things, I'm afraid of them. But I think about Cordelia telling me to do them, not in her scornful voice, in her kind one. I hear her kind voice inside my head. *Do it. Come on.* I would be doing these things to please her.

I consider telling my brother, asking him for help. But

tell him what exactly? I have no black eyes, no bloody noses to report: Cordelia does nothing physical. If it was boys, chasing or teasing, he would know what to do, but I don't suffer from boys in this way. Against girls and their indirectness, their whisperings, he would be helpless.

Also I'm ashamed. I'm afraid he'll laugh at me, he'll despise me for being a sissy about a bunch of girls, for making a fuss about nothing.

I'm in the kitchen, greasing muffin tins for my mother. I see the patterns the grease leaves on the metal, I see the moons of my nails, the raggedy flesh. My fingers go around and around.

My mother makes the batter for the muffins, measuring the salt, sifting the flour. The sifter sounds dry, like sandpaper. "You don't have to play with them," my mother says. "There must be other little girls you can play with instead."

I look at her. Misery washes over me like a slow wind. What has she noticed, what has she guessed, what is she about to do? She might tell their mothers. This would be the worst thing she could do. Also I can't imagine it. My mother is not like the other mothers, she doesn't fit in with the idea of them. She does not inhabit the house, the way the other mothers do; she's airy and hard to pin down. The others don't go skating on the neighborhood rink, or walk in the ravine by themselves. They seem to me grown-up in a way that my own mother is not. I think of Carol's mother in her twin set, her skeptical smile, Cordelia's with her glasses on a chain and her vagueness, Grace's and her hairpins and drooping apron. My mother will turn up on their doorsteps, wearing slacks, carrying a bouquet of weeds, incongruous. They won't believe her.

"When I was little and the kids called names, we used to say, 'Sticks and stones will break my bones but names will never hurt me,' " she says. Her arm goes vigorously around, mixing, efficient and strong.

"They don't call me names," I say. "They're my friends." I believe this.

"You have to learn to stand up for yourself," says my

mother. "Don't let them push you around. Don't be spineless. You have to have more backbone." She dollops the batter into the tins.

I think of sardines and their backbones. You can eat their backbones. The bones crumble between your teeth; one touch and they fall apart. This must be what my own backbone is like: hardly there at all. What is happening to me is my own fault, for not having more backbone.

My mother sets down the bowl and puts her arms around me. "I wish I knew what to do," she says. This is a confession. Now I know what I've been suspecting: as far as this thing is concerned, she is powerless.

I know that muffins have to be baked right away, right after they've been ladled out, or they'll be flat and ruined. I can't afford the distraction of comfort. If I give in to it, what little backbone I have left will crumble away to nothing.

I pull away from her. "They need to go into the oven," I say.

Cordelia brings a mirror to school. It's a pocket mirror, the small plain oblong kind without any rim. She takes it out of her pocket and holds the mirror up in front of me and says, "Look at yourself! Just look!" Her voice is disgusted, fed up, as if my face, all by itself, has been up to something, has gone too far. I look into the mirror but I don't see anything out of the ordinary. It's just my face, with the dark blotches on the lips where I've bitten off the skin.

My parents have bridge parties. They push the furniture in the living room to the walls and unfold two metal bridge tables and eight bridge chairs. In the middle of each table there are two china dishes, one with salted nuts, the other with mixed candies. These candies are called "bridge mixture." There are also two ashtrays on each table.

Then the doorbell begins to ring and the people come in. The house fills with the alien scent of cigarettes, which will still be there in the morning along with a few uneaten candies and salted nuts, and with bursts of laughter that get louder as time passes. I lie in my bed listening to the bursts of laughter. I feel isolated, left out. Also I don't understand why this activity, these noises and smells, is called "bridge." It is not like a bridge.

Sometimes Mr. Banerji comes to these bridge parties. I lurk in the corner of the hallway in my flannelette pajamas, hoping to catch a glimpse of him. I don't have a crush on him or anything like that. My wish to see him is anxiety, and fellow feeling. I want to see how he is managing, how he is coping with his life, with having to eat turkeys, and with other things. Not very well, judging from his dark, haunted-looking eyes and slightly hysterical laughter. But if he can deal with whatever it is that's after him, and something is, then so can I. Or this is what I think.

Princess Elizabeth is coming to Toronto. She's visiting Canada with her husband, who is a Duke. It's a Royal Visit. On the radio there are cheering crowds, and solemn voices describing what color she's wearing, a different color every day. I crouch on the living room floor with the Maritimes fiddle music going on in the background, the *Toronto Star* spread out underneath my elbows, studying the picture of her on the front page. She's older than she should be and more ordinary: no longer in a Girl Guide uniform as in the days of the Blitz, but not in an evening gown and a tiara either, like the Queen at the back of the classroom. She's wearing a plain suit and gloves and carrying a handbag, like anyone, and she has on a ladies' hat. But still she's a Princess. On the inside of the paper there's a full page of her, with women curtsying to her, little girls presenting bouquets of flowers. She smiles down upon them, always the same benevolent smile, and is described as radiant.

Day after day, crouching on the floor, turning the pages of the newspapers, I watch her make her way across the map, by plane, by train, by car, from city to city. I memorize the diagrams of her proposed route through Toronto. I'll get a good chance to see her, because she's supposed to drive right by our house, along the raw, potholed road that runs between the cemetery, with its spindly new trees and heaps of bulldozed earth, and the line of five new mud mountains.

The mud mountains are on our side of the road. They have recently appeared, replacing the strip of weedy field that used to be there before. Each mountain stands beside its own hole, roughly cellar-shaped, with a slop of muddy water at the bottom. My brother has claimed one of them for his own; he plans to excavate it, tunneling down from the top, then in from the side to make a side entrance. What he wishes to do in there is unknown.

I don't know why the Princess is going to be driven past these mud mountains. I don't think of them as anything she would necessarily want to see, but I'm not sure, because she's seeing a lot of other things that don't seem to be of more interest. There's a picture of her outside a city hall, another beside a fish-canning factory. But whether she wants to see them or not, the mud mountains will be a good place to stand.

I am looking forward to this visit. I expect something from it, although I'm not sure what. This is the same Princess that defied the bombs in London, the one that is brave and heroic. I think something will happen for me on that day. Something will change.

The Royal Visit finally reaches Toronto. The day is overcast, with pinpoints of rain; spitting, they call it. I go out early and stand on the top of the middle mud mountain. There's a straggly line of people, adults and children, along the roadside among the draggled weeds. Some of the children have small Union Jacks. I have one as well: they were handed out at school. There's not much of a crowd, because not that many people live around here and some of them have probably gone farther downtown, to where there are sidewalks. I can see Grace and Carol and Cordelia, along the road towards Grace's house. I hope they will not see me.

I stand on the mud mountain with my Union Jack hanging slack from its stick. It gets later and nothing happens. I think maybe I should go back to our house and listen to the radio, to see how far away the Princess is, but suddenly there's a police car, to the left, coming along by the cemetery. It begins to drizzle. In the distance there's cheering.

There are some motorcycles, then some cars. I can see the arms of the people along the road going up into the air, hear scattered hoorays. The cars are going too fast, despite the potholes. I can't see which car is the right one.

Then I do see. It's the car with the pale glove coming out the window, waving back and forth. Already it's opposite me, already it's passing. I don't wave my Union Jack or cheer, because I see that it's too late, I won't have time for what I've been waiting for, which has only now become clear to me. What I must do is run down the mountain with my arms stretched out to either side, for balance, and throw myself in front of the Princess's car. In front of it, or onto it, or into it. Then the Princess will tell them to stop the car. She'll have to, in order to avoid running over me. I don't picture myself being driven away in the royal car, I'm more realistic than that. Anyway I don't want to leave my parents. But things will change, they will be different, something will be done.

The car with the glove is moving away, it's turned the corner, it's gone, and I haven't moved.

31

*M*iss Stuart likes art. She has us bring old shirts of our fathers from home so we can do messier art without getting our clothes dirty. While we scissor and paint and paste she walks the aisles in her nurse's mask, looking over our shoulders. But if anyone, a boy, draws a silly picture on purpose, she holds the page up in mocking outrage. "This lad here thinks he's being smarrut. You've got more between the ears than that!" And she flicks him on the ear with her thumb and fingernail.

For her we make the familiar paper objects, the pumpkins, the Christmas bells, but she has us do other things too. We make complicated floral patterns with a compass, we glue odd substances to cardboard backings: feathers, sequins, pieces of macaroni garishly dyed, lengths of drinking straw. We do group murals on the blackboards or on large rolls of brown paper. We draw pictures about foreign countries: Mexico with cactuses and men in enormous hats, China with cones on the heads and seeing-eye boats, India with what we intend to be graceful, silk-draped women balancing copper urns, and jewels on their foreheads.

I like these foreign pictures because I can believe in them. I desperately need to believe that somewhere else these other, foreign people exist. No matter that at Sunday school I've been told such people are either starving or heathens or both. No matter that my weekly collection goes to convert them, feed them, smarten them up. Miss Lumley saw them as crafty, given to the eating of outlandish or disgusting foods and to acts of treachery against the British, but I prefer Miss Stuart's versions, in which the sun above their heads is a cheerful yellow, the palm trees a clear green, the clothing they wear is floral, their folksongs gay. The women chatter together in quick incomprehensible languages, they laugh, showing perfect, pure-white teeth. If these people exist I can go there sometime. I don't have to stay here.

Today, says Miss Stuart, we are going to draw what we do after school.

The others hunch over their desks. I know what they will draw: skipping ropes, jolly snowmen, listening to the radio, playing with a dog. I stare at my own paper, which remains blank. Finally I draw my bed, with myself in it. My bed has a dark wooden headboard with curlicues on it. I draw the window, the chest of drawers. I color in the night. My hand holding the black crayon presses down, harder and harder, until the picture is almost entirely black, until only a faint shadow of my bed and my head on the pillow remains to be seen.

I look at this picture with dismay. It isn't what I meant to draw. It's unlike everyone else's picture, it's the wrong thing. Miss Stuart will be disappointed in me, she'll tell me I have more between the ears than that. I can feel her standing behind me now, looking over my shoulder; I can smell her smell of hand lotion, and the other smell that is not tea. She moves around so I can see her, her bright blue wrinkly eyes looking at me over the top of her nurse's mask.

For a moment she says nothing. Then she says, not harshly, "Why is your picture so darruk, my dear?"

"Because it's night," I say. This is an idiotic answer, I know that as soon as it's out of my mouth. My voice is almost inaudible, even to me.

"I see," she says. She doesn't say I've drawn the wrong thing, or that surely there's something else I do after school besides going to bed. She touches me on the shoulder, briefly, before continuing down the aisle. Her touch glows briefly, like a blown-out match.

In the schoolroom windows the paper hearts are blossoming. We make a huge Valentine's Day postbox out of a cardboard box covered with pink crêpe paper and red hearts with paper doily edging. Into the slot at the top we slip our valentines, cut from books of them you can buy at Woolworth's, with special, single ones for the people we especially like.

On the day itself the whole afternoon is a party. Miss Stuart loves parties. She's brought dozens of heart-shaped shortbread cookies she's made herself, with pink icing and silver balls on them, and there are tiny cinnamon hearts and pastel hearts with messages on them, messages from some earlier era which is not ours. "Hubba Hubba," they say. "She's My Baby." "Oh You Kid!"

Miss Stuart sits at her desk, supervising, while several girls open the box and deliver the valentines. On my desk the cards pile up. Most of them are from boys. I can tell this because of the sloppy writing, and because a lot of them aren't signed. Others have only initials or *Guess Who?* Some have X's and O's. The cards from girls are all neatly signed, with their full names, so there will be no mistake about who gave what.

On the way home from school Carol giggles and shows off her cards from boys. I have more cards from boys than Carol has, more than Cordelia and Grace have collected in their Grade Six classroom. Only I know this. I've hidden the cards in my desk so they won't be seen on the way home. When questioned I say I didn't get many. I hug my knowledge, which is new but doesn't surprise me: boys are my secret allies.

Carol is only ten and three quarters but she's growing breasts. They aren't very big, but the nipples are no longer flat, they're pointed, and there's a swelling behind them. It's easy to see this because she sticks out her chest, she wears sweaters, pulling them down tightly so the breasts poke out. She complains about these breasts at recess: they hurt, she says. She says she will have to get a bra. Cordelia says, "Oh shut up about your stupid tits." She's older, but she doesn't have any yet.

Carol pinches her lips and cheeks to make them red. She finds a worn-down tube of lipstick in her mother's waste-paper basket and hides it away, and takes it to school in her pocket. Using the tip of her little finger, she rubs some of it on her lips after school. She wipes it off with a Kleenex before we get to her house but she doesn't do a good enough job.

We play upstairs in her room. When we go down to the kitchen for a glass of milk, her mother says, "What's that on your face, young lady?" Right in front of us she scrubs Carol's face with the dirty dishcloth. "Don't let me catch you doing such a cheap thing again! At your age, the idea!" Carol wriggles, cries and screams, abandoning herself. We watch, horrified and thrilled. "Just wait till your father gets home!" her mother says in a cold, furious voice. "Making a spectacle of yourself," as if there's something wrong in the mere act of being looked at. Then she remembers we're still here. "Off you go!"

Two days later Carol says her father has given it to her good, with his belt, buckle end, right across the bare bum. She says she can hardly sit down. She sounds proud of this. She shows us, after school, up in her room: she pulls up her skirt, pulls down her underpants, and sure enough there are the marks, almost like scratch marks, not very red but there.

It's difficult to match this evidence with Carol's father, nice Mr. Campbell, who has a soft mustache and calls Grace Beautiful Brown Eyes and Cordelia Miss Lobelia. It's strange to imagine him hitting anyone with a belt. But fathers and their ways are enigmatic. I know without being told, for instance, that Mr. Smeath lives a secret life of trains and escapes in his head. Cordelia's father is charming to us on the rare occasions when he is seen, he makes wry jokes, his smile is like a billboard, but why is she afraid of him? Because she is. All fathers except mine are invisible in daytime; daytime is ruled by mothers. But fathers come out at night. Darkness brings home the fathers, with their real, unspeakable power. There is more to them than meets the eye. And so we believe the belt.

Carol says she's seen a wet spot on the sheet of her mother's twin bed, in the morning, before the bed was made. We tiptoe into her parents' room. The bed with its tufty chenille bedspread is so neatly made up we're afraid to turn down the covers to look. Carol opens the drawer of her mother's bedside table and we peer in. There's a rubber thing like the top of a mushroom, and a tube of toothpaste that isn't toothpaste. Carol says these things are to keep you from having babies. Nobody giggles, nobody scoffs. Instead we read the label. Somehow the red marks on Carol's bum have given her a credibility she lacked before.

Carol lies on top of her own bed, which has a white ruffled spread that matches the curtains. She's pretending to be sick, with an unspecified illness. We've dampened a washcloth, draped it over her forehead, brought her a glass of water. Illness is now a game we play.

"Oh, I'm so sick, oh, I'm so sick," Carol moans, twisting her body on the bed. "Nurse, do something!"

"We have to listen to her heart," says Cordelia. She pulls up Carol's sweater, then her undershirt. We've all been to the doctor, we know about the brusque humiliations involved. "This won't hurt." There are the breasts, puffy-looking, their nipples bluish, like veins on a forehead. "Feel her heart," Cordelia says to me.

I don't want to. I don't want to touch that swollen, unnatural flesh. "Go on," says Cordelia. "Do as you're told."

"She's being disobedient," says Grace.

I reach out my hand, place it on the left breast. It feels like a balloon half filled with water, or like lukewarm oatmeal porridge. Carol giggles. "Oh, your hand's so cold!" Nausea grips me.

"Her heart, stupid," says Cordelia. "I didn't say her tit. Don't you know the difference?"

An ambulance comes and my mother is carried out to it on a stretcher. I don't see this, Stephen tells me about it. It was in the middle of the night when I was asleep, but Stephen has taken to getting up secretly and looking out of his bedroom window at the stars. He says you can see the stars much better when most of the lights in the city are off. He says that the way to wake up at night without using an alarm clock is to drink two glasses of water before you go to bed. Then you have to concentrate on the hour you want to wake up. This is what the Indians used to do.

So he was awake, and listened, and snuck across to the other side of the house to look out the window there, where he could see what was going on out on the street. He says there were flashing lights but no siren, so it's no wonder I didn't hear anything.

When I get up in the morning my father is in the kitchen frying bacon. He knows how to do this, though he never does it in the city, only over campfires. In my parents' bedroom there's a pile of crumpled sheets on the floor, and the blankets are folded up on a chair; on the mattress there's a huge oval splotch of blood. But when I come home from school the sheets are gone and the bed is made up, and there is nothing more to be seen.

My father says there has been an accident. But how can you have an accident lying in bed asleep? Stephen says it was a baby, a baby that came out too soon. I don't believe him: women who are going to have babies have big fat stomachs, and my mother didn't have one.

My mother comes back from the hospital and is weaker. She has to rest. No one is used to this, she isn't used to it herself. She resists it, getting up as usual, putting her hand on the wall or on the edges of the furniture as she walks, standing hunched over at the kitchen sink, a cardigan over her shoulders. In the middle of something she's doing she has to go and lie down. Her skin is pale and dry. She looks as if she's listening to a sound, outside the house perhaps, but there is no sound. Sometimes I have to repeat things twice before she hears me. It's as if she's gone off somewhere else, leaving me behind; or forgotten I am there.

All of this is more frightening, even, than the splotch of blood. Our father tells us to help out more, which means that he's frightened as well.

After she gets better I find a small knitted sock, pastel green, in my mother's sewing basket. I wonder why she would have knitted only one sock. She doesn't like knitting, so maybe she knitted one and then got tired of it.

I dream that Mrs. Finestein from next door and Mr. Banerji are my real parents.

I dream that my mother has had a baby, one of a set of twins. The baby is gray. I don't know where the other twin is.

I dream that our house has burned down. Nothing of it remains; blackened stumps dot the place where it's been, as if there has been a forest fire. A huge mountain of mud rises beside it.

My parents are dead but also alive. They're lying side by side, in their summer clothes, and sinking down through the earth, which is hard but transparent, like ice. They look up at me sorrowfully as they recede.

32

*I*t's Saturday afternoon. We're going down to the building, to something called a Conversat. I don't know what a Conversat is but I'm relieved to be going to the building, where there are mice and snakes and experiments and no girls. My father asked if I wanted to bring a friend. I said no. My brother is bringing Danny, whose nose runs all the time, who wears knitted vests in diamond patterns, who has a stamp collection. They sit in the back seat—my brother no longer gets carsick—and talk in pig Latin.

"Or-yay ose-nay is-ay unning-ray."

"O-say ut-whay? Awnt-way oo-tay eat-ay ome-say?"

"Um-yay um-yay."

I know that some of this, at least on Danny's part, is for my benefit. He has confused me with other girls, girls who wriggle and shriek. Once I would have replied with something equally disgusting, but I have lost interest in such things as eating snot. I look out the car window, pretending not to hear.

The Conversat turns out to be sort of like a museum. The Zoology Department is throwing itself open to the public, to give people a crack at Science and improve their minds. This is what my father said, grinning the way he does when he's partly joking. He said people's minds could use some improving. My mother said she doesn't think her mind is capable of further improvement, so she's going grocery shopping instead.

There are a lot of people at the Conversat. There isn't that much to do for entertainment on the weekends in Toronto. The building has a festive air: its usual smells of Dustbane and furniture polish and mouse droppings and snakes mingle with other smells, of winter clothing, cigarette smoke, and women's perfume. Streamers of colored paper are taped to the walls, with arrows of construction paper at intervals, along the halls and up and down the stairs and into the different rooms, to show the way. Each room has its own displays, grouped according to what you are supposed to learn.

In the first room there are chicken embryos at various stages of development, from a red dot to a big-headed, bulgy-eyed, pin-feathered chick, looking not fluffy and cute the way they do on Easter cards, but slimy, its claws curled under, its eyelids a slit open, showing a crescent of agate-blue eye. The embryos have been pickled; the scent of formaldehyde is very strong. In another display there's a jar of twins, real dead identical human twins with their placenta attached, gray-skinned, floating in something that looks like dishwater. Their veins and arteries have been injected with colored rubber, blue for the veins, purple for the arteries, so we can see that their blood systems are connected. There's a human brain in a bottle, like a giant flabby gray walnut. I can't believe there is such a thing inside my head.

In another room there's a table where you can get your fingerprints taken, so you can see they aren't the same as anyone else's. There's a large piece of Bristol board with enlarged photographs of people's fingerprints pinned up on it. My brother and Danny and I all get our fingerprints taken. Danny and my brother have made light of the chickens and the twins—"Awnt-way any-nay icken-chay or-fay upper-say?" "Ow-hay about-way ome-say ewed-stay in-tway?"—but they weren't in any hurry to stay in that room. Their enthusiasm for the fingerprints is boisterous. They make fingerprints in the centers of each other's foreheads with their inky fingers, saying, "The Mark of the Black Hand!" in loud, ominous voices, until our father passes nearby and tells them to pipe down. Beautiful Mr. Banerji from India is with him. He smiles nervously at me and says, "How are you, miss?" He always calls me "miss." Among all these winter-white faces he looks darker than usual; his teeth shine and shine.

In the same room with the fingerprints they're handing out pieces of paper; you're supposed to taste them and say whether they taste bitter, like peach pits, or sour, like lemons. This proves that some things are inherited. There's also a mirror where you can do tongue exercises, to see if you can roll your tongue up at the sides or into a cloverleaf shape. Some people can't do either. Danny and my brother hog the mirror and make gruesome faces by sticking their thumbs into the sides of their mouths and pulling the edges of their eyelids down so that the red shows.

Some of the Conversat is less interesting, with too much writing, and some of it is only charts on the wall or looking through microscopes, which we can do whenever we want to anyway.

It's crowded as we shuffle along the halls, following the paper streamers, baby-blue and yellow, in our winter overshoes. We haven't taken our coats off. It's very warm. The clanking radiators are going full-blast, and the air is filling with other people's breath.

We come to a room where there's a cut-open turtle. It's in a white enamel tray, like the ones in butcher shops. The turtle is alive; or it's dead, but its heart is alive. This turtle is an experiment to show how the heart of a reptile can keep on going after the rest of it is dead.

The turtle's bottom shell has a hole sawed into it. The turtle is on its back so you can see down into it, right to the heart, which is beating away slowly, glistening dark red down there in its cave, wincing like the end of a touched worm, lengthening again, wincing. It's like a hand, clenching and unclenching. It's like an eye.

They've attached a wire to the heart, which runs to a loudspeaker, so you can hear the heart beating throughout the entire room, agonizingly slow, like an old man walking up stairs. I can't tell if the heart is going to make it to the next beat, or not. There's a footstep, a pause, then a crackling like the kind of static on the radio that my brother says comes from outer space, then another pulse, a gasp of air sucked in. Life is flowing out of the turtle, I can hear it over the loudspeaker. Soon the turtle will be empty of life.

I don't want to stay in this room but there's a lineup, in front of me and behind. All of the people are grown-ups; I've lost sight of Danny and my brother. I'm hemmed in by tweed coats, my eyes as high as their second buttons. I hear another sound, coming over the sound of the heart like an approaching wind: a rustling, like poplar leaves, only smaller, drier. There's black around the edges of my eyes and it closes in. What I see is like the entrance to a tunnel, rushing away from me; or I am rushing away from it, away from that spot of daylight. After that I'm looking at a lot of overshoes, and the floorboards, stretching into the distance, at eye level. My head hurts.

"She fainted," somebody says, and then I know what I have done.

"It must have been the heat."

I am carried out into the cold gray air; it's Mr. Banerji who carries me, making sounds of distress. My father hurries out and tells me to sit with my head down between my knees. I do this, looking at the tops of my overshoes. He asks if I'm going to be sick and I say no. My brother and Danny come out and stare at me, not saying anything. Finally my brother says, "Eee-shay ainted-fay," and they go back in.

I stay outside until my father brings the car around and we drive home. I'm beginning to feel that I've discovered something worth knowing. There's a way out of places you want to leave, but can't. Fainting is like stepping sideways, out of your own body, out of time or into another time. When you wake up it's later. Time has gone on without you.

Cordelia says, "Think of ten stacks of plates. Those are your ten chances." Every time I do something wrong, a stack of plates comes crashing down. I can see these plates. Cordelia can see them too, because she's the one who says *Crash!* Grace can see them a little, but her crashes are tentative, she looks to Cordelia for confirmation. Carol tries a crash once or twice but is scoffed at: "*That* wasn't a crash!"

"Only four left," says Cordelia. "You better watch yourself. Well?"

I say nothing.

"Wipe that smirk off your face," says Cordelia.

I say nothing.

"Crash!" says Cordelia. "Only three left."

Nobody ever says what will happen if all of the stacks of plates fall down.

I'm standing against the wall, near the GIRLS door, the cold creeping up my legs and in under the edges of my sleeves. I'm not supposed to move. Already I've forgotten why. I've discovered that I can fill my head up with music, *Coming in on a wing and a prayer, Keep happy with the Happy Gang*, and forget almost anything.

It's recess. Miss Lumley patrols the playground with her brass bell, her face clamped against the cold, minding her own business. I'm still just as afraid of her, although she's no longer my teacher. Chains of girls careen past, chanting *We don't stop for anybody*. Other girls promenade more sedately, arms linked two by two. They look at me curiously, then away. It's like the people in cars, on the highway, who slow down and look out the window when there's a car accident by the side of the road. They slow down but they don't stop. They know when there's trouble, they know when to keep out of it.

I'm standing a little out from the wall. I put my head back and stare up into the gray sky and hold my breath. I'm making myself dizzy. I can see a stack of plates as it sways, begin to topple over, into a silent explosion of china shards. The sky closes to a pinpoint and a wave of dry leaves sweeps over my head. Then I can see my own body lying on the ground, just lying there. I can see the girls pointing and gathering, I can see Miss Lumley stalking over, bending with difficulty to look at me. But I'm seeing all this from above, as if I'm in the air, somewhere near the GIRLS sign over the door, looking down like a bird.

I come to with Miss Lumley's face looming inches away from me, scowling more than ever, as if I've made a mess, with a ring of girls around her jostling for a better look.

There's blood, I've cut my forehead. I am taken off to the nurse's office. The nurse wipes off the blood and sticks a wad of gauze onto me with a Band-Aid. The sight of my own blood on the wet white washcloth is deeply satisfying to me.

Cordelia is subdued: blood is impressive, even more impressive than vomit. She and Grace are solicitous on the way home, linking their arms through mine, asking me how I feel. This kind of attention from them makes me tremulous. I'm afraid I will cry, great sopping tears of reconciliation. But I'm far too wary for that by now.

The next time Cordelia tells me to stand against the wall I faint again. Now I can do it almost whenever I want to. I hold my breath and hear the rustling noise and see the blackness and then I slip sideways, out of my body, and I'm somewhere else. But I can't always watch from above, like the first time. Sometimes there's just black.

I begin to be known as the girl who faints.

"She's doing it on purpose," Cordelia says. "Go ahead, let's see you faint. Come on. Faint." But now, when she tells me to, I can't.

I begin to spend time outside my body without falling over. At these times I feel blurred, as if there are two of me, one superimposed on the other, but imperfectly. There's an edge of transparency, and beside it a rim of solid flesh that's without feeling, like a scar. I can see what's happening, I can hear what's being said to me, but I don't have to pay any attention. My eyes are open but I'm not there. I'm off to the side.

VIII
OUR LADY OF PERPETUAL HELP

33

I walk west from Simpsons, still looking for something to eat. Finally I buy a slice of take-out pizza and devour it en route, with my fingers, folding it in two and gnawing. When I'm with Ben I eat at regular times because he does, I eat regular things, but when I'm alone I indulge in junk food and scavenging, my old, singular ways. It's bad for me, but I need to remember what bad for me is like. I could begin to take Ben for granted, with his ties and haircuts and grapefruits for breakfast. It makes me appreciate him more.

Back at the studio I call him, counting the hours backward to the coast. But there's only my own voice on the message, followed by the beep, the Dominion Observatory Official Time Signal, ushering in the future. *Love you*, I say, so he can hear it later. Then I remember: by now he's in Mexico, he won't be back till I am.

By now it's dark outside. I could go out for something more like dinner, or try for a movie. Instead I crawl onto the futon, under the duvet, with a cup of coffee and the Toronto phone book, and start looking up names. There are no more Smeaths, they must have moved or died off, or got married. There are more Campbells than you can shake a stick at. I look up Jon, whose name was once my own. No Josef Hrbik, though there are Hrbeks, Hrens, Hrastniks, Hriczus.

There are no more Risleys.

There is no Cordelia.

It's strange to be lying in Jon's bed again. I haven't thought of this as Jon's bed because I've never seen him in it, but of course it is. It's a lot neater than his beds used to be, and a great deal cleaner. His first bed was a mattress on the floor, with an old sleeping bag on top of it. I didn't mind this, in fact I liked it; it was like camping out. Usually there would be a tide line of empty cups and glasses and plates with scraps of food surrounding it, which I didn't like as much. There was an etiquette about such messes, in those days: there was a line you crossed, from ignoring them to cleaning them up. Whether the man would think you were moving in on him, trying to take him over.

We were lying on that bed one time, right at the beginning, before I'd started to pick up the plates, when the bedroom door opened and a woman I'd never seen before appeared in the doorway. She was wearing dirty jeans and a wan pink T-shirt; her face thin and bleached-looking, with huge pupils. It looked as if she was on some sort of drug, which was beginning to be a possibility, then. She stood there saying nothing, one hand behind her back, her face tight and blank, while I pulled the sleeping bag up over me.

"Hey," said Jon.

She drew her hand out from behind her back then and threw something at us. It was a paper bag full of warm spaghetti, sauce included. It burst when it hit, festooning us. She went out, still without saying anything, and slammed the door.

I was frightened, but Jon started laughing. "What was that?" I said. "How the hell did she get in?"

"Through the door," said Jon, still laughing. He untangled a piece of spaghetti from my hair and leaned over to kiss me. I knew this woman must have been a girlfriend, or an ex-girlfriend, and I was furious with her. It didn't occur to me that she might have reason. I hadn't yet encountered the foreign hairpins left in the bathroom like territorial dog pee on snowy hydrants, the lipstick marks placed strategically on pillows. Jon knew how to cover his tracks, and when he didn't cover them it was for a reason. It didn't occur to me either that she must have had a key.

"She's crazy," I said. "She should be in a bin."

I did not pity her at all. In a way I admired her. I admired her lack of compunction, the courage of her bad manners, the energy of simple rage. Throwing a bag of spaghetti had a simplicity to it, a recklessness, a careless grandeur. It got things over with. I was a long way, then, from being able to do anything like it myself.

Grace says grace. Mr. Smeath says, "Praise the Lord and pass the ammunition," and reaches for the baked beans. Mrs. Smeath says, "Lloyd." Mr. Smeath says, "It's harmless," and shoots me a sideways grin. Aunt Mildred contracts her whiskery mouth. I chew away at the rubber plant Smeath food. Under cover of the tablecloth, I tear at my fingers. Sunday goes on.

After the stewed pineapple Grace wants me to come down into the cellar with her to play school. I do this, but I have to come up the stairs again to go to the bathroom. Grace has given me permission, the same way the teachers in school give you permission. As I come up the cellar stairs I can hear Aunt Mildred and Mrs. Smeath, who are in the kitchen doing the dishes.

"She's exactly like a heathen," says Aunt Mildred. Because she's been a missionary in China, she's an authority. "Nothing you've done has made a scrap of difference."

"She's learning her Bible, Grace tells me," Mrs. Smeath says, and then I know it's me they're discussing. I stop on the top step, where I can see into the kitchen: the kitchen table where the dirty dishes are piled, the back edges of Mrs. Smeath and Aunt Mildred.

"They'll learn all that," says Aunt Mildred. "Till you're blue in the face. But it's all rote learning, it doesn't sink in. The minute your back is turned they'll go right back the way they were."

The unfairness of this hits me like a kick. How can they say that, when I've won a special mention for my essay on Temperance, about drunken men having car accidents and freezing to death in snowstorms because the alcohol dilates their capillaries? I even know what capillaries are, I even spelled it right. I can recite whole psalms, whole chapters, I can sing all the colored-slide white-knight Sunday school songs without looking.

"What can you expect, with that family?" says Mrs. Smeath. She doesn't go on to say what's wrong with my family. "The other children sense it. They know."

"You don't think they're being too hard on her?" says Aunt Mildred. Her voice is relishing. She wants to know how hard.

"It's God's punishment," says Mrs. Smeath. "It serves her right."

A hot wave moves through my body. This wave is shame, which I have felt before, but it is also hatred, which I have not, not in this pure form. It's hatred with a particular shape, the shape of Mrs. Smeath's one breast and no waist. It's like a fleshy weed in my chest, white-stemmed and fat; like the stalk of a burdock, with its rank leaves and little green burrs, growing in the cat piss earth beside the path down to the bridge. A heavy, thick hatred.

I stand there on the top step, frozen with hate. What I hate is not Grace or even Cordelia. I can't go as far as that. I hate Mrs. Smeath, because what I thought was a secret, something going on among girls, among children, is not one. It has been discussed before, and tolerated. Mrs. Smeath has known and approved. She has done nothing to stop it. She thinks it serves me right.

She moves away from the sink and walks to the kitchen table for another stack of dirty plates, into my line of vision. I have a brief, intense image of Mrs. Smeath going through the flesh-colored wringer of my mother's washing machine, legs first, bones cracking and flattening, skin and flesh squeezing up toward her head, which will pop in a minute like a huge balloon of blood. If my eyes could shoot out fatal rays like the ones in comic books I would incinerate her on the spot. She is right, I am a heathen. I cannot forgive.

As if she can feel my stare she turns and sees me. Our eyes meet: she knows I've heard. But she doesn't flinch, she isn't embarrassed or apologetic. She gives me that smug smile, with the lips closed over the teeth. What she says is not to me but to Aunt Mildred. "Little pitchers have big ears."

Her bad heart floats in her body like an eye, an evil eye, it sees me.

We sit on the wooden bench in the church basement, in the dark, watching the wall. Light glints from Grace's glassy eyes as she watches me sideways.

> *God sees the little sparrow fall,*
> *It meets His tender view;*
> *If God so loves the little bird,*
> *I know He loves me too.*

The picture is of a dead bird in an enormous hand, with a shaft of light coming down onto it.

I am moving my lips, but I'm not singing. I am losing confidence in God. Mrs. Smeath has God all sewed up, she knows what things are his punishments. He's on her side, and it's a side from which I'm excluded.

I consider Jesus, who is supposed to love me. But he isn't showing any sign of it, and I don't think he can be of much help. Against Mrs. Smeath and God he can do nothing, because God is bigger. God is not Our Father at all. My image of him now is of something huge, hard, inexorable, faceless and moving forward as if on tracks. God is a sort of engine.

I decide not to pray to God any more. When it's time for the Lord's Prayer I stand in silence, moving my lips only.

Forgive us our trespasses, as we forgive those who trespass against us.

I refuse to say this. If it means I will have to forgive Mrs. Smeath or else go to Hell when I die, I'm ready to go. Jesus must have known how hard it is to forgive, that was why he put this in. He was always putting in things that were impossible to do really, such as giving away all your money.

"You weren't praying," Grace says to me in a whisper.

My stomach goes cold. Which is worse, to contradict her or to admit? Either way there will be penalties.

"Yes I was," I say.

"You weren't. I heard you."

I say nothing.

"You lied," says Grace, pleased, forgetting to whisper.

I still say nothing.

"You should ask God to forgive you," Grace says. "That's what I do, every night."

I sit in the dark, attacking my fingers. I think about Grace asking God to forgive her. But for what? God only forgives you if you're sorry, and she never gives a sign of being sorry. She never thinks she's done anything wrong.

Grace and Cordelia and Carol are up ahead, I am a block behind. They aren't letting me walk with them today because I have been insolent, but they don't want me too far behind either. I am walking along, in time to the music, *Keep happy with the Happy Gang*, my head empty except for these words. I walk head down, scanning the sidewalk, the gutters, for silver cigarette papers, although I no longer collect them as I did long ago. I know that nothing I could make with them would be worthwhile.

I see a piece of paper with a colored picture on it. I pick it up. I know what the picture is: it's the Virgin Mary. The paper is from Our Lady of Perpetual Help, Our Lady of Perpetual Hell. The Virgin Mary is wearing a long blue robe, no feet at all visible below the hem, a white cloth over her head and a crown on top of that, and a yellow halo with light rays coming out of it like nails. She's smiling sadly in a disappointed way; her hands are outstretched as if in welcome, and her heart is on the outside of her chest, with seven swords stuck into it. Or they look like swords. The heart is large, red and tidy, like a satin heart pincushion, or a valentine. Under the picture is printed: *The Seven Sorrows*.

The Virgin Mary is in some of our Sunday school papers, but never with a crown, never with a pincushion heart, never all by herself. She is always more or less in the background. Not much fuss is made over her except at Christmas, and even then Baby Jesus is a lot more important. When Mrs. Smeath and Aunt Mildred speak of Catholics, as they have been known to do at the Sunday dinner table, it's always with contempt. Catholics pray to statues and drink real wine at Communion, instead of grape juice. "They worship the Pope," is what the Smeaths say; or else, "They worship the Virgin Mary," as if this is a scandalous thing to do.

I look at the picture up close. But I know it would be dangerous to keep it, so I throw it away. This is the right impulse, because now the three of them have stopped, they're waiting for me to catch up to them. Anything I do, other than standing, other than walking, attracts their attention.

"What was that thing we saw you pick up?" says Cordelia.

"A paper."

"What sort of a paper?"

"Just a paper. A Sunday school paper."

"Why did you pick it up?"

Once I would have thought about this question, tried to answer it truthfully. Now I say, "I don't know." This is the only answer I can give, to anything, that will not be ridiculed or questioned.

"What did you do with it?"

"I threw it away."

"Don't pick things up off the street," says Cordelia. "They have germs." She lets it go at that.

I decide to do something dangerous, rebellious, perhaps even blasphemous. I can no longer pray to God so I will pray to the Virgin Mary instead. This decision makes me nervous, as if I'm about to steal. My heart beats harder, my hands feel cold. I feel I'm about to get caught.

Kneeling seems called for. In the onion church we don't kneel, but the Catholics are known for it. I kneel down beside my bed and put my hands together, like the children in Christmas cards, except that I'm wearing blue-striped flannelette pajamas and they always have white nightgowns on. I close my eyes and try to think about the Virgin Mary.

I want her to help me or at least show me that she can hear me, but I don't know what to say. I haven't learned the words for her.

I try to picture what she would look like, if I met her on the street for instance: would she be wearing clothes like my mother's, or that blue dress and crown, and if it was the blue dress would a crowd gather? Maybe they would think she was just someone out of a Christmas play; but not if she had her heart on the outside like that, stuck full of swords. I try to think what I would tell her. But she knows already: she knows how unhappy I am.

I pray harder and harder. My prayers are wordless, defiant, dry-eyed, desperate, without hope. Nothing happens. I squeeze my fists into my eyes until they hurt. For an instant I think I see a face, then a splash of blue, but now all I can see is the heart. There it is, bright red, rounded, with a dark light around it, a blackness like luminous velvet. Gold comes out from the center, then fades. It's the heart all right. It looks like my red plastic purse.

35

*I*t's the middle of March. In the schoolroom windows the Easter tulips are beginning to bloom. There's still snow on the ground, a dirty filigree, though the winter is losing its hardness and glitter. The sky thickens, sinks lower.

We walk home under the low thick sky that is gray and bulging with dampness. Moist soft flakes are falling out of it, piling up on roofs and branches, sliding off now and then to hit with a wet cottony *thunk*. There's no wind and the sound is muffled by the snow.

It isn't cold. I undo the ties on my blue knitted wool hat, let it flap loose on my head. Cordelia takes off her mittens and scoops up snowballs, throwing them at trees, at telephone poles, at random. It's one of her friendly days; she puts her arm through my arm, her other arm through Grace's, and we march along the street, singing *We don't stop for anybody*. I sing this too. Together we hop and slide.

Some of the euphoria I once felt in falling snow comes back to me; I want to open my mouth and let the snow fall into it. I allow myself to laugh, like the others, trying it out. My laughter is a performance, a grab at the ordinary.

Cordelia throws herself backward onto a blank front lawn, spreads her arms out in the snow, raises them above her head, draws them down to her sides, making a snow angel. The flakes fall onto her face, into her laughing mouth, melting, clinging to her eyebrows. She blinks, closing her eyes against the snow. For a moment she looks like someone I don't know, a stranger, shining with unknown, good possibilities. Or else a victim of a traffic accident, flung onto the snow.

She opens her eyes and reaches up her hands, which are damp and reddened, and we pull her upward so she won't disturb the image she's made. The snow angel has feathery wings and a tiny pin head. Where her hands stopped, down near her sides, are the imprints of her fingers, like little claws.

We've forgotten the time, it's getting dark. We run along the street that leads to the wooden footbridge. Even Grace runs, lumpily, calling, "Wait up!" For once she is the one left behind.

Cordelia reaches the hill first and runs down it. She tries to slide but the snow is too soft, not icy enough, and there are cinders and pieces of gravel in it. She falls down and rolls. We think she's done it on purpose, the way she made the snow angel. We rush down upon her, exhilarated, breathless, laughing, just as she's picking herself up.

We stop laughing, because now we can see that her fall was an accident, she didn't do it on purpose. She likes everything she does to be done on purpose.

Carol says, "Did you hurt yourself?" Her voice is quavery, she's frightened, already she can tell that this is serious. Cordelia doesn't answer. Her face is hard again, her eyes baleful.

Grace moves so that she's beside Cordelia, slightly behind her. From there she smiles at me, her tight smile.

Cordelia says, to me, "Were you laughing?" I think she means, was I laughing at her because she fell down.

"No," I say.

"She was," says Grace neutrally. Carol shifts to the side of the path, away from me.

"I'm going to give you one more chance," says Cordelia. "Were you laughing?"

"Yes," I say, "but . . ."

"Just yes or no," says Cordelia.

I say nothing. Cordelia glances over at Grace, as if looking for approval. She sighs, an exaggerated sigh, like a grown-up's. "Lying again," she says. "What are we going to do with you?"

We seem to have been standing there for a long time. It's colder now. Cordelia reaches out and pulls off my knitted hat. She marches the rest of the way down the hill and onto the bridge and hesitates for a moment. Then she walks over to the railing and throws my hat down into the ravine. Then the white oval of her face turns up toward me. "Come here," she says.

Nothing has changed, then. Time will go on, in the same way, endlessly. My laughter was unreal after all, merely a gasp for air.

I walk down to where Cordelia stands by the railing, the snow not crunching but giving way under my feet like cotton wool packing. It sounds like a cavity being filled, in a tooth, inside my head. Usually I'm afraid to go so near the edge of the bridge, but this time I'm not. I don't feel anything as positive as fear.

"There's your stupid hat," says Cordelia; and there it is, far down, still blue against the white snow, even in the dimming light. "Why don't you go down and get it?"

I look at her. She wants me to go down into the ravine where the bad men are, where we're never supposed to go. It occurs to me that I may not. What will she do then?

I can see this idea gathering in Cordelia as well. Maybe she's gone too far, hit, finally, some core of resistance in me. If I refuse to do what she says this time, who knows where my defiance will end? The two others have come down the hill and are watching, safely in the middle of the bridge.

"Go on then," she says, more gently, as if she's encouraging me, not ordering. "Then you'll be forgiven."

I don't want to go down there. It's forbidden and dangerous; also it's dark and the hillside will be slippery, I might have trouble climbing up again. But there is my hat. If I go home without it, I'll have to explain, I'll have to tell. And if I refuse to go, what will Cordelia do next? She might get angry, she might never speak to me again. She might push me off the bridge. She's never done anything like that before, never hit or pinched, but now that she's thrown my hat over there's no telling what she might do.

I walk along to the end of the bridge. "When you've got it, count to a hundred," says Cordelia. "Before coming up." She doesn't sound angry any more. She sounds like someone giving instructions for a game.

I start down the steep hillside, holding on to branches

and tree trunks. The path isn't even a real path, it's just a place worn by whoever goes up and down here: boys, men. Not girls.

When I'm among the bare trees at the bottom I look up. The bridge railings are silhouetted against the sky. I can see the dark outlines of three heads, watching me.

My blue hat is out on the ice of the creek. I stand in the snow, looking at it. Cordelia is right, it's a stupid hat. I look at it and feel resentment, because this stupid-looking hat is mine, and deserving of ridicule. I don't want to wear it ever again.

I can hear water running somewhere, down under the ice. I step out onto the creek, reach for the hat, pick it up, go through. I'm up to my waist in the creek, slabs of broken ice upended around me.

Cold shoots through me. My overshoes are filling, and the shoes inside them; water drenches my snowpants. Probably I've screamed, or some noise has come out of me, but I can't remember hearing anything. I clutch the hat and look up at the bridge. Nobody is there. They must have walked away, run away. That's why the counting to a hundred: so they could run away.

I try to move my feet. They're very heavy, because of the water inside my boots. If I wanted to I could just keep standing here. It's true dusk now and the snow on the ground is bluish-white. The old tires and pieces of rusted junk in the creek are covered over; all around me are blue arches, blue caves, pure and silent. The water of the creek is cold and peaceful, it comes straight from the cemetery, from the graves and their bones. It's water made from the dead people, dissolved and clear, and I am standing in it. If I don't move soon I will be frozen in the creek. I will be a dead person, peaceful and clear, like them.

I flounder through the water, the edges of the ice breaking off as I step. Walking with waterlogged overshoes is hard; I could slip, and fall all the way in. I grab a tree branch and haul myself up onto the bank and sit down in the blue snow and take off my overshoes and pour out the water. The arms of my jacket are wet to the elbows, my mittens are soaked. Now there are knives going through my

legs and hands, and tears running down my face from the
pain.

I can see lights along the edges of the ravine, from the
houses there, impossibly high up. I don't know how I'm
going to climb up the hill with my hands and feet hurting
like this; I don't know how I'm going to get home.

My head is filling with black sawdust; little specks of
the darkness are getting in through my eyes. It's as if the
snowflakes are black, the way white is black on a negative. The snow has changed to tiny pellets, more like sleet.
It makes a rustling noise coming down through the
branches, like the shifting and whispering of people in a
crowded room who know they must be quiet. It's the dead
people, coming up invisible out of the water, gathering
around me. *Hush*, is what they say.

I'm lying on my back beside the creek, looking up at the
sky. Nothing hurts any more. The sky has a reddish un-
dercolor. The bridge is different-looking; it seems higher
above me, more solid, as if the railings have disappeared or
been filled in. And it's glowing, there are pools of light along
it, greenish-yellow, not like any light I've ever seen before.
I sit up to get a better look. My body feels weightless, as
it does in water.

There's someone on the bridge, I can see the dark outline. At first I think it's Cordelia, come back for me. Then
I see that it's not a child, it's too tall for a child. I can't see
the face, there's just a shape. One of the yellowish-green
lights is behind it, coming out in rays from around the head.

I know I should get up and walk home, but it seems
easier to stay here, in the snow, with the little pellets of ice
caressing my face gently. Also I'm very sleepy. I close my
eyes.

I hear someone talking to me. It's like a voice calling,
only very soft, as if muffled. I'm not sure I've heard it at
all. I open my eyes with an effort. The person who was
standing on the bridge is moving through the railing, or
melting into it. It's a woman, I can see the long skirt now,
or is it a long cloak? She isn't falling, she's coming down
toward me as if walking, but there's nothing for her to walk

on. I don't have the energy to be frightened. I lie in the snow, watching her with lethargy, and with a sluggish curiosity. I would like to be able to walk on air like that.

Now she's quite close. I can see the white glimmer of her face, the dark scarf or hood around her head, or is it hair? She holds out her arms to me and I feel a surge of happiness. Inside her half-open cloak there's a glimpse of red. It's her heart, I think. It must be her heart, on the outside of her body, glowing like neon, like a coal.

Then I can't see her any more. But I feel her around me, not like arms but like a small wind of warmer air. She's telling me something.

You can go home now, she says. *It will be all right. Go home*.

I don't hear the words out loud, but this is what she says.

36

*T*he lights on the top of the bridge are gone. I make my way in the dark, up the hill, sleet rustling around me, hauling myself up by branches and tree trunks, my shoes slipping on the packed icy snow. Nothing hurts, not even my feet, not even my hands. It's like flying. The small wind moves with me, a warm touch against my face.

I know who it is that I've seen. It's the Virgin Mary, there can be no doubt. Even when I was praying I wasn't sure she was real, but now I know she is. Who else could walk on air like that, who else would have a glowing heart? True, there was no blue dress, no crown; her dress looked black. But it was dark. Maybe the crown was there and I couldn't see it. Anyway she could have different clothes, different dresses. None of that matters, because she came to get me. She didn't want me freezing in the snow. She is still with me, invisible, wrapping me in warmth and painlessness, she has heard me after all.

I am up on the main path now; the lights from the houses are nearer, above me, on either side of me. I can hardly keep my eyes open. I'm not even walking straight. But my feet keep on moving, one in front of the other.

Up ahead is the street. As I reach it I see my mother, walking very fast. Her coat isn't done up, she has no scarf on her head, her overshoes flap, half fastened. When she sees me she begins to run. I stop still, watching her running figure with the coat flying out on either side and the unwieldy overshoes, as if she's just some other person I'm watching, someone in a race. She comes up to me under a streetlamp and I see her eyes, large and gleaming with wet, and her hair dusted with sleet. She has no mittens on. She throws her arms around me, and as she does this the Virgin

Mary is suddenly gone. Pain and cold shoot back into me. I start to shiver violently.

"I fell in," I say. "I was getting my hat." My voice sounds thick, the words mumbled. Something is wrong with my tongue.

My mother does not say, *Where have you been?* or *Why are you so late?* She says, "Where are your overshoes?" They are down in the ravine, covering over with snow. I have forgotten them, and my hat as well.

"It fell over the bridge," I say. I need to get this lie over with as soon as possible. Telling the truth about Cordelia is still unthinkable for me.

My mother takes off her coat and wraps it around me. Her mouth is tight, her face is frightened and angry at the same time. It's the look she used to have when we would cut ourselves, a long time ago, up north. She puts her arm under my armpit and hurries me along. My feet hurt at every step. I wonder if I will be punished for going down into the ravine.

When we reach the house my mother peels off my soggy half-frozen clothes and puts me into a lukewarm bath. She looks carefully at my fingers and toes, my nose, my ear lobes. "Where were Grace and Cordelia?" she asks me. "Did they see you fall in?"

"No," I say. "They weren't there."

I can tell she's thinking about phoning their mothers no matter what I do, but I am too tired to care. "A lady helped me," I say.

"What lady?" says my mother, but I know better than to tell her. If I say who it really was I won't be believed. "Just a lady," I say.

My mother says I'm lucky I don't have severe frostbite. I know about frostbite: your fingers and toes fall off, as punishment for drink. She feeds me a cup of milky tea and puts me into bed with a hot water bottle and flannelette sheets, and spreads two extra blankets on top. I am still shivering. My father has come home and I hear them talking in low, anxious voices out in the hallway. Then my father comes in and puts his hand on my forehead, and fades to a shadow.

.   .   .

I dream I'm running along the street outside the school. I've done something wrong. It's autumn, the leaves are burning. A lot of people are chasing after me. They're shouting.

An invisible hand takes mine, pulls upward. There are steps into the air and I go up them. No one else can see where the steps are. Now I'm standing in the air, out of reach above the upturned faces. They're still shouting but I can no longer hear them. Their mouths close and open silently, like the mouths of fish.

I am kept home from school for two days. The first day I lie in bed, floating in the glassy delicate clarity of fever. By the second day I am thinking about what happened. I can remember Cordelia throwing my blue knitted hat over the bridge, I remember falling through the ice and then my mother running toward me with her sleety hair. All these things are certain, but in between them there's a hazy space. The dead people and the woman in the cloak are there, but in the same way dreams are. I'm not sure, now, that it really was the Virgin Mary. I believe it but I no longer know it.

I'm given a get-well card with violets on it from Carol, shoved through the letter slot. On the weekend Cordelia calls me on the telephone. "We didn't know you fell in," she says. "We're sorry we didn't wait. We thought you were right behind us." Her voice is careful, precise, rehearsed, unrepentant.

I know she's told some story that conceals what really happened, as I have. I know that this apology has been exacted from her, and that I will be made to pay for it later. But she has never apologized to me before. This apology, however fake, makes me feel not stronger but weaker. I don't know what to say. "It's okay," is what I manage. I think I mean it.

When I go back to school, Cordelia and Grace are polite but distant. Carol is more obviously frightened, or interested. "My mother says you almost froze to death," she whispers as we stand in line, two by two, waiting for the bell. "I got a spanking, with the hairbrush. I really *got* it."

• • •

The snow is melting from the lawns; mud reappears on the floors, at school, in the kitchen at home. Cordelia circles me warily. I catch her eyes on me, considering, as we walk home from school. Conversation is artificially normal. We stop at the store for licorice whips, which Carol buys. As we stroll along, sucking in licorice, Cordelia says, "I think Elaine should be punished for telling on us, don't you?"

"I didn't tell," I say. I no longer feel the sinking in my gut, the held-back tearfulness that such a false accusation would once have produced. My voice is flat, calm, reasonable.

"Don't contradict me," Cordelia says. "Then how come your mother phoned our mothers?"

"Yeah, how come?" says Carol.

"I don't know and I don't care," I say. I'm amazed at myself.

"You're being insolent," says Cordelia. "Wipe that smirk off your face."

I am still a coward, still fearful; none of that has changed. But I turn and walk away from her. It's like stepping off a cliff, believing the air will hold you up. And it does. I see that I don't have to do what she says, and, worse and better, I've never had to do what she says. I can do what I like.

"Don't you dare walk away on us," Cordelia says behind me. "You get back here right now!" I can hear this for what it is. It's an imitation, it's acting. It's an impersonation, of someone much older. It's a game. There was never anything about me that needed to be improved. It was always a game, and I have been fooled. I have been stupid. My anger is as much at myself as at them.

"Ten stacks of plates," says Grace. This would once have reduced me. Now I find it silly.

I keep walking. I feel daring, light-headed. They are not my best friends or even my friends. Nothing binds me to them. I am free.

They follow along behind me, making comments on the way I walk, on how I look from behind. If I were to turn I would see them imitating me. "Stuck up! Stuck up!" they cry. I can hear the hatred, but also the need. They need me for this, and I no longer need them. I am indifferent to them. There's something hard in me, crystalline, a kernel of glass. I cross the street and continue along, eating my licorice.

I stop going to Sunday school. I refuse to play with Grace or Cordelia or even Carol after school. I no longer walk home over the bridge, but the long way around, past the cemetery. When they come in a group to the back door to collect me I tell them I'm busy. They try kindness, to lure me back, but I am no longer susceptible to it. I can see the greed in their eyes. It's as if I can see right into them. Why was I unable to do this before?

I spend a lot of time reading comic books in my brother's room when he isn't there. I would like to climb up skyscrapers, fly with a cape, burn holes in metal with my fingertips, wear a mask, see through walls. I would like to hit people, criminals, each fist making a red or yellow lightburst. *Kapow. Krac. Kaboom.* I know that I have the will to do these things. I intend to do them somehow.

At school I make friends with a different girl, whose name is Jill. She is interested in other kinds of games, games of paper and wood. We go to her house and play Old Maid, Snap, Pick Up Sticks. Grace and Cordelia and Carol hang around the edges of my life, enticing, jeering, growing paler and paler every day, less and less substantial. I hardly hear them any more because I hardly listen.

Viiii

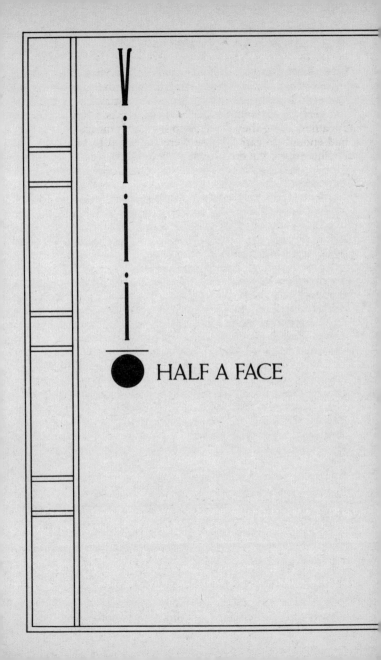 HALF A FACE

37

*F*or a long time, I would go into churches. I told myself I wanted to see the art; I didn't know I was looking for something. I wouldn't seek these churches out, even if they were in a guidebook and of historic significance, and I would never go into them during services, in fact I disliked the idea: it was what was in them, not what went on in them, that interested me. Mostly I would just see them by accident and go into them on impulse.

Once in, I paid little attention to the architecture, although I knew the terms: clerestories and naves were things I'd written papers about. I would look at the stained-glass windows, if any. I preferred Catholic churches to Protestant ones, the more ornate the better, because there was more to look at. I liked the shameless extravaganza: gold leaf and baroque excesses did not put me off.

I would read the inscriptions on walls, and carved into floors, a special foible of rich Anglicans who thought they'd get more points with God by being engraved. Anglicans too went in for tattered military flags, and war memorials of other kinds.

But especially I sought out statues. Statues of saints, and of crusaders on their biers, or those pretending to be crusaders; effigies of all kinds. Statues of the Virgin Mary I would save for last. I would approach them with hope, but I was always disappointed. The statues were of no one I recognized. They were dolls dressed up, insipid in blue and white, pious and lifeless. Then I would not know why I'd been expecting to see something else.

I went to Mexico the first time with Ben. It was also our first trip together, our first time together; I thought it might be only an interlude. I wasn't even sure I wanted a man in my life again; by that time I'd exhausted the notion that the answer to a man is another man, and I was out of breath. But it was a relief to be with someone who was so uncomplicated, and easily pleased.

We were by ourselves, on a two-week excursion that turned out to have something to do with Ben's business. Sarah was staying with her best friend. We began in Veracruz, checking out shrimp and the hotels and cockroaches, then took a car up into the hills, looking as always for something picturesque and undervisited.

There was a small town beside a lake. The place was subdued, for Mexico, which had struck me as visceral, like a body turned inside out so the blood was on the outside. Perhaps it was the coolness, the lake.

While Ben was inspecting the market, searching for things to take pictures of, I went into the church. It wasn't large, and looked poor. There was nobody in it; it smelled of old stone, old neglect, mustiness. I wandered around the outside aisles, looking at the awkward Stations of the Cross, done in grubby oils, almost a paint by numbers. They were bad, but genuine: someone had meant them.

Then I saw the Virgin Mary. I didn't know it was her at first, because she was dressed not in the usual blue or white and gold, but in black. She didn't have a crown. Her head was bowed, her face in shadow, her hands held out open at the sides. Around her feet were the stubs of candles, and all over her black dress were pinned what I thought at first were stars, but which were instead little brass or tin arms, legs, hands, sheep, donkeys, chickens, and hearts.

I could see what these were for: she was a Virgin of lost things, one who restored what was lost. She was the only one of these wood or marble or plaster Virgins who had ever seemed at all real to me. There could be some point in praying to her, kneeling down, lighting a candle. But I didn't do it, because I didn't know what to pray for. What was lost, what I could pin on her dress.

Ben came after a while and found me. "What's the matter?" he said. "What're you doing down on the floor? Are you all right?"

"Yes," I said. "Nothing. Just resting."

I was chilled through from the stone, my muscles were cramped and stiff. I'd forgotten how I got down there.

My daughters, both of them, went through a phase when they would say *So?* Meaning *So what*. It was when the first one hit twelve or thirteen. They'd fold their arms and stare at me, or at their friends, or at each other. *So?*

"Don't do that," I'd say. "It's driving me crazy."

"So?"

Cordelia did the same thing, at the same age. The same folded arms, the same immobile face, the blank-eyed stare. Cordelia! Put on your gloves, it's cold out. *So?* I can't come over, I have to finish my homework. *So?*

Cordelia, I think. You made me believe I was nothing. *So?*

To which there is no answer.

38

*T*he summer comes and goes and then it's fall and then winter, and the King dies. I hear it on the news at lunchtime. I walk back to school along the snowy street, thinking, *The King is dead*. Now all the things that happened when he was alive are over and done with: the war, the planes with only one wing, the mud outside our house, a lot of things. I think of those heads of his, thousands of them, on the money, which are now the heads of a dead person instead of a living one. The money will have to be changed, and the postage stamps; they will have the Queen on them instead. The Queen used to be Princess Elizabeth. I remember seeing her in photos, when she was much younger. I have some other memory of her, but it's indistinct and makes me faintly uneasy.

Cordelia and Grace have both skipped a grade. They're now in Grade Eight, even though they're only eleven and the other ones in Grade Eight are thirteen. Carol Campbell and I are merely in Grade Six. All of us are at a different school now, one that's finally been built on our side of the ravine, so we don't have to take the school bus in the mornings or eat our lunches in the cellar or walk home over the collapsing footbridge after school. Our new school is a modern one-story yellow brick building that looks like a post office. It has soft-textured, eye-saving green blackboards instead of screechy black ones, and tiled pastel floors instead of the old creaky wooden ones in Queen Mary. There are no BOYS and GIRLS doors, there are no separate playgrounds. Even the teachers are different: younger, more casual. Some of them are young men.

· · · · ·

I've forgotten things, I've forgotten that I've forgotten them. I remember my old school, but only dimly, as if I was last there five years ago instead of five months. I remember going to Sunday school, but not the details. I know I don't like the thought of Mrs. Smeath, but I've forgotten why. I've forgotten about fainting and about the stacks of plates, and about falling into the creek and also about seeing the Virgin Mary. I've forgotten all of the bad things that happened. Although I see Cordelia and Grace and Carol every day, I remember none of those things; only that they used to be my friends, when I was younger, before I had other friends. There's something to do with them, something like a sentence in tiny dry print on a page, flattened out, like the dates of ancient battles. Their names are like names in a footnote, or names written in spidery brown ink in the fronts of Bibles. There is no emotion attached to these names. They're like the names of distant cousins, people who live far away, people I hardly know. Time is missing.

Nobody mentions anything about this missing time, except my mother. Once in a while she says, "That bad time you had," and I am puzzled. What is she talking about? I find these references to bad times vaguely threatening, vaguely insulting: I am not the sort of girl who has bad times, I have good times only. There I am, in the Grade Six class picture, smiling broadly. *Happy as a clam*, is what my mother says for happy. I am happy as a clam: hardshelled, firmly closed.

My parents labor away at our house. Rooms are being constructed in the cellar, gradually and with a lot of hammering and sawing, in my father's spare time: a darkroom, a storeroom for jars and jellies and jams. The lawn is a lawn now. In the garden they've planted a peach tree, a pear tree, an asparagus bed, rows and rows of vegetables. The borders are lush with flowers: tulips and daffodils, irises, peonies, pinks, chrysanthemums, something for every season. Occasionally I have to help, but mostly I watch with detachment as they upend themselves in the mud, digging and weeding, clay stains on the knees of their pants. They're like kids in a sandpile. I like the flowers, but know I would not go to such lengths, make such efforts, get myself so dirty to produce them.

The wooden footbridge over the ravine is torn down. Everyone says it's about time, it was getting so unsafe. They're going to replace it with a bridge made of concrete. I go one day and stand at the top of the hill on our side of the ravine, watching the bridge come down. There's a pile of rotten boards down by the creek. The vertical piles are still standing, like the trunks of dead trees, and part of the cross-planking is attached to them, but the railings are gone. I have an uneasy feeling, as if something's buried down there, a nameless, crucial thing, or as if there's someone still on the bridge, left by mistake, up in the air, unable to get to the land. But it's obvious there's no one.

Cordelia and Grace graduate and go elsewhere; Cordelia, it's rumored, to St. Sebastian's, a private school for girls, Grace to a high school farther north which emphasizes math. She's good at adding things up in neat little rows. She still has her long braids when she graduates. Carol hangs around near the boys at recess, and is often chased by two or three of them. They like to throw her into snowbanks and rub snow into her face, or, when snow is lacking, to tie her up with skipping ropes. When she runs away from them she flings her arms around a lot. She runs in a funny wiggling way, slow enough to be caught, and screams loudly when she is. She wears a training bra. She isn't much liked by the other girls.

For Social Studies I do a project on Tibet, where there are prayer wheels and reincarnation and women have two husbands, and for Science I do different kinds of seeds. I have a boyfriend, as is the fashion. Occasionally he sends me a note across the aisle, written in very black pencil. Sometimes there are parties, with awkward dancing and clumsy guffaws and horseplay by the boys, and wet, inexpert, toothy kisses. My boyfriend carves my initials into the top of his new school desk and gets the strap for it. He gets the strap for other things too. This is admired. I see my first television set, which is like a small black-and-white puppet show of no great interest.

Carol Campbell moves away and I hardly notice. I skip Grade Seven and go straight into Grade Eight, missing the Kings of England in chronological order, missing the circulatory system, leaving my boyfriend behind. I get my hair cut. I want to do this. I'm tired of having long wavy hair that has to be held back by barrettes or hairbands, I'm tired of being a child. I watch with satisfaction as my hair falls away from me like fog and my head emerges, sharper-featured, more clearly defined. I'm ready for high school, I want to go there right away.

I reorganize my room in preparation. I clear old toys out of my cupboard, I empty out all the drawers in my bureau. I find a solitary cat's eye marble rolling around at the back of the drawer, and some old dried-up chestnuts. Also a red plastic purse, which I remember getting for Christmas once. It's a babyish purse. It rattles when I pick it up; inside there's a nickel. I take the nickel out to spend, and put the marble inside the purse. I throw out the chestnuts.

I find my photo album with the black pages. I haven't taken any pictures with my Brownie camera for a long time, so this album has slipped from view. Stuck into it with the black triangles there are pictures I can't recall taking. For instance, there are several pictures of what look like large boulders, beside a lake. Underneath is printed, in white pencil: *Daisy. Elsie.* It's my writing, but I don't remember printing this.

I take these things down to the cellar and put them into the trunk, where old things go that are not thrown out. My mother's wedding dress is in there, several pieces of ornate silver, some sepia-toned portraits of people I don't know, a packet of bridge tallies with silk tassels on them, left over from before the war. Some of our old drawings are in there, my brother's spaceships and red and gold explosions, my delicate, old-fashioned little girls. I look at their pinafores and hair bows and their rudimentary faces and hands with distaste. I don't like looking at things connected so closely with my life as a child. I think these drawings are inept: I can do much better now.

The day before the first day of high school the telephone rings. It's Cordelia's Mummie; she wants to speak to my mother. I assume it's boring grown-ups' business and go back to reading the newspaper on the living room floor. But after she puts the phone down, my mother comes into the room.

"Elaine," she says. This is unusual, as she doesn't often use my name. She sounds solemn.

I look up from *Mandrake the Magician*. She looks down. "That was Cordelia's mother," she says. "Cordelia will be going to your high school. Cordelia's mother wonders whether you girls would like to walk to school together."

"Cordelia?" I say. I haven't seen or spoken to Cordelia for a whole year. She has vanished completely. I've chosen that school because I can walk to it, instead of going on a bus; so why not walk with Cordelia? "Okay," I say.

"Are you sure you want to?" my mother says, a little anxiously. She doesn't say why Cordelia will be coming to my school now and I don't ask.

"Why wouldn't I?" I say. I'm already sliding into flippancy, which goes with high school, but also I can't see what she's getting at. I'm being asked to do Cordelia, or Cordelia's mother, some kind of a minor favor. My mother's usual line is that you should do these favors when asked, so why is she hedging on this one?

She doesn't answer this. Instead she hovers. I go back to reading the comics. "Shall I call her mother back, then, or would you like to speak with Cordelia yourself?" she says.

"You can call her," I say. I add, "Please." I have no particular wish to speak to Cordelia right now.

The next morning I go to Cordelia's house, which is on the way to school, to pick her up. The door opens and Cordelia is there, but she is no longer the same. She's no longer angular and rangy; she's grown full breasts and is heavier in the hips and face. Her hair is longer now, not a pageboy.

She wears it in a ponytail with small white cloth lilies of the valley wired around the elastic band. She's bleached a peroxide streak into the bangs. She has orange lipstick, and orange nail polish to match. My own lipstick is pale pink. Seeing Cordelia, I realize that I don't look like a teenager, I look like a kid dressed up as one. I am still thin, still flat. I have a ferocious desire to be older.

We walk to school together, not saying much at first, past a gas station, a funeral parlor, then a mile along a strip of shops, a Woolworth's, an I.D.A. drugstore, a fruit and vegetable shop, a hardware store, all of them side by side in two-story flat-roofed yellow brick buildings. We hold our schoolbooks up against our chests, our full cotton skirts brushing against our bare legs. Right now it's the end of summer, when all the lawns are dull green or yellow and used up.

I've assumed Cordelia would be a grade ahead of me. But she isn't, she's in the same grade now. She's been expelled from St. Sebastian's for drawing a penis on a bat. Or this is what she says. She says there was a large drawing of a bat on the blackboard, with its wings outspread and just a tiny bump between its legs. So she went up to the blackboard when the teacher was out of the room and rubbed out the little bump and made a bigger, longer one— "Not that much bigger"—and the teacher came into the room and caught her doing it.

"Is that all?" I say.

Not exactly. She also printed *Mr. Malder*, neatly, underneath the bump. Mr. Malder was the teacher's name.

Probably this wasn't all she did, but it's all she's telling about. As an afterthought she mentions that she failed her year. "I was too young for it," she says. This sounds like something she's been told by other people, her mother most likely. "I was only twelve. They shouldn't have skipped me."

Now she's thirteen. I am twelve. I too have been skipped. I begin to wonder if I'll end up the same way she has, drawing penises on bats, failing my year.

39

*T*he school we go to is called Burnham High School. It's recently built, oblong in shape, flat-roofed, undecorated, unrevealing, sort of like a factory. It's the latest thing in modern architecture. Inside, it has long corridors with mottled floors of something that looks like granite but is not. The yellowish walls are lined with dark-green lockers, and there's an auditorium and a P.A. system.

Every morning we have announcements over the P.A. system. First we have a Bible reading and prayers. I bow my head during the prayers but I refuse to pray, though I don't know why I do this. After the prayers the principal tells us of coming events, and he also warns us to pick up our chewing gum wrappers and not to moon around in the halls like old married couples. His name is Mr. MacLeod, although everyone calls him Chrome Dome behind his back because he's bald on top; and he's a Scot by affiliation. Burnham High has a school plaid, a school crest with a thistle and a couple of those Scottish knives they stick in their socks, and a Gaelic motto. The plaid, the crest, the motto, and the school colors all belong to Mr. MacLeod's personal clan.

In the front hall, alongside the Queen, hangs a portrait of Dame Flora MacLeod with her two bagpipe-playing grandsons, posed outside Dunvegan Castle. We are encouraged to think of this castle as our ancestral home, and of Dame Flora as our spiritual leader. In choir we learn "The Skye Boat Song," about Bonnie Prince Charlie escaping the genocidal English. We learn "Scots Wha' Hae," and a poem about a mouse, which causes some snickering as it contains the word *breast*. I think all this Scottishness is normal for high schools, never having gone to one before; and even the several Armenians, Greeks, and Chinese in our school lose the edges of their differences, immersed as we all are in a mist of plaid.

I don't know many people at this school and neither does Cordelia. In my graduating class from public school there were only eight people, and in Cordelia's there were four. So it's a school full of strangers. In addition to that, we're in different homerooms, so we don't even have each other to rely on.

Everyone in my homeroom is bigger than I am. This is to be expected, because everyone is also older. The girls have breasts and a drowsy, powdery, hot-day smell; the skin of their faces is slippery-looking, slick with oily juice. I'm wary of them and dislike the changing room where we have to put on the blue cotton bloomer-bottomed gym suits with our names embroidered on the pockets. In there I feel skinnier than ever; when I catch sight of myself in the mirror I can see the ribs below my collarbone. During volleyball games, these other girls lollop and thunder around me, their voices outsized and raucous, their new, extra flesh wobbling. I take care to keep out of their way, simply because they are bigger and might knock me over. But I'm not really afraid of them. In a way I despise them, because they are so much like Carol Campbell, squealing and flinging themselves around.

Among the boys there are a few pipsqueaks whose voices have not yet changed, but many of the boys are gigantic. Some are fifteen, almost sixteen. They have hair that's long at the sides and greased back into ducktails, and they shave. Some of them look as if they shave a lot. They sit at the back of the room and stick their long legs out into the aisle. They've already failed a grade, at least once; they've given up and been given up on, and they're doing time until they can leave. Although they call remarks at other girls in the halls and make kissing sounds at them, or dangle around their lockers, they pay no attention at all to me. To them I'm just a child.

But I don't feel younger than these people. In some ways I feel older. In our Health book there's a chapter on teenage emotions. According to this book, I'm supposed to be caught in a whirlwind of teenage emotions, laughing one minute, crying the next, zooming around on a roller coaster, which is their term. However, this description does not apply to me. I am calm; I regard the antics of my fellow students, who act like the textbook, with a combination of scientific curiosity and almost matronly indulgence. When Cordelia says, "Don't you think he's a dreamboat?" I have a hard time understanding what she means. Occasionally I do cry for no reason, as it says you're supposed to. But I can't believe in my own sadness, I can't take it seriously. I watch myself crying in the mirror, intrigued by the sight of tears.

At lunchtime I sit with Cordelia in the cafeteria, which is pale-colored, with long whitish tables. We eat the lunches that have been sweltering in our school lockers all morning and which taste faintly of gym shoes, and drink chocolate milk through straws, and make what we consider to be witty, sarcastic remarks to each other, about the other kids at the school, about the teachers. Cordelia has been to high school for a year already and knows how to do this. She wears the collar of her blouse turned up and affects a derisive laugh. "He's a pill," she says; or, "What a creep." These are words that apply only to boys. Girls can be tough, stuck-up or cheap, mousy or boy-crazy; or they can be brains and sucks and brownnosers, like boys, if they are thought to study too much. But they can't be pills and creeps. I like the word *pill*. I think it refers to the little balls of wool that form on sweaters. Boys who are pills have sweaters like that. I take care to pick all such woolen balls off of my own sweaters.

Cordelia collects glossy photos of movie stars and singers, which she sends away for, finding the addresses of the fan clubs in movie magazines that advertise Frederick's of Hollywood peekaboo lingerie at the back, and chocolate-flavored tablets you chew to lose weight. She thumbtacks the photos to the bulletin board over her desk and Scotch tapes them to the walls of her room. Whenever I'm in there I feel as if there's a crowd watching me, their glossy black-and-white eyes following me around the room. Some of these pictures have signatures on them, and we examine them under the light to see if the pen has dented the paper.

If not, they're only printed on. Cordelia likes June Allyson, but she also likes Frank Sinatra and Betty Hutton. Burt Lancaster is the sexiest, according to Cordelia.

On the way home from school we go to the record store and try out 78 records in the tiny cork-lined booth. Sometimes Cordelia will buy a record with her allowance, which is larger than mine, but most of the time she just tries them out. She expects me to roll my eyes in ecstasy, the way she does; she expects me to groan. She knows the rituals, she knows how we're supposed to be behaving, now that we're in high school. But I think these things are impenetrable and fraudulent, and I can't do them without feeling I'm acting.

We take the records back to Cordelia's house and put them on the record player in the living room, and turn up the sound. Frank Sinatra appears, a disembodied voice, sliding around on the tune like someone slipping on a muddy sidewalk. He slithers up to a note, hits it, flails, recovers, oozes in the direction of another note.

"Don't you just love the way he does that?" says Cordelia. She flings herself onto the chesterfield, legs across the arm, head hanging upside-down. She's eating a doughnut covered with powdered sugar; the sugar has come off on her nose. "I feel as if he's right here, running his hand up and down my spine."

"Yeah," I say.

Perdie and Mirrie come in, and Perdie says, "Not mooning over *him* again," and Mirrie says, "Cordelia dear, would you mind turning down the sound?" These days she speaks to Cordelia in tones of extra sweetness and calls her *dear* a lot.

Perdie is in university now. She goes to frat parties. Mirrie's in the last year of high school, though not our high school. They are both more charming and beautiful and sophisticated than ever. They wear cashmere sweaters and pearl button earrings, and smoke cigarettes. They call them ciggie-poos. They call eggs eggie-poos, and breakfast brekkers. If someone is pregnant they say preggers. They call their mother Mummie, still. They sit and smoke their cigarettes and talk casually and with amused, semi-contemptuous irony about their friends, who have names like Mickie and Bobbie and Poochie and Robin. It's hard to figure out from the names whether these people are boys or girls.

"Are you sufficiently sophonsified?" Perdie asks Cordelia. This is a new thing they've taken to saying. It means, have you had enough to eat? "Those were supposed to be for dinner." She means the doughnuts.

"There's a lot left," says Cordelia, still upside-down, wiping her nose.

"Cordelia," says Perdie. "Don't turn your collar up like that. It's cheap."

"It's not cheap," says Cordelia. "It's sharp."

"Sharp," says Perdie, rolling her eyes, blowing smoke from her nose. Her mouth is little and plump and curly at the edges. "That sounds like a hair oil ad."

Cordelia sits around right side up and sticks her tongue in the corner of her mouth and looks at Perdie. "So?" she says at last. "What do you know? You're already over the hill."

Perdie, who's old enough to drink cocktails with the grown-ups before dinner although she's not supposed to do it in bars, curls up her mouth. "I think high school's bad for her," she says to Mirrie. "She's turning into a hardrock." She pronounces this word in a mocking drawl, to show that it's the sort of word she herself has outgrown. "Pull up your socks, Cordelia, or you'll flunk your year again. You know what Daddy said last time."

Cordelia flushes, and can't think what to say back.

Cordelia begins to pinch things from stores. She doesn't call it stealing, she calls it pinching. She pinches tubes of lipstick from Woolworth's, packets of licorice Nibs from the drugstore. She goes in and buys some small item, such as bobby pins, and when the salesgirl has her back turned getting the change out of the till she slips something off the counter and hides it under her coat or in her coat pocket. By this time it's autumn, and we have long coats which flap against the backs of our legs, coats with baggy, outsize patch pockets, good for pinching. Outside the store she shows me what she's gotten away with. She seems to think there's nothing wrong in what she's doing; she laughs with delight, her eyes sparkle, her cheeks are flushed. It's as if she's won a prize.

The Woolworth's has old wooden floors, stained from years of winter slush on people's boots, and dim overhead lights that hang down from the ceiling on metal stems. Nothing in it is anything we would really want, except maybe the lipsticks. There are photo frames with strangely tinted pictures of movie stars in them to show what the frame would look like with a photo in it; these stars have names like Raymon Novarro and Linda Darnell, stars from some remote period several years ago. There are cheesy hats, old-lady hats with veiling around them, and hair combs stuck with imitation rhinestones. Just about everything in here is imitation something else. We walk up and down the aisles, spraying ourselves from the cologne testers, rubbing the sample lipsticks on the backs of our hands, fingering the merchandise and disparaging it in loud voices, while the middle-aged salesladies glare at us.

Cordelia pinches a pink nylon scarf and thinks she's been seen by one of the glaring salesladies, so we don't go back there for a while. We go into the drugstore and buy Cream-sicles, and while I'm paying for them Cordelia pinches two horror comics. As we walk the rest of the way home from school we take turns reading them out loud, dramatizing the parts like radio plays, pausing to shriek with laughter. We sit on the low stone wall in front of the funeral parlor so we can both see the pictures, reading and laughing.

The comic books are drawn in great detail and garishly colored, with green and purple and sulfur-yellow prevailing. Cordelia reads a story about two sisters, a pretty one and one who has a burn covering half her face. The burn is maroon-colored and wrinkled like a dead apple. The pretty one has a boyfriend and goes to dances, the burned one hates her and loves the boyfriend. The burned one hangs herself in front of a mirror, out of jealousy. But her spirit goes into the mirror, and the next time the pretty one is brushing her hair in front of that mirror, she looks up and there's the burned one looking back at her. This is a shock and she faints, and the burned one gets out of the mirror and into the pretty one's body. She takes over the body and fools the boyfriend, she even gets him to kiss her, but although her face is now perfect, her reflection in that one mirror still shows her real, ruined face. The boyfriend sees it. Luckily he knows what to do. He breaks the mirror.

"Sob, sob," says Cordelia. "Oh, Bob . . . it was . . . horrible. Never mind, my darling, it's all over now. She's gone . . . back . . . to where she came from . . . forever. Now we can truly be together, without fear. Clinch. The End. Oh, puke!"

I read one about a man and a woman who drown at sea but find they aren't dead exactly. Instead they are enormously bloated and fat, and living on a desert island. They don't love each other any more because of being so fat. Along comes a ship and they wave to it. "They don't see us! They're passing right through us! Oh no . . . that must mean . . . we're condemned to be this way *forever!* Is there no way out?"

In the next picture they've hanged themselves. The fat bodies are dangling from one of the palm trees, and their previous thin bodies, wispy-looking and dressed in falling-apart bathing suits, are holding hands and walking into the ocean. "Clinch. The End."

"Oh, double puke," says Cordelia.

Cordelia reads one about a dead man coming back out of a swamp, covered with dripping, peeling-off flesh, to strangle the brother who pushed him into the swamp in the first place, and I read one about a man picking up a beautiful girl hitchhiker who turns out to have been dead for ten years. Cordelia reads one about a man who gets cursed by a voodoo witch doctor and grows a big red lobster claw on his hand, which turns on him and attacks him.

When we get to Cordelia's house, Cordelia doesn't want to take the horror comics inside with her. She says someone might find them and wonder where she got them. Even if they think she bought them, she'll be in trouble. So I end up taking them home with me. It doesn't occur to either of us to throw them out.

Once I get them home, I realize I don't want them in the same room with me at night. It's one thing to laugh at them in the daylight, but I don't like the idea of them lying there, right in my bedroom, while I'm asleep. I think of them glowing in the dark, with a lurid sulfur-yellow light; I think of curling wisps of mist coming out of them and materializing on top of my bureau. I'm afraid I'll find out that there's someone else trapped inside my body; I'll look into the bathroom mirror and see the face of another girl, someone who looks like me but has half of her face darkened, the skin burned away.

I know these things won't really happen, but I don't like the thought. Nor do I want to throw the comics away: that would be letting them loose, they might go out of control. So I take them into Stephen's room and slide them in among his own old comic books, which are still there, stacked up under his bed. He never reads them any more, so he won't find these ones. Whatever emanations may seep from them at night, he will be impervious to them. In my opinion he is up to things, which includes things of this kind.

40

*I*t's Sunday evening. There's a fire in the fireplace; the drapes are drawn against the heavy November darkness. My father sits in the easy chair marking drawings of spruce budworms cut open to show their digestive systems, my mother has made grilled cheese squares with bacon on them. We're listening to "The Jack Benny Show" on the radio, which is punctuated by singing commercials for Lucky Strike cigarettes. On this show there is a man who talks in a raspy voice and another one who says "Pickle in the middle and the mustard on top." I have no idea that the first one is supposed to be black and the second one Jewish; I think they just have funny voices.

Our old radio with the green eye has vanished, and a new, blond one has appeared, in a smooth unornamented cabinet that holds a long-playing record player as well. We have little wooden nesting tables for our plates with the cheese squares; these tables are blond also, with legs that are wide at the top and taper down without a bump or curlicue, no dust catchers. They look like the legs of fat women as they appear in comic books: no knees, no ankles. All this blond wood is from Scandinavia. Our silverware has descended to the steamer trunk. In its place there is new silverware, which is not silver but stainless steel.

These items have been chosen, not by my mother, but by my father. He picks out my mother's dressing-up clothes as well; my mother, laughing, says that all her taste is in her mouth. As far as she is concerned a chair is there to sit down on, and she couldn't care less whether it has pink petunias on it or purple polka dots, as long as it doesn't collapse. It's as if, like a cat, she cannot see things unless they are moving. She is becoming even more indifferent to fashion, and strides around in improvised getups, a ski jacket, an old scarf, mitts that don't match. She says she doesn't care what it looks like as long as it keeps out the wind.

Worse, she's taken up ice dancing; she goes to classes at the local indoor rink, and tangos and waltzes in time to tinny music, holding hands with other women. This is mortifying but at least she does it indoors, where no one can see her. I can only hope she won't take to practicing, later when it's really winter, on the outdoor rink, where somebody I might know could see her. But she isn't even aware of the chagrin this could cause. She never says, *What will people think?* the way other mothers do, or are supposed to. She says she doesn't give a hoot.

I think this is irresponsible of her. At the same time, the word *hoot* pleases me. It makes my mother into a nonmother, a sort of mutant owl. I have become picky about my own clothes, and given to looking at myself from behind with the aid of a hand mirror: although I may appear all right from the front, treachery could sneak up on me: a loose thread, a dropped hem. *Not giving a hoot* would be a luxury. It describes the fine, irreverent carelessness I myself would like to cultivate, in these and other matters.

My brother sits in one of the taper-legged blond chairs that go with the tables. He has become bigger and older, all of a sudden, when I wasn't looking. He has a razor now. Because it's the weekend and he hasn't shaved, he has a line of fine bristles poking out of the skin around his mouth. He's got on his moccasins, old ones he wears around the house, with holes worn under the big toes, and his V-neck maroon sweater with the ravels coming off the elbows. He resists my mother's efforts to mend this sweater or replace it. My mother says frequently that she doesn't give a hoot about clothes, but this indifference does not extend to holes, frayed edges, or dirt.

My brother's ragged sweater and sievelike moccasins are the clothes he studies in. On weekdays he has to wear a jacket and tie and gray flannels, all of which are required at his school. He can't have a ducktail, like the boys at my school, or even a crewcut: his hair is shaved up the back of the neck and parted at one side, like the hair of English choirboys. This too is a school requirement. With his hair cut this way he looks like an illustration from an adventure book of the 1920s or earlier, of which there are a number in our cellar, or like an Allied air officer from a comic book. He has that kind of nose, that kind of chin, although thinner: clean-cut, good-looking, old-fashioned. His eyes are like that too, a piercing, slightly fanatical blue. His scorn for boys who give a hoot about how they look is devastating. He calls them fruity clothes horses.

His school is a private school for brainy boys, though not an expensive one: you get in by passing tough exams. My parents asked me, a little anxiously, if I wanted to go to a private school for girls; they thought I'd feel left out if they didn't make the effort for me too. I know about these schools, where you have to wear kilts and play field hockey. I said they were for snobs and had low academic standards, which was true. But in fact I wouldn't be caught dead in a girls' school. The idea fills me with claustrophobic panic: a school with nothing in it but girls would be like a trap.

My brother is listening to Jack Benny too. As he listens, he stuffs the cheese squares into his mouth with his left hand, but his right hand holds a pencil, and this hand is never still. He hardly looks at the scrap pad on which he's doodling, but once in a while he tears off a sheet and crumples it up. These crumpled notes land on the floor. When I gather them up to put them into the wastebasket after the show, I see that they're covered with numbers, long lines of numbers and symbols that go on and on, like writing, like a letter in code.

My brother sometimes has friends over. They sit in his room with the chess table between them, not moving except for their hands, which lift, hover over the board, plunge down. Sometimes they grunt or say "Aha" or "Trade you" or "Got you back"; or they exchange new, obscure good-natured insults: "You surd!" "You square root!" "You throwback!" The captured chess pieces, knights and pawns and bishops, line up on the outskirts of the board. Once in a while, to see how the game is going, I bring in glasses of milk and vanilla-chocolate pinwheel cookies which I've made out of the *Betty Crocker Picture Cookbook*. This is a form of showing-off on my part, but it doesn't get much response. They grunt, drink the milk with their left hands, stuff in the cookies, their eyes never leaving the board. The bishops topple, the queen falls, the king is encircled. "Mate in two," they say. A finger comes down, knocks over the king. "Best of five." And they start again.

In the evenings my brother studies. Sometimes he does this in a curious way. He stands on his head, to improve the circulation to his brain, or he throws spitballs at the ceiling. The area around his ceiling light fixture is pimply with little wads of once-chewed paper. At other times he indulges in manic bouts of physical activity: he splits huge piles of kindling, much more than is needed, or goes running down in the ravine, wearing disgraceful baggy pants and a forest-green sweater even more unraveled than his maroon one, and frayed gray running shoes that look like the kind you see one of in vacant lots. He says he's training for the marathon.

A lot of the time my brother doesn't seem aware of me. He's thinking about other things, solemn things that are important. He sits at the dinner table, his right hand moving, pinching a breadcrust into pellets, staring at the wall behind my mother's head, on which there is a picture of three milkweed pods in a vase, while my father explains why the human race is doomed. This time it's because we've discovered insulin. All the diabetics aren't dying the way they used to, they're living long enough so that they're passing the diabetes on to their children. Soon, by the law of geometric progression, we'll all be diabetics, and since insulin is made from cows' stomachs the whole world will be covered with insulin-producing cows, the parts that aren't covered with human beings, who are reproducing much too rapidly for their own good anyway. The cows burp methane gas. Far too much methane gas is entering the atmosphere already, it will choke out the oxygen and perhaps cause the entire earth to become a giant greenhouse. The polar seas will melt and New York will be under six feet of water, not to mention many another coastal city.

Also we have to worry about deserts, and erosion. If we don't get burped to death by the cows we'll end up like the Sahara Desert, says my father cheerfully, finishing up the meatloaf.

My father has nothing against diabetics, or cows either. He just likes following chains of thought to their logical conclusions. My mother says it's coffee soufflé for dessert.

Once my brother would have been more interested in the fate of the human race. Now he says that if the sun went supernova it would be eight minutes before we'd see it. He's taking the long-range view. Sooner or later we're going to be a cinder anyway, he implies, so why worry about a few cows more or less? Although he still collects butterfly sightings, he's moving farther and farther away from biology. In the larger picture, we're just a little green scuzz on the surface, says my brother.

My father eats his coffee soufflé, frowning a little. My mother tactfully pours him a cup of tea. I see that the future of the human race is a battleground, that Stephen has won a point and my father has lost one. Whoever cares the most will lose.

I know more about my father than I used to know: I know he wanted to be a pilot in the war but could not, because the work he did was considered essential to the war effort. How spruce budworms could be essential to the war effort I have not yet figured out, but apparently they were. Maybe this is why he always drives so fast, maybe he's heading for takeoff.

I know he grew up on a farm in the backwoods of Nova Scotia, where they didn't have running water or electricity. This is why he can build things and chop things: everyone there could use an ax and a saw. He did his high school courses by correspondence, sitting at the kitchen table and studying by the light of a kerosene lamp; he put himself through university by working in lumber camps and cleaning out rabbit hutches, and was so poor that he lived in a tent in the summers to save money. He used to play country fiddle at square dances and was twenty-two before he heard an orchestra. All this is known, but unimaginable. Also I wish I did not know it. I want my father to be just my father, the way he has always been, not a separate person with an earlier, mythological life of his own. Knowing too much about other people puts you in their power, they have a claim on you, you are forced to understand their reasons for doing things and then you are weakened.

I harden my heart toward the fate of the human race, and calculate in my head how much money I'll need to save to buy a new lamb's-wool sweater. In Home Economics, which really means cooking and sewing, I've learned how to install a zipper and make a flat-fell seam, and now I make a lot of my clothes myself because it's cheaper, although they don't always turn out exactly like the picture on the front of the pattern. I get very little help from my mother on the fashion front, because whatever I wear she says it looks lovely, as long as it has no visible rents.

For advice I turn to Mrs. Finestein next door, for whom I baby-sit on weekends. "Blue is your color, honey," she says. "Very gorgeous. And cerise. You'd look stunning in cerise." Then she goes out for the evening with Mr. Finestein, her hair upswept, her mouth vivid, teetering in her tiny shoes with high heels, jingling with bracelets and dangly gold earrings, and I read *The Little Engine That Could* to Brian Finestein and tuck him into bed.

Sometimes Stephen and I still get stuck doing the dishes together, and then he remembers he's my brother. I wash, he dries, and he asks me benign, avuncular, maddening questions, such as how do I like Grade Nine. He is in Grade Eleven, stairways and stairways above me; he doesn't have to rub it in.

But on some of these dish-drying nights he reverts to what I consider to be his true self. He tells me the nicknames of the teachers at his school, all of which are rude, such as The Armpit or The Human Stool. Or we invent new swear-words together, words that suggest an unspecific dirtiness. "Frut," he says. I counter with "pronk," which I tell him is a verb. We lean against the kitchen counter, doubled over with laughter, until our mother comes into the kitchen and says, "What are you two kids up to?"

Sometimes he decides that it's his duty to educate me.

He has a low opinion of most girls, it seems, and doesn't want me turning into one of the ordinary kind. He doesn't want me to be a pin-headed fuzzbrain. He thinks I'm in danger of becoming vain. In the mornings he stands outside the bathroom door and asks if I can bear to unstick myself from the mirror.

He thinks I should develop my mind. In order to help me do this, he makes a Möbius strip for me by cutting out a long slip of paper, twisting it once and gluing the ends together. This Möbius strip has only one side, you can prove it by running your finger along the surface. According to Stephen, this is a way of visualizing infinity. He draws me a Klein bottle, which has no outside and no inside, or rather the outside and the inside are the same. I have more trouble with the Klein bottle than the Möbius strip, probably because it's a bottle, and I can't think of a bottle that isn't intended to contain something. I can't see the point of it.

Stephen says he's interested in the problems of two-dimensional universes. He wants me to imagine what a three-dimensional universe would look like to someone who was perfectly flat. If you stood in a two-dimensional universe you would only be perceived at the point of intersection, you'd be perceived as two oblong discs, two two-dimensional cross sections of your own feet. Then there are five-dimensional universes, seven-dimensional ones. I try very hard to picture these but I can't seem to get past three.

"Why three?" says Stephen. This is a favorite technique of his, asking me questions to which he knows the answers, or other answers.

"Because that's how many there are," I say.

"That's how many we *perceive*, you mean," he says. "We're limited by our own sensory equipment. How do you think a fly sees the world?" I know how a fly perceives the world, I've seen many flies' eyes, through microscopes. "In facets," I say. "But each facet would still have only three dimensions."

"Point taken," he says, which makes me feel grown-up, worthy of this conversation. "But actually we perceive four."

"Four?" I say.

"Time is a dimension," he says. "You can't separate it from space. Space-time is what we live in." He says there are no such things as discrete objects which remain unchanged, set apart from the flow of time. He says space-time is curved and that in curved space-time the shortest distance between two points is not a straight line but a line following the curve. He says that time can be stretched or shrunk, and that it runs faster in some places than in others. He says that if you put one identical twin in a high-speed rocket for a week, he'd come back to find his brother ten years older than he is himself. I say I think that would be sad.

My brother smiles. He says the universe is like a dot-covered balloon that's being blown up. The dots are the stars; they're moving farther and farther away from one another all the time. He says that one of the really interesting questions is whether the universe is infinite and unbounded, or infinite but bounded, like the balloon idea. All I can think of in connection with a balloon is the explosion when it breaks.

He says that space is mostly empty and that matter is not really solid. It's just a bunch of widely spaced atoms moving at greater or lesser speeds. Anyway, matter and energy are aspects of each other. It's as if everything is made of solid light. He says that if we knew enough we could walk through walls as if they were air, if we knew enough we could go faster than light, and at that point space would become time and time would become space and we would be able to travel through time, back into the past.

This is the first of these ideas of his that has really interested me. I'd like to see dinosaurs and a good many other things, such as the Ancient Egyptians. On the other hand there's something menacing about this notion. I'm not so sure I want to travel back into the past. I'm not so sure I want to be that impressed, either, by everything he says. It gives him too much of an advantage. Anyway it isn't a sensible way to talk. A lot of it sounds like comic books, the kind with ray guns.

So I say, "What good would that be?"

He smiles. "If you could do it, you'd know you could do it," is what he says.

I tell Cordelia that Stephen says we could walk through walls if we knew enough. This is the only one of his latest ideas I can trust myself to expound, at the moment. The rest are too complicated, or bizarre.

Cordelia laughs. She says that Stephen is a brain and that if he weren't so cute he'd be a pill.

Stephen has a summer job this summer, teaching canoeing at a boys' camp, but I don't, because I'm only thirteen. I go with my parents up to the north, near Sault Ste. Marie, where my father is overseeing an experimental colony of tent caterpillars in screened-in cages.

Stephen writes me letters, in pencil, on pages torn from lined workbooks, in which he ridicules everything he can get his hands on, including his fellow camp instructors and the girls they go drooling around after on their days off. He describes these instructors with pimples popping from their skins, fangs sprouting in their mouths, their tongues hanging out like those of dogs, their eyes crossed in permanent, girl-inspired imbecility. This makes me think I have power, of a sort. Or will have it: I too am a girl. I go fishing by myself, mostly so I'll have something to put in my letters to him. Other than that I don't have much to tell.

Cordelia's letters are in real ink, black in color. They are full of superlatives and exclamation marks. She dots her I's with little round circles, like Orphan Annie eyes, or bubbles. She signs them with things like, "Yours till Niagara Falls," "Yours till the cookie crumbles," or "Yours till the sea wears rubber pants to keep its bottom dry."

"I am *so bored!!!*" she writes, with triple underlining. She sounds enthusiastic even about boredom. And yet her burbly style does not ring true. I have seen her, sometimes, when she thinks I'm not looking: her face goes still, remote, unreflecting. It's as if she's not inside it. But then she'll turn and laugh. "Don't you just *love* it when they roll up their sleeves and tuck the cigarette pack inside?" she'll say. "That takes biceps!" And she will be back to normal.

I feel as if I'm marking time. I swim in the lake provided, and eat raisins and crackers spread thickly with peanut butter and honey while reading detective stories, and sulk because there's no one my age around. My parents' relentless cheer is no comfort. It would almost be better if they could be as surly as I am, or surlier; this would make me feel more ordinary.

IX
LEPROSY

41

*I*n late morning the phone wakes me. It's Charna. "Hey," she says. "We made the front page of Entertainment, and three, count them, three pictures! It's a real rave!"

I shudder at her idea of a rave; and what does she mean, *we*? But she's pleased: I've graduated from Living to Entertainment; this is a good sign. I remember when I had ideas about eternal greatness, when I wanted to be Leonardo da Vinci. Now I'm in with the rock groups and the latest movie. Art is what you can get away with, said somebody or other, which makes it sound like shoplifting or some other minor crime. And maybe that's all it ever was, or is: a kind of stealing. A hijacking of the visual.

I know it will be bad news. Still, I can't resist. I pull on my clothes, go down in search of the nearest paper box. I do have the decency to wait until I get upstairs before I open the paper.

The bold print says: CROTCHETY ARTIST STILL HAS POWER TO DISTURB. I take note: *artist* instead of *painter*, the foreboding *still*, sign-pointing the way to senility. Andrea the acorn-headed ingénue getting her own back. I'm surprised she'd use an old-fashioned word like *crotchety*. It manages to suggest both crotches and crocheting, both of which seem appropriate. But probably she didn't write the headline.

There are indeed three photos. One is of my head, shot a little from beneath so it looks as if I have a double chin. The other two are of paintings. One is of Mrs. Smeath, bare-naked, flying heavily through the air. The church spire with the onion on it is in the distance. Mr. Smeath is stuck to her back like an asparagus beetle, grinning like a maniac; both of them have shiny brown insect wings, done to scale and meticulously painted. *Erbug, The Annunciation*, it's called. The other is of Mrs. Smeath by/herself, with a sickle-moon paring knife and a skinless potato, unclad from the waist up and the thighs down. This is from the *Empire Bloomers* series. The newspaper photos don't do these paintings justice, because there's no color. They look too much like snapshots. I know that in real life the bloomers on Mrs. Smeath are an intense indigo blue that took me weeks to get right, a blue that appears to radiate a dark and stifling light.

I scan the first paragraph: "Eminent artist Elaine Risley returns to hometown Toronto this week for a long overdue retrospective." *Eminent*, the mausoleum word. I might as well climb onto the marble slab right now and pull the bedsheet over my head. There are the usual misquotations, nor does my blue jogging suit escape comment. "Elaine Risley, looking anything but formidable in a powder-blue jogging suit that's seen better days, nevertheless can come out with a few pungent and deliberately provocative comments on women today."

I suck in some coffee, skip to the last paragraph: the inevitable *eclectic*, the obligatory *post-feminist*, a *however* and a *despite*. Good old Toronto bet hedging and qualification. A blistering attack would be preferable, some flying fur, a little fire and brimstone. That way I would know I'm still alive.

I think savagely of the opening. Perhaps I should be deliberately provocative, perhaps I should confirm their deepest suspicions. I could strap on some of Jon's axmurder special effects, the burnt face with its one peeled bloodshot eye, the plastic blood-squirting arm. Or slip my feet into the hollow casts of feet and lurch in like something from a mad scientist movie.

I won't do these things, but thinking about them is soothing. It distances the entire thing, reduces it to a farce or prank, in which I have no involvement aside from mockery.

Cordelia will see this piece in the paper, and maybe she will laugh. Even though she's not in the phone book, she must still be around here somewhere. It would be like her to have changed her name. Or maybe she's married; maybe she's married more than once. Women are hard to keep track of, most of them. They slip into other names, and sink without a trace.

At any rate she'll see this. She'll know it's Mrs. Smeath, she'll get a kick out of that. She'll know it's me, and she'll come. She'll come in the door and she'll see herself, titled, framed, and dated, hanging on the wall. She will be unmistakable: the long line of jaw, the slightly crooked lip. She appears to be in a room, alone; a room with walls of a pastel green.

This is the only picture I ever did of Cordelia, Cordelia by herself. *Half a Face*, it's called: an odd title, because Cordelia's entire face is visible. But behind her, hanging on the wall, like emblems in the Renaissance, or those heads of animals, moose or bear, you used to find in northern bars, is another face, covered with a white cloth. The effect is of a theatrical mask. Perhaps.

I had trouble with this picture. It was hard for me to fix Cordelia in one time, at one age. I wanted her about thirteen, looking out with that defiant, almost belligerent stare of hers. *So?*

But the eyes sabotaged me. They aren't strong eyes; the look they give the face is tentative, hesitant, reproachful. Frightened.

Cordelia is afraid of me, in this picture.

I am afraid of Cordelia.

I'm not afraid of seeing Cordelia. I'm afraid of being Cordelia. Because in some way we changed places, and I've forgotten when.

42

*A*fter the summer I'm in Grade Ten. Although I'm still shorter, still younger, I have grown. Specifically, I've grown breasts. I have periods now, like normal girls; I too am among the knowing, I too can sit out volleyball games and go to the nurse's for aspirin and waddle along the halls with a pad like a flattened rabbit tail wadded between my legs, sopping with liver-colored blood. There are satisfactions in this. I shave my legs, not because there's much to shave but because it makes me feel good. I sit in the bathtub, scraping away at my calves, which I wish were thicker, bulgier, like the calves of cheerleaders, while my brother mutters outside.

"Mirror, mirror on the wall, who is the most beautiful of all?" he says.

"Go away," I say tranquilly. I now have that privilege.

In school I am silent and watchful. I do my homework. Cordelia plucks her eyebrows into two thin lines, thinner than mine, and paints her nails Fire and Ice. She loses things, such as combs and also her French homework. She laughs raucously in the halls. She comes up with new, complicated swearwords: *excrement of the ungulate*, she says, meaning *bullshit*, and *great flaming blue-eyed bald-headed Jesus*. She takes up smoking and gets caught doing it in the girls' washroom. It must be hard for the teachers, looking, to figure out why we are friends, what we're doing together.

Today on the way home it snows. Big soft caressing flakes fall onto our skin like cold moths; the air fills with feathers. Cordelia and I are elated, we racket along the sidewalk through the twilight while the cars drift past us, hushed and slowed by the snow. We sing:

> *Remember the name*
> *Of Lydia Pinkham,*
> *Whose remedies for women brought her FAME!*

This is a singing commercial from the radio. We don't know what Lydia Pinkham's remedies are, but anything that says "for women" on it has to do with monthly blood or some equally unspeakable female thing, and so we think it's funny. Also we sing:

> *Leprosy,*
> *Night and day you torture me,*
> *There goes my eyeball*
> *Into my highball . . . .*

Or else:

> *Part of your heart,*
> *That's what I'm eating now,*
> *Too bad we had to part . . . .*

We sing these, and other parodies of popular songs, all of which we think are very witty. We run and slide, in our rubber boots with the tops turned down, and make snowballs which we throw at lampposts, at fire hydrants, bravely at passing cars, and as close as we dare at people walking on the sidewalk, women most of them, with shopping bags or dogs. We have to set our schoolbooks down to make the snowballs. Our aim is poor and we don't hit much of anything, though we hit a woman in a fur coat, from behind, by mistake. She turns and scowls at us and we run away, around a corner and up a side street, laughing so much with terror and embarrassment we can hardly stand up. Cordelia throws herself backward onto a snow-covered lawn. "The evil eye!" she shrieks. For some reason I don't like the sight of her lying there in the snow, arms spread out.

"Get up," I say. "You'll catch pneumonia."

"So?" says Cordelia. But she gets up.

The streetlights come on, though it isn't yet dark. We reach the place where the cemetery begins, on the other side of the street.

"Remember Grace Smeath?" Cordelia says.

I say yes. I do remember her, but not clearly, not continuously. I remember her from the time I first knew her, and later, sitting in the apple orchard with a crown of flowers on her head; and much later, when she was in Grade Eight and about to go off to high school. I don't even know what high school she went to. I remember her freckles, her little smile, her coarse horsehair braids.

"They rationed their toilet paper," Cordelia says. "Four squares a time, even for Number Two. Did you know that?"

"No," I say. But it seems to me that I did know it, once.

"Remember that black soap they had?" says Cordelia. "Remember? It smelled like tar."

I know what we're doing now: we're making fun of the Smeath family. Cordelia remembers all kinds of things: the greying underwear dripping on the clothesline in the cellar, the kitchen paring knife that was worn right down to a sliver, the winter coats from the *Eaton's Catalogue*. Simpsons is the right place to shop, according to Cordelia. That's where we go now on Saturday mornings, bareheaded, jerking downtown stop by stop on the streetcar. And shopping from the *Eaton's Catalogue* is much worse than shopping at Eaton's.

"The Lump-lump Family!" Cordelia shouts into the snowy air. It's cruel and appropriate; we snort with laughter. "What does the Lump-lump Family have for dinner? Plates of gristle!"

Now it's a full-blown game. What color is their underwear? Grunt color. Why did Mrs. Lump-lump have a Band-Aid on her face? Cut herself shaving. Anything can be said about them, invented about them. They're defenseless, they're at our mercy. We picture the two adult Lump-lumps making love, but this is too much for us, it can't be done, it's too vomit-making. Vomit-making is a new word, from Perdie.

"What does Grace Lump-lump do for fun? Pops her pimples!" Cordelia laughs so hard she doubles over and almost falls down. "Stop, stop, you'll make me pee," she says. She says that Grace started to grow pimples in Grade Eight: by now they must have increased in number. This is not made up but true. We relish the thought.

The Smeaths in our rendition of them are charmless, miserly, heavy as dough, boring as white margarine, which we claim they eat for dessert. We ridicule their piety, their small economies, the size of their feet, their rubber plant, which sums them up. We speak of them in the present tense, as if we still know them.

This for me is a deeply satisfying game. I can't account for my own savagery; I don't question why I'm enjoying it so much, or why Cordelia is playing it, insists on playing it, whips it to life again when it seems to be flagging. She looks at me sideways, as if estimating how far, how much farther I'll go in what we both know, surely, is base treachery. I have a fleeting image of Grace once more, disappearing into her house through the front door, in her skirt with the straps, her pilly sweater. She was adored, by all of us. But she is not any more. And in Cordelia's version, now, she never was.

We run across the street in the falling snow, open the small wrought-iron gate in the cemetery fence, go in. We've never done this before.

This is the raw end of the cemetery. The trees are only saplings; they look even more temporary without their leaves. Much of the ground is untouched, but there are scars like giant claw marks, diggings, earthworks going on. The gravestones are few and recent: blockish oblongs of granite polished to a Presbyterian gloss, the letters cut plainly and without any attempt at prettiness. They remind me of men's overcoats.

We walk among these gravestones, pointing out which ones—particularly gray, particularly oafish—the Lumplump Family would choose to bury one another beneath. From here we can look through the chain-link fence and see the houses on the other side of the street. Grace Smeath's is one of them. It's strange and oddly pleasant to think that she might be inside it at this very moment, inside that ordinary-looking brick box with the white porch pillars, not knowing a thing about what we've just been saying about her. Mrs. Smeath might be in there, lying on the velvet chesterfield, the afghan spread over her; I remember this much. The rubber plant will be on the landing, not much bigger. Rubber plants grow slowly. We are bigger though, and the house looks small.

The cemetery stretches out before us, acres and acres. Now the ravine is on our left, with the new concrete bridge just visible. I have a quick memory of the old bridge, of the creek beneath it: under our feet the dead people must be dissolving, turning to water, cold and clear, flowing down-hill. But I forget about this immediately. Nothing about the cemetery is frightening, I tell myself. It's too pragmatic, too ugly, too neat. It's only like a kitchen shelf, where you put things away.

We walk for a while without speaking, not knowing where we're going, or why. The trees are taller, the tomb-stones older. There are Celtic crosses now, and the occa-sional angel.

"How do we get out of here?" says Cordelia, laughing a little.

"If we keep going we'll hit a road," I say. "Isn't that the traffic?"

"I need a ciggie-poo," Cordelia says. We find a bench and sit down so Cordelia can free her hands for the cigarette, cupping it against the air, lighting it. She isn't wearing gloves, or a scarf on her head. She has a tiny black and gold lighter.

"Look at all the little dead people houses," she says.

"Mausoleums," I say knowingly.

"The Lump-lump Family Mausoleum," she says, giving the joke one last push.

"They wouldn't have one," I say. "Too ritzy."

"Eaton," Cordelia reads. "That must be the store, it's the same lettering. The *Eaton's Catalogues* are buried in there."

"Mr. and Mrs. Catalogue," I say.

"I wonder if they're wearing foundation garments," says Cordelia, inhaling. We're trying for a return to our hilarity, but it isn't working. I think of the Eatons, both of them or maybe more, tucked away for storage as if they're fur coats or gold watches, in their private tomb, which is all the stranger for being shaped like a Greek temple. Where exactly are they, inside there? On biers? In cobwebby stone-lidded coffins, as in the horror comics? I think of their jewels, glinting in the dark—of course they would have jewels—and of their long dry hair. Your hair grows after you're dead, also your fingernails. I don't know how I know this.

"Mrs. Eaton is really a vampire, you know," I say slowly. "She comes out at night. She's dressed in a long white ballgown. That door creaks open and she comes out."

"To drink the blood of Lump-lumps out too late," says Cordelia hopefully, stubbing out her cigarette.

I refuse to laugh. "No, seriously," I say. "She does. I happen to know."

Cordelia looks at me nervously. The snow is falling, it's twilight, there's nobody here but us. "Yeah?" she says, waiting for the joke.

"Yes," I say. "We sometimes go together. Because I'm a vampire too."

"You're not," says Cordelia, standing up, brushing off the snow. She's smiling uncertainly.

"How do you know?" I say. "How do you *know*?"

"You walk around in the daytime," Cordelia says.

"That's not me," I say. "That's my twin. You've never known, but I'm one of a twins. Identical ones, you can't tell us apart by looking. Anyway it's just the sun I have to avoid. On days like this it's perfectly safe. I have a coffin full of earth where I sleep; it's down in, down in"—I search for a likely place—"the cellar."

"You're being silly," Cordelia says.

I stand up too. "Silly?" I say. I lower my voice. "I'm just telling you the truth. You're my friend, I thought it was time you knew. I'm really dead. I've been dead for years."

"You can stop playing that," says Cordelia sharply. I'm surprised at how much pleasure this gives me, to know she's so uneasy, to know I have this much power over her.

"Playing what?" I say. "I'm not playing. But *you* don't

have to worry. I won't suck any of your blood. You're my friend."

"Don't be a brat," says Cordelia.

"In a minute," I say, "we're going to be locked in." It strikes both of us that this may be the truth. We run along the roadway, gasping and laughing, and find a large gateway, which is luckily still open. Beyond it is Yonge Street, lined with rush-hour traffic.

Cordelia wants to point out Lump-lump Family cars, but I'm tired of this. I have a denser, more malevolent little triumph to finger: energy has passed between us, and I am stronger.

43

Now I'm in Grade Eleven, and as tall as many other girls, which is not very tall. I have a charcoal-gray pencil skirt that's hard to walk in despite the kick pleat, and a bat wing sweater, a red one with modulated gray horizontal stripes across it. I have a wide black elastic cinch belt with an imitation gold clasp buckle, and flat ballerina shoes of velveteen that scuff as I walk and bulge out at the sides. I have a shortie coat to go with the pencil skirt. This is the look: boxy and flared at the top, with a long skinny stem of thighs and legs coming out the bottom. I have a mean mouth.

I have such a mean mouth that I become known for it. I don't use it unless provoked, but then I open my mean mouth and short, devastating comments come out of it. I hardly have to think them up, they're just there suddenly, like thought balloons with light bulbs in them. "Don't be a pain" and "Takes one to know one" are standard repartee among girls, but I go much farther than that. I'm willing to say *pain in the ass*, which skirts good taste, and to go in for crushing inventions, such as The Walking Pimple and The Before Part of an Arrid Armpit Ad. If any girl calls me a brain, I say, "Better a brain than a pin-headed moron like you." "Use much hair grease?" I will say, or "Suck much?" I know where the weak spots are. "Suck" is an especially satisfying word, especially annihilating. Boys say it mostly, to one another; it suggests thumbs and babies. I haven't yet considered what else might be sucked, or under what circumstances.

Girls at school learn to look out for my mean mouth and avoid it. I walk the halls surrounded by an aura of potential verbal danger, and am treated with caution, which suits me fine. Strangely enough, my mean behavior doesn't result in fewer friends, but, on the surface, more. The girls are afraid of me but they know where it's safest: beside me, half a step behind. "Elaine is a riot," they say, without conviction. Some of them are already collecting china and housewares, and have Hope Chests. For this kind of thing I feel amused disdain. And yet it disturbs me to learn I have hurt someone unintentionally. I want all my hurts to be intentional.

I don't have occasion to use my mean mouth on boys, since they don't say provoking things to me. Except for Stephen, of course. These days we trade verbal meannesses as a kind of game, like badminton. *Got you. Got you back.* I can usually silence him with, "Where'd you get that haircut? Lawnmowerville?" He's sensitive about the haircut. Or, when he's all spiffed up in his private school gray flannels and jacket: "Hey, you look like a Simpsons Rep." Simpsons Reps are sucky kids who appear in high school yearbooks wearing blazers with crests on the pockets, looking clean-cut, and advertising Simpsons.

My father says, "Your sharp tongue will get you in trouble some day, young lady." *Young lady* is a sign that I've gone too daringly close to some edge or other, but although it silences me for the moment it doesn't tone me down. I've come to enjoy the risk, the sensation of vertigo when I realize that I've shot right over the border of the socially acceptable, that I'm walking on thin ice, on empty air.

The person I use my mean mouth on the most is Cordelia. She doesn't even have to provoke me, I use her as target practice. We sit on the hill overlooking the football field, wearing our jeans, which are only allowed at school on the days of football games. We have our overlong pant cuffs pinned up with blanket pins, the latest thing. The cheerleaders leap around in their mid-thigh skirts, waving their paper pom-poms; they don't look long-legged and golden, like the cheerleaders at the back of *Life* magazine, but ill-assorted, dumpy, and dark. However I still envy their calves. The football team jogs on. Cordelia says, "That Gregory! What a hunk," and I say, "Of cheese." Cordelia gives me a hurt look. "I think he's a doll." "If you like them covered with corn oil," I say. When she says it's a bad idea to sit down on the high school toilet seats without wiping them off first because you might get a disease, I say, "Who told you that? Your Mummie?"

I make fun of her favorite singers. "Love, love, love," I say. "They're always moaning." I have developed a searing contempt for gushiness and schmaltz. Frank Sinatra is The Singing Marshmallow, Betty Hutton is The Human Grindstone. Anyway, these people are out of date, they are sentimental mushballs. The real truth is to be found in rock and roll: "Hearts Made of Stone" is more like it.

Sometimes Cordelia can think of things to say back, but sometimes she can't. She says, "That's cruel." Or she sticks her tongue in the side of her mouth and changes the subject. Or she lights a cigarette.

I sit in History class, doodling on the side of the page. We are taking the Second World War. The teacher is an enthusiast, he's hopping around at the front of the room, waving his arms and his pointer. He's a short man with an unruly strand of hair and a limp, who may have been in the war himself, or so rumor goes. On the board he's drawn a large map of Europe, in white, with yellow dotted lines for the borders between countries. Hitler's armies invade, by means of pink chalk arrows. Now it's the *Anschluss*, and now Poland falls, and now France. I draw tulips and trees, putting a line for the ground and including the root systems in every case. Submarines appear in the English Channel, in green. I draw the face of the girl sitting across the aisle from me. The Blitz is on, bombs drift down through the air like sinister silver angels, London is disintegrating block by block, house by house, mantelpieces, chimneys, double beds hand-carved and passed down through the generations blasted into burning splinters, history reduced to shards. "It was the end of an era," says the teacher. It's hard for us to understand, he says, but nothing will ever be the same again. He is deeply moved by this, you can tell, it's embarrassing. The same as what? I think.

It's incredible to me that I myself was alive when all those chalk things were going on, all those statistical deaths.

I was alive when women wore those ridiculous clothes with the big shoulder pads and the nipped-in waists, with peplums over their bums like backward aprons. I draw a woman with wide shoulders and a picture hat. I draw my own hand. Hands are the hardest. It's difficult to keep them from looking like clumps of sausages.

I go out with boys. This is not part of a conscious plan, it just happens. My relationships with boys are effortless, which means that I put very little effort into them. It's girls I feel awkward with, it's girls I feel I have to defend myself against; not boys. I sit in my bedroom picking the pilly fuzzballs off my lambswool sweaters and the phone will ring. It will be a boy. I take the sweater into the hall, where the phone is, and sit on the hall chair with the receiver cradled between my ear and shoulder and continue to pick off the fuzzballs, while a long conversation goes on that is mostly silence.

Boys by nature require these silences; they must not be startled by too many words, spoken too quickly. What they actually say is not that important. The important parts exist in the silences between the words. I know what we're both looking for, which is escape. They want to escape from adults and other boys, I want to escape from adults and other girls. We're looking for desert islands, momentary, unreal, but there.

My father paces the living room, jingling his keys and small change in his pockets. He's impatient, he can't help hearing these monosyllables, these murmurs, these silences. He walks into the hall and makes snipping motions with his fingers, meaning I'm to cut it short. "I have to go now," I say. The boy makes a sound like air coming out of an inner tube. I understand it.

I know things about boys. I know what goes on in their heads, about girls and women, things they can't admit to other boys, or to anyone. They're fearful about their own bodies, shy about what they say, afraid of being laughed at. I know what kind of talk goes on among them as they horse around in the locker room, sneak cigarettes behind the field house. *Stunned broad*, *dog*, *bag* and *bitch* are words they apply to girls, as well as worse words. I don't hold these words against them. I know these words are another version of pickled ox eyes and snot eating, they're prove-it words boys need to exchange, to show they are strong and not to be taken in. The words don't necessarily mean they don't like real girls, or one real girl. Sometimes real girls are an alternative to these words and sometimes they're an incarnation of them, and sometimes they're just background noise.

I don't think any of these words apply to me. They apply to other girls, girls who walk along the high school halls in ignorance of them, swinging their hair, swaying their little hips as if they think they're seductive, talking too loudly and carelessly to one another, fooling nobody; or else acting pastel, blank, daisy-fresh. And all the time these clouds of silent words surround them, *stunned broad*, *dog*, *bag* and *bitch*, pointing at them, reducing them, cutting them down to size so they can be handled. The trick with these silent words is to walk in the spaces between them, turn sideways in your head, evade. Like walking through walls.

This is what I know about boys in general. None of it has to do with individual boys by themselves, the boys I go out with. These boys are usually older than I am, although they aren't the kind with greasy ducktails and a lot of leather, they're nicer than that. When I go out with them I'm supposed to be home on time. If I'm not, my father has long conversations with me in which he explains that being home on time is like being on time for a train. If I were to be late for a train, I would miss the train, wouldn't I? "But this house isn't a train," I say. "It's not going anywhere." My father is exasperated; he jingles his keys in his pocket. "That's not the point," he says.

What my mother says is, "We worry." "What about?" I say. There's nothing to worry about, as far as I can see.

My parents are a liability in this as in other matters. They won't buy a television, like everyone else, because my father says it turns you into a cretin and emits harmful radiation and subliminal messages as well. When the boys come to pick me up, my father emerges from the cellar wearing his old gray felt hat and carrying a hammer or a saw, and grips their hands in his bear paw handshake. He assesses them with his shrewd, twinkly, ironic little eyes and calls them "sir," as if they're his graduate students. My mother goes into her nice lady act and says almost nothing. Or else she tells me I look sweet, right in front of the boys.

In the spring they appear around the corner of the house in their baggy gardening pants, smudged with mud, to see me off. They drag the boys out to the backyard, where there is now a large pile of cement blocks accumulated by my father for some future contingency. They want the boys to see their display of irises, as if these boys are old ladies; and the boys have to say something about the irises, although irises are the last thing on their minds. Or else my father attempts to engage them in improving conversation about current topics, or asks them if they've read this book or that one, pulling books from the bookshelves while the boys shift on their feet. "Your father's a card," the boys say uneasily, later.

My parents are like younger, urchinlike brothers and sisters whose faces are dirty and who blurt out humiliating things that can neither be anticipated nor controlled. I sigh and make the best of it. I feel I'm older than they are, much older. I feel ancient.

What I do with the boys is nothing to worry about. It's normal. We go to movies, where we sit in the smoking section and neck, or we go to drive-ins and eat popcorn and neck there as well. There are rules for necking, which we observe: approach, push away, approach, push away. Garter belts are going too far and so are brassieres. No zippers. The boys' mouths taste of cigarettes and salt, their skin smells like Old Spice after-shave. We go to dances and twirl around during the rock numbers, or shuffle in the blue light, surrounded by the shuffling of the other couples. After formal dances we go to someone's house or to the St. Charles Restaurant, and after that we neck, though not for long because the time has usually run out. For formal dances I have dresses which I sew myself because I can't afford to buy them. They have layers of tulle and are propped up underneath with crinolines, and I worry about the hooks coming unfastened. I have shoes in matching satin or silver straps, I have earrings which pinch like hell. For these dances the boys send corsages, which I press afterward and keep in my bureau drawer: squashed carnations and brown-edged rosebuds, wads of dead vegetation, like a collection of floral shrunken heads.

My brother Stephen treats these boys with scorn. As far as he's concerned they are dimwits and unworthy of my serious consideration. He laughs at them behind their backs and makes fun of their names. They are not George but Georgie-Porgie, not Roger but Rover. He makes bets as to how long each one will last. "Three months for him," he'll say, after seeing the boy for the first time; or, "When are you going to throw him over?"

I don't dislike my brother for this. I expect it of him, because he's partly right. I don't feel about these boys the way girls do in true romance comic books. I don't sit around wondering when they'll call. I like them but I don't fall in love with them. None of the teenage magazine descriptions of girls moping, one tear on each cheek like pearl earrings, applies to me. So partly the boys are not a serious matter. But at the same time they are.

The serious part is their bodies. I sit in the hall with the cradled telephone, and what I hear is their bodies. I don't listen much to the words but to the silences, and in the silences these bodies re-create themselves, are created by me, take form. When I am lonely for boys it's their bodies I miss. I study their hands lifting the cigarettes in the darkness of the movie theaters, the slope of a shoulder, the angle of a hip. Looking at them sideways, I examine them in different lights. My love for them is visual: that is the part of them I would like to possess. *Don't move*, I think. *Stay like that. Let me have that.* What power they have over me is held through the eyes, and when I'm tired of them it's an exhaustion partly physical, but also partly visual.

Only some of this has to do with sex; although some of it does. Some of the boys have cars, but others do not, and with them I go on buses, on streetcars, on the newly opened

Toronto subway that is clean and uneventful and looks like a long pastel-tiled bathroom. These boys walk me home, we walk the long way around. The air smells of lilac or mown grass or burning leaves, depending on the season. We walk over the new cement footbridge, with the willow trees arching overhead, the sound of running water from the creek beneath. We stand in the dim light coming from the lampposts on the bridge and lean back against the railing, their arms around me and mine around them. We lift each other's clothing, run our hands over each other's backbones, and I feel the backbone tensed and strung to breaking. I feel the length of the whole body, I touch the face, amazed. The faces of the boys change so much, they soften, open up, they ache. The body is pure energy, solidified light.

44

A girl is found murdered, down in the ravine. Not the ravine near our house, but a larger branch of it, farther south, past the brickworks, where the Don River, willow-bordered, junk-strewn and dingy, winds sluggishly toward the lake. Such things are not supposed to happen in Toronto, where people leave their back doors unlocked, their windows unlatched at night; but they do happen, it seems. It's on the front pages of all the papers.

This girl is our age. Her bicycle has been found near her. She has been strangled, and also molested. We know what *molested* means. There are photos of her when alive, which already have that haunted look such photos usually take years to acquire, the look of vanished time, unrecoverable, unredeemed. There are extensive descriptions of her clothing. She was wearing an angora sweater, and a little fur collar with pom-poms, of the sort that is currently fashionable. I don't have a collar like this, but would like one. Hers was white but you can get them in mink. She was wearing a pin on the sweater, in the shape of two birds with red glass jewels for eyes. It's what anyone would wear to school. All these details about her clothing strike me as unfair, although I devour them. It doesn't seem right that you can just walk out one day, wearing ordinary clothes, and be murdered without warning, and then have all those people looking at you, examining you. Murder ought to be a more ceremonial occasion.

I have long since dismissed the idea of bad men in the ravine. I've considered them a scarecrow story, put up by mothers. But it appears they exist, despite me.

This murdered girl troubles me. After the first shock, nobody at school says much about her. Even Cordelia does not want to talk about her. It's as if this girl has done something shameful, herself, by being murdered. So she goes to that place where all things go that are not mentionable, taking her blond hair, her angora sweater, her ordinariness with her. She stirs up something, like dead leaves. I think of a doll I had once, with white fur on the border of her skirt. I remember being afraid of this doll. I haven't thought about that in years.

Cordelia and I sit at the dining table doing our homework. I am helping Cordelia, I'm trying to explain the atom to her, but she's refusing to take it seriously. The diagram of the atom has a nucleus, with electrons circling it. The nucleus looks like a raspberry, the electrons and their rings look like the planet Saturn. Cordelia sticks her tongue in the side of her mouth and frowns at the nucleus. "This looks like a raspberry," she says.

"Cordelia," I say. "The exam is tomorrow." Molecules do not interest her, she doesn't seem able to grasp the Periodic Table. She refuses to understand mass, she refuses to understand why atom bombs blow up. There's a picture of one blowing up in the Physics book, mushroom cloud and all. To her it's just another bomb. "Mass and energy are different aspects," I tell her. "That's *why* $E = mc^2$."

"It would be easier if Percy the Prude weren't such a creep," she says. Percy the Prude is the Physics teacher. He has red hair that stands up at the top like Woody Woodpecker's, and he lisps.

Stephen walks through the room, looks over our shoulders. "So they're still teaching you kiddie Physics," he says indulgently. "They've still got the atom looking like a raspberry."

"See?" says Cordelia.

I feel subverted. "This is the atom that's going to be on the exam, so you'd better learn it," I say to Cordelia. To Stephen I say, "So what does it really look like?"

"A lot of empty space," Stephen says. "It's hardly there at all. It's just a few specks held in place by forces. At the subatomic level, you can't even say that matter exists. You can only say that it has a tendency to exist."

"You're confusing Cordelia," I say. Cordelia has lit a cigarette and is looking out the window, where several squirrels are chasing one another around the lawn. She is paying no attention to any of this.

Stephen considers Cordelia. "Cordelia has a tendency to exist," is what he says.

Cordelia doesn't go out with boys the way I do, although she does go out with them. Once in a while I arrange double dates, through whatever boy I'm going out with. Cordelia's date is always a boy of lesser value, and she knows this and refuses to approve of him.

Cordelia can't seem to decide what kind of boy she really does approve of. The ones with haircuts like my brother's are drips and pills, but the ones with ducktails are sleazy greaseballs, although sexy. She thinks the boys I go out with, who go no further than crewcuts, are too juvenile for her. She's abandoned her ultrared lipstick and nail polish and her turned-up collars and has taken up moderate pinks and going on diets, and grooming. This is what magazines call it: Good Grooming, as in horses. Her hair is shorter, her wardrobe more subdued.

But something about her makes boys uneasy. It's as if she's too attentive to them, too polite, studied and overdone. She laughs when she thinks they've made a joke and says, "That's very witty, Stan." She will say this even when they haven't intended to be funny, and then they aren't sure whether or not she's making fun of them. Sometimes she is, sometimes she isn't. Inappropriate words slip out of her. After we've finished our hamburgers and fries she turns to the boys and says brightly, "Are you sufficiently sophonsified?" and they gape at her. They are not the kind of boys who would have napkin rings.

She asks them leading questions, tries to draw them into conversations, as a grown-up would do, not appearing to know that the best thing, with them, is to let them exist in their own silences, to look at them only out of the corners of the eyes. Cordelia tries to look at them sincerely, head-on; they are blinded by the glare, and freeze like rabbits in a headlight. When she's in the back seat with them I can tell, from the breathing and gasps, that she's going too far in that direction as well. "She's kind of strange, your friend," the boys say to me, but they can't say why. I decide it's because she has no brother, only sisters. She thinks that what matters with boys is what you say; she's never learned the intricacies, the nuances of male silence.

But I know Cordelia isn't really interested in anything the boys themselves have to say, because she tells me so. Mostly she thinks they're dim. Her attempts at conversation with them are a performance, an imitation. Her laugh, when she's with them, is refined and low, like a woman's laugh on the radio, except when she forgets herself. Then it's too loud. She's mimicking something, something in her head, some role or image that only she can see.

The Earle Grey Players come to our high school, as they do every year. They go from high school to high school, they are well known for this. Every year they do one play by Shakespeare; it's always the play that's on the province-wide Grade Thirteen Examinations, the ones you have to pass in order to get into university. There aren't many theaters in Toronto, in fact there are only two, so many people go to these plays. The kids go to them because it's on the exam and the parents go because they don't often get a chance to see plays.

The Earle Grey Players are Mr. Earle Grey, who always plays the leads, Mrs. Earle Grey who plays the lead woman, and two or three other actors who are thought to be Earle Grey cousins and who are likely to double up and do two or more parts. The rest of the parts are played by students in whatever high school they're performing at that week. Last year the play was *Julius Caesar*, and Cordelia got to be part of the crowd. She had to smear burnt cork on her face for dirt and wrap herself up in a bedsheet from home, and say *rabble rabble* during the crowd scene when Mark Antony was making his Ears speech.

This year the play is *Macbeth*. Cordelia is a serving woman, and also a soldier in the final battle scene. This time she has to bring a plaid car-rug from home. She's lucky because she also has a kilt, an old one of Perdie's from when she went to her girls' private school. In addition to her parts, Cordelia is the props assistant. She's in charge of tidying up the props after each performance, setting them in order, always the same order, so that the actors can grab them backstage and run on without a moment's thought.

During the three days of rehearsal Cordelia is very excited. I can tell by the way she chain-smokes on the way home and acts bored and nonchalant, referring, every once in a while, to the real, professional actors by their first names. The younger ones make such an effort to be funny, she says. They call the Witches The Three Wired Sisters; they call Cordelia a cream-faced loon, and they threaten to put eye of newt and toe of frog into her coffee. They say that when Lady Macbeth says, "Out, damned spot," during the mad scene, she's referring to her dog Spot, who has poo'ed on the carpet. She says real actors will never say the name *Macbeth* out loud, because it's bad luck. They call it "The Tartans" instead.

"You just said it," I say.

"What?"

"Macbeth," I say.

Cordelia stops short in the middle of the sidewalk. "Oh God," she says. "I did, didn't I?" She pretends to laugh it off, but it bothers her.

At the end of the play Macbeth's head gets cut off and Macduff has to bring it onto the stage. The head is a cabbage wrapped up in a white tea towel; Macduff throws it onto the stage, where it hits with an impressive, flesh-and-bone thud. Or this is what has happened in rehearsal. But the night before the first performance—there are to be three—Cordelia notices that the cabbage is going bad, it's getting soft and squooshy and smells like sauerkraut. She replaces it with a brand-new cabbage.

The play is put on in the school auditorium, where the school assemblies are, and the choir practice. Opening night is packed. Things go without much mishap, apart from the sniggers in the wrong places and the anonymous voice that says, "Go on, do it!" when Macbeth is hesitating outside Duncan's chamber, and the catcalls and whistles from the back of the auditorium when Lady Macbeth appears in her nightgown. I watch for Cordelia in the battle scene, and there she is, running across backstage in her kilt with a wooden sword, her car rug thrown over her shoulder. But when Macduff comes in at the end and tosses down the cabbage in the tea towel, it doesn't hit once and lie still. It bounces, bumpity-bump, right across the stage like a rubber ball, and falls off the edge. This dampens the tragic effect, and the curtain comes down on laughter.

It's Cordelia's fault, for replacing the cabbage. She is mortified. "It was *supposed* to be rotten," she wails backstage, where I have gone to congratulate her. "So now they tell me!" The actors have made light of it; they tell her it's a novel effect. But although Cordelia laughs and blushes and tries to pass it off lightly, I can see she is almost in tears.

I ought to feel pity, but I do not. Instead, on the way home from school the next day, I say "Bumpity bumpity bump, plop," and Cordelia says, "Oh, don't." Her voice is toneless, leaden. This is not a joke. I wonder, for an instant, how I can be so mean to my best friend. For this is what she is.

Time passes and we are older, we are the oldest, we are in Grade Thirteen. We can look down on the incoming students, those who are still mere children as we were once. We can smile at them. We're old enough to take Biology, which is taught in the Chemistry lab. For this we leave our homeroom group and meet with students from other homerooms. This is why Cordelia is my Biology lab partner, at the Chemistry lab table, which is black and has a sink. Cordelia doesn't like Biology any more than she liked Physics, which she barely squeaked through, but she has to take something in the sciences and it's easier, to her mind, than a number of the things she might have to take otherwise.

We are given dissecting kits with scalpel-like knives that could be sharper, and trays with a coating of wax at the bottom, and a package of pins, as in sewing classes. First we have to dissect a worm. Each of us is given one of these. We look at the diagram of the inside of the worm, in the Zoology textbook: this is what we're supposed to see once we get the worm open. The worms wriggle and twine in the wax-bottomed trays, and snout their way along the sides, trying to get out. They smell like holes in the ground.

I pin my worm at either end and make a slick vertical cut; the worm twists as they do on fishhooks. I pin the worm's skin out to the sides. I can see its worm heart, which isn't the shape of a heart, its central artery pumping worm blood, its digestive system, which is full of mud. "Oh," says Cordelia. "How can you." Cordelia is becoming mushier and mushier, I think. She is becoming a drip. I do her worm for her, when the teacher isn't looking. Then I draw a diagram of the worm, cut open, beautifully labeled.

After that comes the frog. The frog kicks and is more difficult than the worm, it looks a little too much like a person swimming. I conk the frog out with chloroform as directed and dissect it with flair, sticking in the pins. I make a drawing of the inside of the frog, with all its curlicues and bulbs, its tiny lungs, its cold-blooded amphibian heart.

Cordelia can't do the frog either. She says she feels sick to her stomach just thinking about putting her dissecting knife through its skin. She looks at me, pale, her eyes big. The frog smell is getting to her. I do her frog for her. I'm good at this.

I memorize the statocysts of the crayfish, its gills and mouth parts. I memorize the circulatory system of the cat. The teacher, who is usually the boys' football coach but who has recently taken a summer course in Zoology so he can teach us this, orders a dead cat for us, with its veins and arteries pumped full of blue and pink latex. He's disappointed when it arrives, because the cat is definitely rancid, you can smell it even through the formaldehyde. So we don't have to dissect it, we can just use the diagram in the book.

But worms, frogs, and cats aren't enough for me. I want more. I go down to the Zoology Building on Saturday afternoons to use the microscopes in the empty labs. I look at slides, planaria worms in section with their triangular heads and cross-eyes, bacteria colored with vivid dyes, hot pinks, violent purples, radiant blues. These are lit up from beneath, they're breathtaking, like stained-glass windows.

I draw them, delineating the structures with different colored pencils; though I can never get the same luminous brilliance.

Mr. Banerji, who is now Dr. Banerji, discovers what I'm doing. He brings me slides he thinks I would like to see and offers them to me shyly and eagerly, with a conspiratorial giggle, as if we are sharing a delicious, esoteric secret, or something religious. "Parasite of the tent caterpillar," he says, depositing the slide with reverence on a clean piece of paper at my table. "Egg of the budworm."

"Thank you," I say, and he looks at my drawings, picking them up by the corners with his deft, bitten fingers. "Very good, very good, miss," he says. "Soon you will take over my job."

He has a wife now, who has come from India, and a little boy. I see them sometimes, looking in through the doorway of the lab, the child gentle and dubious, the wife anxious. She wears gold earrings and a scarf with spangles on it. Her red sari shows beneath her brown Canadian winter coat, her overshoes poking out beneath it.

Cordelia comes to my house and I help her with her Zoology homework, and she stays to dinner. My father, dishing out the beef stew, says that a species a day is becoming extinct. He says we are poisoning the rivers and ruining the gene pools of the planet. He says that when a species becomes extinct, some other species moves in to fill up the ecological niche, because Nature abhors a vacuum. He says that the things that move in are common weeds, and cockroaches and rats: soon all flowers will be dandelions. He says, waving his fork, that if we continue to overbreed as a species, a new epidemic will arise to redress the balance. All this will happen because people have neglected the basic lessons of Science, they have gone in for politics and religion and wars instead, and sought out passionate excuses for killing one another. Science on the other hand is dispassionate and without bias, it is the only universal language. The language is numbers. When at last we are up to our ears in death and garbage, we will look to Science to clean up our mess.

Cordelia listens to all of this, smirking a little. She thinks my father is quaint. I hear him the way she must: this is not what people are supposed to talk about at the dinner table.

I go to dinner at Cordelia's house. Dinners at Cordelia's house are of two kinds: those when her father is there and those when he isn't. When he isn't there, things are slapdash. Mummie comes to the table absent-mindedly still in her painting smock, Perdie and Mirrie and also Cordelia appear in blue jeans with a man's shirt over top and their hair in pin curls. They jump up from the table, saunter into the kitchen for more butter, or the salt, which has been forgotten. They talk all at once, in a languid, amused way, and groan when it's their turn to clear the table, while Mummie says "Now girls," but without conviction. She is losing the energy for disappointment.

But when Cordelia's father is there, everything is different. There are flowers on the table, and candles. Mummie has on her pearls, the napkins are neatly rolled in the napkin rings instead of crumpled in under the edges of the plates. Nothing is forgotten. There are no pin curls, no elbows on the table, even the spines are straighter.

Today is one of the candle days. Cordelia's father sits at the head of the table, with his craggy eyebrows, his wolvish look, and bends upon me the full force of his ponderous, ironic, terrifying charm. He can make you feel that what he thinks of you matters, because it will be accurate, but that what you think of him is of no importance.

"I'm hag-ridden," he says, pretending to be mournful. "The only man in a houseful of women. They won't let me into the bathroom in the morning to shave." Mockingly, he invites my sympathy and collusion. But I can think of nothing to say.

Perdie says, "He should consider himself lucky that we put up with him." She can get away with a little impertinence, with coltish liberties. She has the haircut for it. Mirrie, when hard-pressed, looks reproachful. Cordelia is not good at either of these things. But they all play up to him.

"What are you studying these days?" he says to me. It's a usual question of his. Whatever I say amuses him.

"The atom," I say.

"Ah, the atom," he says. "I remember the atom. And what does the atom have to say for itself these days?"

"Which one?" I say, and he laughs.

"Which one, indeed," he says. "That's very good." This may be what he wants: a give and take, of sorts. But Cordelia can never come up with it, because she's too frightened of him. She's frightened of not pleasing him. And yet he is not pleased. I've seen it many times, her dithering, fumble-footed efforts to appease him. But nothing she can do or say will ever be enough, because she is somehow the wrong person.

I watch this, and it makes me angry. It makes me want to kick her. How can she be so abject? When will she learn?

Cordelia fails the mid-year Zoology test. She doesn't seem to care. She has spent half the exam time drawing surreptitious cartoons of various teachers in the school, which she shows to me on the way home, laughing her exaggerated laugh.

Sometimes I dream about boys. These are wordless dreams, dreams of the body. They stay with me for minutes after I wake up and I luxuriate in them, but I forget them soon.

I have other dreams as well.

I dream that I can't move. I can't talk, I can't even breathe. I'm in an iron lung. The iron is clenched around my body like a hard cylindrical skin. It's this iron skin that is doing my breathing for me, in and out. I'm dense and heavy, I feel nothing other than this heaviness. My head sticks out the end of the iron lung. I'm looking up at the ceiling, on which there is a light fixture like yellowish cloudy ice.

I dream that I'm trying on a fur collar, in front of the mirror on my bureau. There's someone standing behind me. If I move so that I can see into the mirror, I'll be able to look over my own shoulder without turning around. I'll be able to see who it is.

I dream that I've found a red plastic purse, hidden in a drawer or trunk. I know there is treasure inside it, but I can't get it open. I try and try and finally it bursts, like a balloon. It's full of dead frogs.

I dream that I've been given a head wrapped up in a white tea towel. I can see the outlines of the nose, the chin, the lips through the white cloth. I could unwrap the cloth to see whose head it is, but I don't want to, because I know that if I do the head will come alive.

45

Cordelia tells me that when she was younger she broke
a thermometer and ate some of the mercury in it to make
herself sick so she wouldn't have to go to school. Or she'd
stick her finger down her throat and throw up, or she'd hold
the thermometer near a light bulb to make it look as if she
had a temperature. Her mother caught her doing that because she left it near the light bulb too long and the mercury
shot up to a hundred and ten. After that her other deceptions
were harder to pull off.

"How old were you then?" I ask her.

"Oh, I don't know. Before high school," she tells me.
"You know, the age when you do those things."

It's Tuesday, in the middle of May. We're sitting in a
booth at Sunnysides. Sunnysides has a soda fountain
counter, which is speckled bloodstone red with chrome trim
and has a row of round swivel-seat stools screwed to the
floor along beside it. The black tops of the seats, which may
not be leather, make a gentle farting sound when you sit
down on them, so Cordelia and I and all girls prefer the
booths. They're dark wood, and the tabletop between the
two facing benches is red like the soda fountain counter.
This is where the Burnham students go after school to smoke
and to drink glasses of Coca-Cola with maraschino cherries
in them. If you drink a Coke and mix two aspirins in with
it, it's supposed to make you drunk. Cordelia says she has
tried this; she says it's nothing like being really drunk.

Instead of Cokes, we're drinking vanilla milkshakes, with
two straws each. We ease the paper covers off the straws
so that they pleat up into short caterpillars of paper. Then
we drop water onto them out of our water glasses, and the
paper caterpillars expand and look as if they're crawling.

The tables at Sunnysides are littered with strips of soggy paper.

"What did the chickens say when the hen laid an orange?" Cordelia says, because there is a wave of corny chicken jokes sweeping the school. Chicken jokes, and moron jokes. *Why did the moron throw the clock out the window? To see time fly.*

"Look at the orange marmalade," I say in a bored voice. "What did the moron say when he saw the three holes in the ground?"

"What?" says Cordelia, who has trouble remembering jokes even when she's heard them.

"Well, well, well," I say.

"Ha ha," says Cordelia. Part of this ritual is mild derision, of other people's jokes.

Cordelia doodles on the table, using our spilled water. "Remember those holes I used to dig?" she says.

"What holes?" I say. I don't remember any holes.

"Those holes in my backyard. Boy, did I want a hole out there. I started one, but the ground was too hard, it was full of rocks. So I dug another one. I used to work away at it after school, day after day. I got blisters on my hands from the shovel." She smiles a pensive, reminiscent smile.

"What did you want it for?" I ask.

"I wanted to put a chair in it and sit down there. By myself."

I laugh. "What for?"

"I don't know. I guess I wanted someplace that was all mine, where nobody could bug me. When I was little, I used to sit on a chair in the front hall. I used to think that if I kept very still and out of the way and didn't say anything, I would be safe."

"Safe from what?" I say.

"Just safe," she says. "When I was really little, I guess I used to get into trouble a lot, with Daddy. When he would lose his temper. You never knew when he was going to do it. 'Wipe that smirk off your face,' he would say. I used to stand up to him." She squashes out her cigarette, which has been smoldering in the ashtray. "You know, I hated moving to that house. I hated the kids at Queen Mary's, and those boring things like skipping. I didn't really have any good friends there, except for you."

Cordelia's face dissolves, re-forms: I can see her nine-year-old face taking shape beneath it. This happens in an eyeblink. It's as if I've been standing outside in the dark and a shade has snapped up, over a lighted window, revealing the life that's been going on inside in all its clarity and detail. There is that glimpse, during which I can see. And then not.

A wave of blood goes up to my head, my stomach shrinks together, as if something dangerous has just missed hitting me. It's as if I've been caught stealing, or telling a lie; or as if I've heard other people talking about me, saying bad things about me, behind my back. There's the same flush of shame, of guilt and terror, and of cold disgust with myself. But I don't know where these feelings have come from, what I've done.

I don't want to know. Whatever it is, it's nothing I need or want. I want to be here, in Tuesday, in May, sitting in the red-topped booth at Sunnysides, watching Cordelia as she delicately slurps the last of her milkshake up through the straws. She's noticed nothing.

"I've got one," I say. "Why did the unwashed chicken cross the road twice?"

"Why?" says Cordelia.

"Because it was a dirty double-crosser," I say.

Cordelia rolls her eyes, like Perdie. "Very funny," she says.

I close my eyes. In my head there's a square of darkness, and of purple flowers.

46

I begin to avoid Cordelia. I don't know why.

I no longer arrange double dates with her. I tell her that the boy I'm going out with doesn't have any suitable friends. I say I have to stay after school, which is true: I'm painting the decorations for the next dance, palm trees and girls in hula skirts.

Some days Cordelia waits for me, so I have to walk home with her anyway. She talks and talks as if there's nothing wrong, and I say little; but then I've never said a lot anyway. After a while she'll say, overly brightly, "But here I've been going on and on about me. What's doing with you?" and I smile and say "Nothing much." Sometimes she makes a joke of it and says, "But that's enough about me. What do *you* think of me?" and I add to the joke by saying, "Nothing much."

Cordelia is failing more and more tests. It doesn't seem to bother her, or at any rate she doesn't want to talk about it. I no longer help her with her homework, because I know she won't pay attention even if I do. She has trouble concentrating on anything. Even when she's just talking, on the way home, she changes the subject in the middle of a sentence so it's hard to follow what she's saying. She's slipping up on the grooming, too, reverting to her old sloppy ways of years ago. She's let her bleached strip grow out, so it's disconcertingly two-toned. There are runs in her nylons, buttons popped off her blouses. Her lipstick doesn't seem to fit her mouth.

It is decided that it would be best for Cordelia to change schools again, so she does. After this she phones me frequently, but then less frequently. She says we should get together soon. I never deny this, but I never set a time either. After a while I say, "I have to go now."

Cordelia's family moves to a different, larger house, in a ritzier neighborhood farther north. Some Dutch people move into her old house. They plant a lot of tulips. That seems to be the end of her.

I write the final Grade Thirteen exams, subject after subject, day after day, sitting at a desk in the gymnasium. The leaves are fully out, the irises are in bloom, there's a heat wave; the gymnasium heats up like an oven and we all sit in there, superheated, writing away, while the gymnasium exudes its smell of bygone athletes. The teachers police the aisles. Several girls faint. One boy keels over and is found afterward to have drunk a pitcher of tomato juice out of the refrigerator which was really Bloody Marys for his mother's bridge club. As the bodies are carried out I scarcely look up from the page.

I know I'll do well in the two Biology exams. I can draw anything: the insides of crayfish ears, the human eye, frogs' genitalia, the blossom of the snapdragon (*Antirrhinum majus*) in cross section. I know the difference between a raceme and a rhizome, I explicate photosynthesis, I can spell *Scrofulariaciae*. But in the middle of the Botany examination it comes to me, like a sudden epileptic fit, that I'm not going to be a biologist, as I have thought. I am going to be a painter. I look at the page, where the life cycle of the mushroom from spore to fruiting body is taking shape, and I know this with absolute certainty. My life has been changed, soundlessly, instantaneously. I continue my explication of tubers, bulbs, and legumes, as if nothing has happened.

One night, just after the exams have finished, the phone rings. It's Cordelia. I realize I've been expecting this.

"I'd like to see you," she says. I don't want to see her, but I know I will. What I hear is not *like* but *need*.

The next afternoon I take the subway and then the bus, northward through the heated city, to where Cordelia now lives. I've never been up here before. The streets wind in and around, the houses are large, ponderous, Georgian, set off with weighty shrubbery. I see or think I see Cordelia's face, pale and indistinct, behind the front window as I come up the walk. She opens the door before I have time to ring.

"Well, hi there," she says. "Long time no see." This is false heartiness and we both know it, because Cordelia is a wreck. Her hair is lusterless, the flesh of her face pasty. She's gained a lot of weight, not solid-muscled weight, but limp weight, bloated and watery. She's gone back to the too-vivid orange-red lipstick, which turns her yellowish. "I know," she says. "I look like Haggis McBaggis."

The house is cool inside. The front hall floor is white and black squares; there's a graceful central staircase. A flower arrangement with gladioli sits on a polished table beside it. The house is silent, except for a clock chiming in the living room. Nobody else seems to be home.

We don't go into the living room but back past the stairs and through a door into the kitchen, where Cordelia makes me a cup of instant coffee. The kitchen is beautiful, perfectly arranged, pale-colored and peaceful. The refrigerator and stove are white. Some people now have colored refrigerators, pale-green or pink, but I don't like these colors and I'm pleased that Cordelia's mother doesn't either. There's a lined school notebook open on the kitchen table, which I recognize as the dining table from their other house with the two middle leaves taken out. That means they must have a new dining table. It appals me to discover that I want to see this new dining table more than I want to see Cordelia.

Cordelia rummages in the fridge and brings out an opened package of store doughnuts. "I've been waiting for an excuse to eat the rest of these," she says. But as soon as she's taken her first bite she lights a cigarette.

"So," she says. "What are you up to these days?" It's her too-bright voice, the one she used to use on boys. Right now it frightens me.

"Oh, just the usual," I say. "You know. Finishing exams." We look at each other. Things are bad for her, that much is clear. I don't know whether she wants me to ignore this or not. "What about you?" I say.

"I have a tutor," she says. "I'm supposed to be studying.

For summer courses." We both know without mentioning it that she must have failed her year, despite the new school. She must have failed badly. Unless she passes whatever subjects she failed, at the next set of exams or sometime or other, she'll be locked out of university forever.

"Is the tutor nice?" I say, as if I'm asking about a new dress.

"I guess so," says Cordelia. "Her name is Miss Dingle. It really is. She blinks all the time, she has watery eyes. She lives in this squalid apartment. She has salmon-colored lingerie, I see it hanging over the shower curtain rod in her squalid bathroom. I can always get her off the subject by asking about her health."

"Off what subject?" I ask.

"Oh, any subject," says Cordelia. "Physics, Latin. Any of it." She sounds a little ashamed of herself, but proud and excited too. It's like the time when she used to pinch things. This is her accomplishment these days: deluding the tutor. "I don't know why they all think I spend the days studying," she says. "I sleep a lot. Or else I drink coffee and smoke and listen to records. Sometimes I have a little nip out of Daddy's whisky decanter. I fill it up with water. He hasn't found out!"

"But, Cordelia," I say. "You have to do *something!*"

"Why?" she says, with a little of her old belligerence. She isn't only joking.

And I have no reason to give her. I can't say, "Because everyone does." I can't even say, "You have to earn a living," because she obviously doesn't, she's here in this large house and she isn't earning a living at all. She could just go on like this, like a woman from old-fashioned times, a maiden aunt, some aging perennial girl who never leaves home. It isn't likely that her parents would kick her out.

So I say, "You'll get bored."

Cordelia laughs, too loudly. "So what if I study?" she says. "I pass my exams. I go to university. I learn it all. I turn into Miss Dingle. No thanks."

"Don't be a cretin," I say. "Who says you have to be Miss Dingle?"

"Maybe I am a cretin," she says. "I can't concentrate on

that stuff, I can hardly look at the page, it all turns into little black dots."

"Maybe you could go to secretarial school," I say. I feel like a traitor as soon as I've said it. She knows what we both think of girls who would go to secretarial school, with their spidery plucked eyebrows and pink nylon blouses.

"Thanks a bundle." There's a pause. "But let's not talk about all that," she says, returning to her ultrabright voice. "Let's talk about fun things. Remember that cabbage? The bouncy one?"

"Yes," I say. It occurs to me that she could be pregnant, or that she might have been. It's natural to wonder that about girls who drop out of school. But I decide this is unlikely.

"I was so mortified," she says. "Remember when we used to go downtown and take our pictures at Union Station? We thought we were so sharp!"

"Right before the subway was built," I say.

"We used to throw snowballs at old ladies. We used to sing those silly songs."

"Leprosy," I say.

"Part of your heart," she says. "We thought we were the cat's ass. I see kids that age now and I think: *Brats!*"

She's looking back on that time as if it was her golden age; or maybe it seems that way to her because it's better than now. But I don't want her to remember any more. I want to protect myself from any further, darker memories of hers, get myself out of here gracefully before something embarrassing happens. She's balanced on the edge of an artificial hilarity that could topple over at any moment into its opposite, into tears and desperation. I don't want to see her crumple up like that, because I have nothing to offer her in the way of solace.

I harden toward her. She's acting like a jerk. She doesn't have to stay locked into place, into this mournful, drawn-out, low-grade misery. She has all kinds of choices and possibilities, and the only thing that's keeping her away from them is lack of willpower. *Smarten up*, I want to tell her. *Pull up your socks*.

I say I have to get back, that I'm going out later. This

isn't true and she suspects it. Although she's a mess, her instinct for social fraud has sharpened. "Of course," she says. "That's entirely understandable." It's her distant, grown-up voice.

Now that I'm hurrying, making a show of bustle, it strikes me that one of my reasons for escape is that I don't want to meet her mother coming back, from wherever she's been. Her mother would look at me with reproach, as if I am responsible for Cordelia in her present shape, as if she's disappointed, not in Cordelia, but in me. Why should I have to undergo such a look, for something that is not my fault?

"Goodbye, Cordelia," I say in the front hall. I squeeze her arm briefly, move back before she can kiss me on the cheek. Kissing on the cheek is what they do in her family. I know she has expected something from me, some connection to her old life, or to herself. I know I have failed to provide it. I am dismayed by myself, by my cruelty and indifference, my lack of kindness. But also I feel relief.

"Call you soon," I say. I'm lying, but she chooses not to acknowledge this.

"That would be nice," she says, shielding us both with politeness.

I go down the walk toward the street, turn to look back. There's her face again, a blurred reflection of a moon, behind the front window.

LIFE DRAWING

47

*T*here are several diseases of the memory. Forgetfulness of nouns, for instance, or of numbers. Or there are more complex amnesias. With one, you can lose your entire past; you start afresh, learning how to tie your shoelaces, how to eat with a fork, how to read and sing. You are introduced to your relatives, your oldest friends, as if you've never met them before; you get a second chance with them, better than forgiveness because you can begin innocent. With another form, you keep the distant past but lose the present. You can't remember what happened five minutes ago. When someone you've known all your life goes out of the room and then comes back in, you greet them as if they've been gone for twenty years; you weep and weep, with joy and relief, as if at a reunion with the dead.

I sometimes wonder which of these will afflict me, later; because I know one of them will.

For years I wanted to be older, and now I am.

I sit in the harsh ultrablack of the Quasi, drinking red wine, staring out the window. On the other side of the glass, Cordelia drifts past; then melts and reassembles, changing into someone else. Another mistaken identity.

Why did they name her that? Hang that weight around her neck. Heart of the moon, jewel of the sea, depending on which foreign language you're using. The third sister, the only honest one. The stubborn one, the rejected one, the one who was not heard. If she'd been called Jane, would things have been different?

My own mother named me after her best friend, as women did in those days. Elaine, which I once found too plaintive. I wanted something more definite, a monosyllable:

Dot or Pat, like a foot set down. Nothing you could make a mistake about; nothing watery. But my name has solidified around me, with time. I think of it as tough but pliable now, like a well-worn glove.

There's a lot of neo-black in here, some of it leather, some shiny vinyl. I've come prepared this time, I have my black cotton turtleneck and my black trenchcoat with the button-on hood, but I'm not the right texture. Also not the right age: everyone in here is twelve. This place was Jon's suggestion. Trust him to cling to the surfboard as it upends in the froth of the latest wave.

He always made a fetish of lateness, to indicate that his life was crammed with many things, all of them more important than I was, and today is no exception. Thirty minutes later than agreed he breezes in. This time however he apologizes. Has he learned something, or does his new wife run a tighter ship? Funny I still think of her as new.

"That's all right, I programmed for it," I say. "I'm glad you could come out to play." A small preliminary kick at the wife.

"Having lunch with you hardly qualifies as play," he says, grinning.

He's still up to it. We look each other over. In four years he's achieved more wrinkles, and the sideburns and mustache are graying further. "Don't mention the bald spot," he says.

"What bald spot?" I say, meaning I'll overlook his physical degeneration if he'll overlook mine. He's up to that one, too.

"You're looking better than ever," he says. "Selling out must agree with you."

"Oh, it does," I say. "It's so much better than licking bums and hacking up women's bodies in screw-and-spew movies." Once this would have drawn blood, but he must have accepted his lot in life by now. He shrugs, making the best of it; but he looks tired.

"Live long enough and the licker becomes the lickee," he says. "Ever since the exploding eyeball I can do no wrong. Right now I'm head-to-toe saliva."

The possibility for crude sexual innuendo is there, but

I duck it. Instead I think, he's right: we are the establishment now, such as it is. Or that's what we must look like. Once the people I knew died of suicide and motorcycle crashes and other forms of violence. Now it's diseases: heart attacks, cancer, the betrayals of the body. The world is being run by people my age, men my age, with falling-out hair and health worries, and it frightens me. When the leaders were older than me I could believe in their wisdom, I could believe they had transcended rage and malice and the need to be loved. Now I know better. I look at the faces in newspapers, in magazines, and wonder: what greeds, what furies drive them on?

"How's your real work going?" I say, relenting, letting him know I still take him seriously.

This bothers him. "All right," he says. "I haven't been able to get to it much lately."

We are silent, considering shortfalls. There's not much time left, for us to become what we once intended. Jon had potential, but it's not a word that can be used comfortably any more. Potential has a shelf life.

We talk about Sarah, easily and without competing, as if we are her aunt and uncle. We talk about my show.

"I guess you saw that hatchet job in the paper," I say.

"Was that a hatchet job?" he says.

"It's my fault. I was rude to the interviewer," I say, with what I try to pass off as penitence. "I'm well on my way to becoming a cantankerous old witch."

"I'd be disappointed in you if you weren't," he says. "Make 'em sweat, it's what they're paid for." We both laugh. He knows me. He knows what a shit I can be.

I look at him with the nostalgic affection men are said to feel for their wars, their fellow veterans. I think, I once threw things at this man. I threw a glass ashtray, a fairly cheap one which didn't break. I threw a shoe (his) and a handbag (mine), not even snapping the handbag shut first, so that he was showered with a metal rain of keys and small change. The worst thing I threw was a small portable television set, standing on the bed and heaving it at him with the aid of the bouncy springs, although the instant I let fly

I thought, *Oh God, let him duck!* I once thought I was capable of murdering him. Today I feel only a mild regret that we were not more civilized with each other at the time. Still, it was amazing, all those explosions, that recklessness, that Technicolor wreckage. Amazing and agonizing and almost lethal.

Now that I'm more or less safe from him, and him from me, I can recall him with fondness and even in some detail, which is more than I can say for several others. Old lovers go the way of old photographs, bleaching out gradually as in a slow bath of acid: first the moles and pimples, then the shadings, then the faces themselves, until nothing remains but the general outlines. What will be left of them when I'm seventy? None of the baroque ecstasy, none of the grotesque compulsion. A word or two, hovering in the inner emptiness. Maybe a toe here, a nostril there, or a mustache, floating like a little curl of seaweed among the other flotsam.

Across from me at the night-black table, Jon, though diminishing, still moves and breathes. There's a sliver of pain, of longing in me: *Don't go yet! It's not time! Don't go!* It would be stupid, as always, to reveal my own sentimentality, my weakness to him.

What we eat is vaguely Thai: chicken, spicy and succulent, a salad of exotic foliage, red leaves, tiny splinters of purple. Gaudy food. This is the kind of thing people eat now, people who eat in places like this: Toronto is no longer the land of chicken pot pie, beef stew, overboiled vegetables. I recall my first avocado, when I was twenty-two. It was like my father's first symphony orchestra. Perversely I long for the desserts of my childhood, the desserts of war, simple and inexpensive and bland: tapioca pudding, with its gelatinous fish eyes, Jell-O caramel pudding, Junket. Junket was made with white tablets that came out of a tube, and served with a dollop of grape jelly on the top. Probably it's vanished by now.

Jon has ordered a bottle, no glass-by-glass for him. It's a hint of the old bombast, the old peacock tail, and reassuring.

"How's your wife?" I ask him.

"Oh," he says, looking down, "Mary Jean and I have decided to try it apart for a while."

This may explain the herbal tea: some younger, more vegetarian influence, in the studio, on the sly. "I suppose you've got some little number," I say. "They say 'he goes' instead of 'he says,' have you noticed?"

"As a matter of fact," he says, "Mary Jean was the one who left."

"I'm sorry," I say. And immediately I am, I'm indignant, how could she do that to him, the cold unfeeling bitch. I side with him, despite the fact that I did the same thing to him myself, years ago.

"I guess I'm partly to blame," he says. This is not something he ever would have admitted before. "She said she couldn't get through to me."

I bet that isn't all she said. He's lost something, some illusion I used to think was necessary to him. He's come to realize he too is human. Or is this a performance, for my benefit, to show me he's up-to-date? Maybe men shouldn't have been told about their own humanity. It's only made them uncomfortable. It's only made them trickier, slier, more evasive, harder to read.

"If you hadn't been so crazy," I say, "it could have worked out. With us, I mean."

That perks him up. "Who was crazy?" he says, grinning again. "Who drove who to the hospital?"

"If it hadn't been for you," I say, "I wouldn't have needed to be driven to the hospital."

"That's not fair and you know it," he says.

"You're right," I say. "It's not fair. I'm glad you drove me to the hospital."

Forgiving men is so much easier than forgiving women.

"I'll walk you where you're going," he says when we're out on the sidewalk. I would like that. We're getting along so well, now there's nothing at stake. I can see why I fell in love with him. But I don't have the energy for it now.

"That's okay," I say. I don't want to admit that I don't know where I'm going. "Thanks for the studio. Let me know if you need anything out of it." Though I know he won't come over while I'm there, it's still too awkward, and hazardous, for us to be together behind a door that locks.

"Maybe we could have a drink, later," he says.

I say, "Maybe we could."

After leaving Jon I walk east along Queen, past the street dealers selling risqué T-shirts, past the garter belts and satin underpants in the windows. What I'm thinking about is a picture I painted, years ago now. *Falling Women*, it was called. A lot of my paintings then began in my confusion about words.

There were no men in this painting, but it was about men, the kind who caused women to fall. I did not ascribe any intentions to these men. They were like the weather, they didn't have a mind. They merely drenched you or struck you like lightning and moved on, mindless as blizzards. Or they were like rocks, a line of sharp slippery rocks with jagged edges. You could walk with care along between the rocks, picking your steps, and if you slipped you'd fall and cut yourself, but it was no use blaming the rocks.

That must be what was meant by fallen women. Fallen women were women who had fallen onto men and hurt themselves. There was some suggestion of downward motion, against one's will and not with the will of anyone else. Fallen women were not pulled-down women or pushed women, merely fallen. Of course there was Eve and the Fall; but there was nothing about falling in that story, which was only about eating, like most children's stories.

Falling Women showed the women, three of them, falling as if by accident off a bridge, their skirts opened into bells by the wind, their hair streaming upward. Down they fell, onto the men who were lying unseen, jagged and dark and without volition, far below.

48

I'm staring at a naked woman. In a picture she would be a nude, but she is not in a picture. This is the first live naked woman I've ever seen, apart from myself in the mirror. The girls in the high school locker room always had their underwear on, which is not the same thing, and neither are the women in stretch Lycra one-piece bathing suits with modesty panels, in magazine ads.

Even this woman is not entirely naked, as she has a sheet draped over her left thigh and tucked in between her legs: no hair shows. She's sitting on a stool, her buttocks squashing out sideways; her stocky back is curved, her right leg is crossed over her left at the knee, her right elbow rests on her right knee, her left arm is placed behind her with the hand on the stool. Her eyes are bored, her head droops forward, the way it has been put. She looks cramped and uncomfortable, and also cold: I can see the goose bumps on her upper arms. She has a thick neck. Her hair is frizzly and short, red with darker roots, and I suspect she is chewing gum: every once in a while there is a slow, furtive, sideways motion of her jaw. She is not supposed to move.

I am trying to draw this woman, with a piece of charcoal. I am trying for fluidity of line. This is how the teacher has arranged her: for fluidity of line. I would rather be using a hard pencil; the charcoal gets on my fingers and smears, and is no good for hair. Also this woman frightens me. There is a lot of flesh to her, especially below the waist; there are folds across her stomach, her breasts are saggy and have enormous dark nipples. The harsh fluorescent light, falling straight down on her, turns her eye sockets to caverns, emphasizes the descending lines from nose to chin; but the massiveness of her body makes her head look like an afterthought. She is not beautiful, and I am afraid of turning into that.

This is a night class. It's called Life Drawing, and is held on Tuesdays at the Toronto College of Art, in a large bare room, beyond which is a utilitarian stairway, then McCaul Street, then Queen with its drunks and streetcar tracks, and beyond that square, boxy Toronto. There are a dozen of us in the room, with our hopeful, almost-new Bristol drawing boards and our black-tipped fingers; two older women, eight young men, another girl my own age, and me. I am not a student here, but even those who aren't students can sign up for this class, under certain circumstances. The circumstances are that you have to convince the teacher you are serious. It's not clear however how long I will last.

The teacher is Mr. Hrbik. He is in his mid-thirties, with dark thickly curled hair, a mustache, an eagle nose, and eyes that look almost purple, like mulberries. He has a habit of staring at you without saying anything, and, it seems, without blinking.

It was the eyes I noticed first, when I went for my interview with him. He was sitting in his tiny paper-covered office at the college, leaning back in his chair and chewing the end of a pencil. When he saw me he put the pencil down.

"How old are you?" he said.

"Seventeen," I said. "Almost eighteen."

"Ah," he said, and sighed as if this was bad news. "What have you done?"

This made it sound as if he was accusing me of something. Then I saw what he meant: I was supposed to bring something called "a portfolio of recent work," which meant pictures, so he could judge me. But I didn't have much work to bring. About the only contact I'd had with art was in high school, in the Art Appreciation class we had to take in Grade Nine, where we listened to the *Moonlight Sonata* and interpreted it with wavy crayon lines, or drew a tulip in a vase. I'd never been to an art gallery, though I'd read an article on Picasso, in *Life* magazine.

Over the past summer, when I'd had a job making beds

and cleaning toilets at a resort in Muskoka to earn extra money, I'd bought a small oil painting set in one of the tourist shops. The names on the little tubes were like passwords: Cobalt Blue, Burnt Umber, Crimson Lake. On my time off I'd take this set out along the shore, and sit with my back against a tree with the pine needles digging into me from underneath, and mosquitoes collecting around me, looking out across the flat sheet metal water, the varnished mahogany inboards moving across it, little flags at their sterns. In these boats were sometimes other chambermaids, the kind who went to illegal parties in people's rooms to drink rye and ginger ale out of paper cups, and were rumored to go all the way. There had been tearful confrontations in the laundry room, over the folded sheets.

I did not know how to paint or even what to paint, but I knew I had to begin. After a while I'd painted a picture of a beer bottle minus the label, and a tree shaped like a damaged whisk, and several uncertain, sludge-colored pictures of rocks, with a violently blue lake in the background. Also a sunset, which came out looking like something you might spill on yourself.

I produced these from the black file folder in which I'd been carrying them. Mr. Hrbik frowned and twiddled his pencil and said nothing. I was discouraged, and also in awe of him, because he had power over me, the power to shut me out. I could see he thought my paintings were bad. They were bad.

"Any more?" he said. "Any drawings?"

Out of desperation I'd included some of my old Biology drawings, in hard lead pencil with colored shadings. I knew I could draw better than I could paint, I'd been doing it longer. I had nothing to lose and so I brought them out.

"What is it you call this?" he said, holding the top one upside-down.

"It's the inside of a worm," I said.

He did not show surprise. "This?"

"It's a planaria. In stained section."

"And this?"

"It's the reproductive system of a frog. A male frog," I added.

Mr. Hrbik stared up at me with his shining purple eyes. "Why do you want to take this class?" he said.

"It's the only one I can get into," I said. Then I realized how bad that sounded. "It's my only hope. I don't know anyone else who can teach me."

"Why do you want to learn?"

"I don't know," I said.

Mr. Hrbik picked up his pencil and stuck the end of it into the side of his mouth, like a cigarette. Then he took it out again. He twirled his fingers in his hair. "You are a complete amateur," he said. "But sometimes this is better. We can begin from nothing." He smiled at me, the first time. He had uneven teeth. "We will see what we can make of you," he said.

Mr. Hrbik paces the room. He despairs of us, all of us, including the model, whose surreptitious gum chewing maddens him. "Keep still," he says to her, tugging at his hair. "Enough gum." The model shoots him a malevolent look and clenches her jaw. He takes her arms and her sulky-faced head and rearranges them, as if she is a mannequin. "We will try again."

He strides up and down among us, looking over our shoulders and grunting to himself, as the room fills with the sandy rasping noise of charcoal on paper. "No, no," he says to a young man. "This is a *body*." He pronounces it "bowdy." "This is not an automobile. You must think of the fingers, touching this flesh, or the running of the hand over. This must be tactile." I try to think the way he wants me to, but recoil. I have no wish to run my fingers over this woman's goose-pimply flesh.

To one of the older women he says, "We do not want pretty. The bowdy is not pretty like a flower. Draw what is there." He stops behind me, and I cringe, waiting. "We are not making a medical textbook," he says to me. "What you have made is a corpse, not a woman." He pronounces it "voman."

I look at what I have drawn, and he is right. I am careful and accurate, but I have drawn a person-shaped bottle, inert and without life. Courage, which has brought me here, flows out of me. I have no talent.

But at the end of the class, when the model has risen stiffly to her feet and has clutched her sheet around her and padded off to dress, when I am putting away my charcoal, Mr. Hrbik comes to stand beside me. I rip out the drawings I have made, intending to crumple them up, but he puts his hand quickly on mine. "Save these," he says.

"Why?" I say. "They're no good."

"You will look at them later," he says, "and you will see how far you have come. You can draw objects very well. But as yet you cannot draw life. God first made the bowdy out of dirt, and after he breathed in the soul. Both are necessary. Dirt and soul." He gives me a brief smile, squeezes my upper arm. "There must be passion."

I look at him uncertainly. What he says is a trespass: people don't talk about bodies unless they're discussing illnesses, or about souls except in church, or about passion unless they mean sex. But Mr. Hrbik is a stranger, and can't be expected to know this.

"You are an unfinished voman," he adds in a lower voice, "but here you will be finished." He doesn't know that *finished* means over and done with. He intends to be encouraging.

49

I sit in the darkened auditorium, downstairs at the Royal Ontario Museum, leaning back in the hard seat covered with scratchy plush and breathing in the smell of dust and air-lessness and stale upholstery and the sweetish face powder of the other students. I feel my eyes getting rounder and rounder, the pupils enlarging like an owl's: for an hour I've been looking at slides, yellowy, sometimes unfocused slides of white marble women with flat-topped heads. These heads are holding up stone entablatures, which look very heavy; no wonder the tops of their heads are flat. These marble women are called caryatids, which originally referred to the priestesses of Artemis at Caryae. But they are no longer priestesses; they are now ornamental devices doubling as supporting columns.

There are many slides of columns as well, various kinds of columns from various periods: Doric, Ionic, Corinthian. Doric columns are the strongest and simplest, Corinthian ones are the lightest and most ornate, adorned with rows of acanthus leaves giving rise to graceful volutes and helices. A long pointer, emerging from the area of no light beside the screen, rests on the volutes and helices, indicating which is which. I will need these words later, when I have to regurgitate them for exams, so I attempt to write them in my notebook, bending my head down close to the paper in order to see. I spend a lot of time now writing obscure words in the dark.

I expect things to be better next month, when we'll get away from the Greeks and Romans and into Mediaeval and Renaissance. *Classical* has come to mean, for me, bleached-out and broken. Most of the Greek and Roman things have body parts missing, and the general armlessness, leglessness, and noselessness is getting to me, not to mention the snapped-off penises. Also the grayness and whiteness, although I have learned to my surprise that all these marble statues used to be painted, in bright colors, with yellow hair and blue eyes and flesh tones, and sometimes dressed up in real clothing, like dolls.

This class is a survey course. It's supposed to orient us in time, in preparation for later, more specialized courses. It's part of Art and Archaeology at the University of Toronto, which is the only sanctioned pathway that leads anywhere close to art. Also the only thing I can afford: I have won a scholarship to university, which was no more than expected. "You should use the brains God gave you," my father is in the habit of saying, though we both know he thinks this gift was really bestowed by him. If I left university, threw over my scholarship, he would not see his way clear to putting up the cash for anything else.

When I first told my parents I was not going into Biology after all but was going to be an artist, they reacted with alarm. My mother said that was fine if it was what I really wanted to do, but they were worried about how I would make a living. Art was not something that could be depended on, though all right for a hobby, like shellwork or wood carving. But Art and Archaeology was reassuring to them: I could veer off in the archaeology direction and take to digging things up, which was more serious.

At the very least I will come out of it with a degree, and with a degree you can always teach. I have private reservations about this: I think of Miss Creighton, the Art Appreciation teacher at Burnham High, pudgy and beleaguered, who got routinely locked into the supply closet where the paper and paints were kept by some of the greasier and more leathery boys.

One of my mother's friends tells her that art is something you can always do at home, in your spare time.

The other students in Art and Archaeology are all girls but one, just as the professors are all men but one. The student who is not a girl and the professor who is not a man are considered strange; the first has an unfortunate skin condition, the second a nervous stammer. None of the girl students wants to be an artist; instead they want to be teachers of art in high schools, or, in one case, a curator in a gallery. Or else they are vague about their wants, which means they intend to get married before any of these other things becomes necessary.

What they wear is cashmere twin sets, camel's-hair coats, good tweed skirts, pearl button earrings. They wear tidy medium-heel pumps and tailored blouses, or jumpers, or little weskits with matching skirts and buttons. I wear these things too, I try to blend in. Between classes I drink cups of coffee with them and eat doughnuts, sitting in various common rooms and butteries and coffee shops. They discuss clothes, or talk about the boys they are going out with, licking the doughnut sugar off their fingers. Two of them are already pinned. Their eyes during these conversations look dewy, blurred, pulpy, easily hurt, like the eyes of blind baby kittens; but also sly and speculative, and filled with greed and deceit.

I feel ill at ease with them, as if I am here under false pretences. Mr. Hrbik and the tactility of the body do not fit into Art and Archaeology; my botched attempts at drawing naked women would be seen as a waste of time. Art has been accomplished, elsewhere. All that remains to be done with it is the memory work. The entire Life Drawing class would be viewed as pretentious, and also ludicrous.

But it is my lifeline, my real life. Increasingly I begin to eliminate whatever does not fit in with it, paring myself down. To the first class I made the mistake of wearing a plaid jumper and a white blouse with a Peter Pan collar, but I learn quickly. I switch to what the boys wear, and the other girl: black turtlenecks and jeans. This clothing is not a disguise, like other clothing, but an allegiance, and in time I work up the courage to wear these things even in the daytime, to Art and Archaeology; all except the jeans, which nobody wears. Instead I wear black skirts. I grow out my high school bangs and pin my hair back off my face, hoping to look austere. The girls at university, in their cashmere and pearls, make jokes about arty beatniks and talk to me less.

The two older women in Life Drawing notice my transformation as well. "So who died?" they ask me. Their names are Babs and Marjorie, and they are professionals. They both do portraits, Babs of children, Marjorie of dog owners and their dogs; they are doing Life Drawing as a refresher course, they say. They themselves do not wear black turtlenecks, but smocks, like pregnant women. They call each other "kid" and make raucous comments about their work, and smoke in the washroom, as if it's naughty. Because they are my mother's age, it embarrasses me to be in the same room with them and the naked model both together. At the same time I find them undignified. They remind me less of my mother, however, than of Mrs. Finestein from next door.

Mrs. Finestein has taken to wearing fitted red suits and jaunty pillbox hats trimmed with matching cherries. She catches sight of me in my new getup and is disappointed. "She looks like an Italian widow," she tells my mother. "She's letting herself go. Such a shame. With a good haircut and a little makeup, she could be stunning." My mother reports this to me, smiling as if it's funny, but I know it's her way of expressing concern. I am verging on grubbiness. *Letting yourself go* is an alarming notion; it is said of older women who become frowzy and fat, and of things that are sold cheap.

Of course there is something to it. I am letting myself go.

50
·

I'm in a beer parlor, drinking ten-cent draft beer, with the other students from Life Drawing. The grumpy waiter comes, balancing a circular tray on one hand, and plonks down the glasses, which are like ordinary water glasses only full of beer. Froth slops over. I don't like the taste of beer much, but by now I know how to drink it. I even know enough to sprinkle salt on the top, to cut down the foam.

This beer parlor has a dingy red carpet and cheesy black tables and plastic-upholstered chairs and scant lighting, and reeks of car ashtray; the other beer parlors we drink in are similar. They are called things like Lundy's Lane and The Maple Leaf Tavern, and they're all dark, even in daytime, because they aren't allowed to have windows you can see in through from the street. This is to avoid corrupting minors. I am a minor myself—the legal drinking age is twenty-one—but none of the waiters ever asks for my I.D. Jon says I look so young they think I'd never have the nerve to try it unless I was really overage.

The beer parlors are divided into two sections. The Men Only sections are where the rowdy drunks and rubby-dubs hang out; they're floored with sawdust, and the smell of spilled beer and old urine and sickness wafts out from them. Sometimes you can hear shouts and the crash of glass from within, and see a man being ejected by two wrestler-sized waiters, his nose bleeding, his arms flailing.

The Ladies and Escorts sections are cleaner and quieter and more genteel, and smell better. If you're a man you can't go into them without a woman, and if you're a woman you can't go into the Men Onlys. This is supposed to keep prostitutes from bothering men, and to keep the male hard drinkers from bothering women. Colin, who is from England, tells us about pubs, where there are fireplaces and you can play darts and stroll around and even sing, but none of that is allowed in beer parlors. They are for drinking beer, period. If you laugh too much you can be asked to leave.

The Life Drawing students prefer Ladies and Escorts, but they need a woman to get in. This is why they invite me: they even buy me free beers. I am their passport. Sometimes I'm the only one available after class, because Susie, the girl my age, frequently begs off, and Marjorie and Babs go home. They have husbands, and are not taken seriously. The boys call them "lady painters."

"If they're lady painters, what does that make me?" I say.

"A girl painter," Jon says, joking.

Colin, who has manners of a sort, explains: "If you're bad, you're a lady painter. Otherwise you're just a painter." They don't say "artist." Any painter who would call himself an artist is an asshole, as far as they're concerned.

I've given up on going out on dates in the old way: somehow it's no longer a serious thing to do. Also I haven't been asked that often since the advent of the black turtlenecks: boys of the blazer-and-white-shirt variety know what's good for them. In any case they are boys, not men. Their pink cheeks and group sniggering, their good-girl and bad-girl categories, their avid, fumbling attempts to push back the frontiers of garter belt and brassiere no longer hold my attention. Mustaches of long standing do, and nicotine-stained fingers; experienced wrinkles, heavy eyelids, a world-weary tolerance; men who can blow cigarette smoke out through their mouths and breathe it in through their nostrils without a second thought. I'm not sure where this picture has come from. It seems to have arrived fully formed, out of nowhere.

The Life Drawing students aren't like this, though they don't wear blazers either. With their deliberately shoddy and paint-stained clothing, their newly sprouted facial hair, they are a transitional form. Although they talk, they distrust words; one of them, Reg from Saskatchewan, is so inarticulate he's practically mute, and this wordlessness of his gives him a special status, as if the visual has eaten up part of his brain and left him an idiot saint. Colin the Englishman is distrusted because he talks not too much but too well. Real painters grunt, like Marlon Brando.

But they can make their feelings known. There are shrugs, mutterings, half-finished sentences, hand movements: jabs, fists, openings of the fingers, jerky sculptings of the air. Sometimes this sign language is about other people's painting: "It sucks," they say, or very occasionally, "Fan-fuckin'-tastic." They don't approve of much. Also they think Toronto is a dump. "Nothing's happening here," is what they say, and many of their conversations revolve around their plans for escape. Paris is finished, and even Colin the Englishman doesn't want to go back to England. "They all paint yellowy-green there," he says. "Yellowy-green, like goose turds. Bloody depressing." Nothing but New York will do. That's where everything is happening, that's where the action is.

When they've had several beers they might talk about women. They refer to their girlfriends, some of whom live with them; these are called "my old lady." Or they make jokes about the models in Life Drawing, who change from night to night. They speak of going to bed with them, as if this depends only on their inclination or lack of it. There are two possible attitudes to this: lip smacking or nauseated revulsion. "A cow," they say. "A bag." "What a discard." Sometimes they do this with an eye toward me, looking to see how I will take it. When the descriptions of body parts get too detailed—"Cunt like an elephant's arse," "How would you know, eh, screw elephants much?"—they shush one another, as if in front of mothers; as if they haven't decided who I am.

I don't resent any of this. Instead I think I am privileged: I am an exception, to some rule I haven't even identified.

I sit in the dankness and beer fug and cigarette smoke, getting a little dizzy, keeping my mouth shut, my eyes open. I think I can see them clearly because I expect nothing from them. In truth I expect a lot. I expect to be accepted.

There's one thing they do that I don't like: they make fun of Mr. Hrbik. His first name is Josef and they call him Uncle Joe, because he has a mustache and an Eastern European accent and is authoritarian in his opinions. This is unfair, since I know—all of us know by now—that he was shunted around in four different countries, because of the upheavals of the war, and got trapped behind the Iron Curtain and lived on garbage and almost starved, and escaped during the Hungarian Revolution, probably with danger to his life. He has never mentioned the exact circumstances. In fact he has mentioned none of this, in class. Nevertheless it is known.

But it cuts no ice with the boys. Drawing sucks and Mr. Hrbik is a throwback. They call him a D.P., which means *displaced person*, an old insult I remember from high school. It was what you called refugees from Europe, and those who were stupid and uncouth and did not fit in. They mimic his accent, and the way he talks about the body. They only take Life Drawing because it's a requirement. Life Drawing is not what's happening, Action Painting is, and for that you sure as hell don't need to know how to draw. In particular you don't need to know how to draw a cow with no clothes on. Nevertheless they sit in Life Drawing, scratching away with the charcoal and turning out rendering after rendering of breasts and buttocks, thighs and necks, and some nights nothing but feet, as I do, while Mr. Hrbik strides up and down, tugging at his hair and despairing.

The faces of the boys are impassive. To me their contempt is obvious, but Mr. Hrbik doesn't notice. I feel sorry for him, and grateful to him, for letting me into the class. Also I admire him. The war is far enough away now to be romantic, and he has been through it. I wonder if he has any bullet holes in him, or other marks of grace.

Tonight, in the Ladies and Escorts of the Maple Leaf Tavern, it isn't just the boys and me. Susie is here too.

Susie has yellow hair, which I can tell she rolls and sets and then dishevels, and tips ash-blond at the ends. She wears jeans and black turtlenecks too, but her jeans are skintight and she's usually got something around her neck, a silver chain or a medallion. She does her eyes with a heavy black line over the lid like Cleopatra, and black mascara and smoky dark-blue eye shadow, so her eyes are blue-rimmed, bruise-colored, as if someone's punched her; and she uses white face powder and pale pink lipstick, which makes her look ill, or as if she's been up very late every night for weeks. She has full hips, and breasts that are too large for her height, like a rubber squeaky toy that's been pushed down on the top of the head and has bulged out in these places. She has a little breathless voice and a startled little laugh; even her name is like a powder puff. I think of her as a silly girl who's just fooling around at art school, too dumb to get into university, although I don't make judgments like this about the boys.

"Uncle Joe was raving tonight," says Jon. Jon is tall, with sideburns and big hands. He has a denim jacket with a lot of snap fasteners on it. Besides Colin the Englishman, he's the most articulate one. He uses words like *purity* and *the picture plane*, but only among two or three, never with the whole group.

"Oh," says Susie, with a tiny, gaspy laugh, as if the air is going into her instead of out, "that's mean! You shouldn't call him that!"

This irritates me: because she's said something I should have said myself and didn't have the guts to, but also because she's made even this defense come out like a cat rubbing against a leg, an admiring hand on a bicep.

"Pompous old fart," says Colin, to get some of her attention for himself.

Susie turns her big blue-rimmed eyes on him. "He's not old," she says solemnly. "He's only thirty-five." Everyone laughs.

But how does she know? I look at her and wonder. I remember the time I went early to class. The model wasn't there yet, I was in the room by myself, and then Susie walked in with her coat already off, and right after that Mr. Hrbik.

Susie came over to where I was sitting and said, "Don't you just hate the snow!" Ordinarily she didn't talk to me. And I was the one who'd been out in the snow: she looked warm as toast.

*I*n the daytime it's February. The gray museum auditorium steams with wet coats and the slush melting from winter boots. There's a lot of coughing.

We've finished the Mediaeval period, with its reliquaries and elongated saints, and are speeding through the Renaissance, hitting the high points. Virgin Marys abound. It's as if one enormous Virgin Mary has had a whole bunch of daughters, most of which look something like her but not entirely. They've shed their gold-leaf halos, they've lost the elongated, flat-chested look they had in stone and wood, they've filled out more. They ascend to Heaven less frequently. Some are dough-faced and solemn, sitting by fireplaces or in chairs of the period, or by open windows, with roof work going on in the background; some are anxious-looking, others are milk-fed and pinky-white, with wire-thin halos and fine gold tendrils of hair escaping from their veils and clear Italian skies in the distance. They bend over the cradle of the Nativity, or they hold Jesus on their laps.

Jesus has trouble looking like a real baby because his arms and legs are too long and spindly. Even when he does look like a baby, he's never newborn. I've seen newborn babies, with their wizened dried-apricot look, and these Jesuses are nothing like them. It's as if they've been born at the age of one year, or else are shriveled men. There's a lot of red and blue in these pictures, and a lot of breast-feeding.

The dry voice from the darkness concentrates on the formal properties of the compositions, the arrangement of cloth in folds to accentuate circularity, the rendering of textures, the uses of perspective in archways and in the tiles underfoot. We skim over the breast-feeding: the pointer emerging from nowhere never alights on these bared breasts, some of which are an unpleasant pinky-green or veiny, or have a hand pressing the nipple and even real milk. There is some shifting in the seats at this: nobody wants to think about breast-feeding, not the professor and certainly not the girls. Over coffee they shiver: they themselves are fastidious, they will bottle-feed, which is anyway more sanitary.

"The point of the breast-feeding," I say, "is that the Virgin is humble enough to do it. Most women then got their kids wet-nursed by somebody else, if they could afford it." I have read this in a book, dug up from the depths of the stacks, in the library.

"Oh, Elaine," they say. "You're such a brain."

"The other point is that Christ came to earth as a mammal," I say. "I wonder what Mary did for diapers? Now that would be a relic: the Sacred Diaper. How come there are no pictures of Christ on the potty? I know there's a piece of the Holy Foreskin around, but what about the Holy Shit?"

"You're awful!"

I grin, I cross my ankle over my knee, I put my elbows on the table. I enjoy pestering the girls in this minor, trivial way: it shows I am not like them.

This is one life, my life of daytimes. My other, my real life, takes place at night.

I've been watching Susie closely, and paying attention to what she does. Susie is not in fact my age, she is two years older and more, she's almost twenty-one. She doesn't live at home with her parents, but in a bachelorette apartment in one of the new high-rise buildings on Avenue Road, north of St. Clair. It is thought her parents pay for this. How else could she afford it? These buildings have elevators in them, and wide foyers with plants, and are called things like The Monte Carlo. Living in them is a daring and sophisticated thing to do, though scoffed at by the painters: trios of nurses live there. The painters themselves live on Bloor Street or Queen, above hardware stores and places that sell suitcases wholesale, or on side streets where there are immigrants.

Susie stays after class, she turns up early, she hangs around; during the class itself she looks at Mr. Hrbik only sideways, furtively. I meet her coming out of his office and she jumps and smiles at me, then turns and calls, artificially and too loudly, "Thank you, Mr. Hrbik! See you next week!" She gives a little wave, although the door is partly closed and he can't possibly see her: the wave is for me. I now guess what I should have spotted right away: she is having a love affair with Mr. Hrbik. Also, she thinks nobody has figured this out.

In this she is wrong. I overhear Marjorie and Babs discussing it in an oblique way: "Listen, kid, it's one way to pass the course," is what they say. "Wish I could do it just by flipping on my back." "Don't you wish! Those days are long gone, eh?" And they laugh in a comfortable way, as if what is going on is nothing at all, or funny.

I don't think this love affair is at all funny. Love affair is how I think of it; I can't detach the word *affair* from the word *love*, although which of them loves the other is not clear. I decide that it's Mr. Hrbik who loves Susie. Or he doesn't really love her: he's besotted by her. I like this word *besotted*, suggestive as it is of sogginess, soppiness, flies drunk on syrup. Susie herself is incapable of love, she's too shallow. I think of her as the conscious one, the one in control: she's toying with him, in a hard, lacquered way straight out of forties movie posters. Hard as nails, and I even know what color of nails: Fire and Ice. This, despite her easily hurt look, her ingratiating ways. She throws off guilt like a sweet aroma, and Mr. Hrbik staggers besotted toward his fate.

After she realizes the people in the class know—Babs and Marjorie have a way of conveying their knowledge—Susie becomes bolder. She starts referring to Mr. Hrbik by his first name, and popping him into sentences: Josef thinks, Josef says. She always knows where he is. Sometimes he is in Montreal for the weekend, where they have much better restaurants and decent wines. She's definite about this, although she's never been there. She throws out inside tidbits of information about him: he was married in Hungary, but his wife didn't come with him and now he's divorced. He has two daughters whose pictures he keeps in his wallet. It kills him to be separated from them—"It just *kills* him," she says softly, her eyes misting.

Marjorie and Babs gobble this up. Already she's losing her floozie status with them, she's entering the outskirts of domesticity. They egg her on: "Listen, I don't blame you! I think he's just cute as a button!" "I could eat him up! But that would be robbing the cradle, eh?" In the washroom the two of them sit side by side in separate cubicles, talking over the noise of gushing pee, while I stand in front of the mirror, listening in. "I just hope he knows what he's doing. A nice kid like her." What they mean is that he should marry her. Or perhaps they mean that he should marry her if she gets pregnant. That would be the decent thing.

The painters, on the other hand, turn rough on her. "Jeez, will you shut up about Josef! You'd think the sun shines out of his ass!" But she can't shut up. She resorts to craven, apologetic giggling, which annoys them further, and me also. I've seen that saturated, brimming look before.

I feel that Mr. Hrbik needs protecting, or even rescuing. I don't yet know that a man can be admirable in many ways but a jerk in others. Also I haven't yet learned that chivalry in men is idiocy in women: men can get out of a rescue a lot more easily, once they get into it.

52

I am still living at home, which is humiliating; but why should I pay extra to live in a dormitory, when the university is in the same city? This is my father's view, and the reasonable one. Little does he know it isn't a dormitory I have in mind, but a crumbling walk-up above a bakery or cigar store, with streetcars rumbling by outside and the ceilings covered with egg cartons painted black.

But I no longer sleep in my childhood room with the vanilla-colored light fixture and the window curtains. I've retreated to the cellar, claiming I can study better. Down there in a dim storage room adjacent to the furnace I've set up a realm of ersatz squalor. From the cupboardful of old camping equipment I've excavated one of the army surplus cots and a lumpy khaki sleeping bag, short-circuiting my mother's plans to move my bed down to the cellar so I can have a proper mattress. On the walls I've taped theater posters, from local productions—Beckett's *Waiting for Godot*, Sartre's *No Exit*—with deliberate fingerprints and inkstain-black lettering on them, and shadowy figures that look as if they've run in the wash; also several of my careful drawings of feet. My mother thinks the theater posters are gloomy, and doesn't understand the feet at all: feet should have a body. I narrow my eyes at her, knowing better.

As for my father, he thinks my talent for drawing is impressive, but wasted. It would have been better applied to cross sections of stems and the cells of algae. For him I am a botanist manqué.

His view of life has darkened since Mr. Banerji returned to India. There is some obscurity around this: it is not talked of much. My mother says he was homesick, and hints at a nervous breakdown, but there was more to it than that. "They wouldn't promote him," says my father. There's a lot behind *they* (not *we*), and *wouldn't* (not *didn't*). "He wasn't properly appreciated." I think I know what this means. My father's view of human nature has always been bleak, but scientists were excluded from it, and now they aren't. He feels betrayed.

My parents' footsteps pace back and forth above my head; the sounds of the household, the Mixmaster and the telephone and the distant news, filter down to me as if in illness. I emerge, blinking, for meals and sit in stupor and demi-silence, picking at my chicken fricassee and mashed potatoes, while my mother comments on my lack of appetite and pallor and my father tells me useful and interesting things as if I am still young. Do I realize that nitrogen fertilizers are destroying fish life by fostering an overgrowth of algae? Have I heard of the new disease which will turn us all into deformed cretins unless the paper companies are forced to stop dumping mercury into the rivers? I do not realize, I have not heard.

"Are you getting enough sleep, dear?" says my mother.

"Yes," I say untruthfully.

My father has noticed an ad in the paper, for an atomic radiation monster insect movie. "As you know," he says, "those giant grasshoppers could never actually exist. At that size their respiratory systems would collapse."

I do not know.

In April, while I'm studying for exams and before the buds come out, my brother Stephen gets arrested. This happens the way it would.

Stephen has not been here as he should have been to help me out at the dinner table, he hasn't been home all year. Instead he's running around loose in the world. He's studying Astrophysics at a university in California, having finished his undergraduate degree in two years instead of four. Now he is doing graduate work.

I have no clear picture of California, having never been there, but I think of it as sunny, and warm all the time. The sky is a vibrant aniline blue, the trees a preternatural green. I populate it with tanned, handsome men in sunglasses and sports shirts with palm trees on them, and with real palm trees, and with blond, long-legged women, also tanned, with white convertibles.

Among these sunglassed, fashionable people my brother is an anomaly. After he left his boys' school he reverted to his old, unkempt ways, and goes around in his moccasins and his sweaters with the worn-through elbows. He doesn't get a haircut unless reminded, and who is there to remind him? He walks among the palm trees, oblivious, whistling, his head sheathed in a halo of invisible numbers. What do the Californians make of him? They think he is a kind of tramp.

On this particular day he takes his binoculars and his butterfly book and heads out into the countryside on his secondhand bicycle, to look for Californian butterflies. He comes to a promising field, descends, locks up the bike: he is prudent enough within limits. He heads into the field, which must have tall grass in it and some smallish bushes. He sees two exotic Californian butterflies and starts in pursuit of them, pausing to scan them with his binoculars; but at this distance he can't identify them, and every time he moves forward they take off.

He follows them to the end of the field, where there is a chain-link fence. They fly through it, he climbs over. On the other side there's another field, a flatter one with less vegetation. There's a dirt road crossing it, but he disregards this and follows the butterflies, red and white and black in color, with an hourglass pattern, something he's never seen before. At the other side of this field there's another fence, a higher one, and he scales this too. Then, when the butterflies have finally stopped, on a low tropical bush with pink flowers, and he's down on one knee focusing his binoculars, three uniformed men in a jeep drive up.

"What're you doing in here?" they say.

"In where?" says my brother. He's impatient with them, they've disturbed the butterflies, which have flown off again.

"Didn't you see the signs?" they say. "The ones that said, DANGER, KEEP OUT?"

"No," says my brother. "I was chasing those butterflies."

"Butterflies?" says one. The second one makes a twirling motion beside his ear, with his finger, denoting craziness. "Wacko," he says. The third one says, "You expect us to believe that?"

"What you believe is your own concern," says my brother. Or something of the sort.

"Wise guy," they say, because this is what Americans say in comic books. I add some cigarettes, in the sides of their mouths, a few pistols and other hardware, and boots.

It turns out they are the military and this is a military testing zone. They take my brother back to their headquarters and lock him up. Also they confiscate his binoculars. They don't believe he's a graduate student in Astrophysics out chasing butterflies, they think he's a spy, although they can't figure out why he would have been so open about it. Spy novels, as I and the military know but my brother does not, are crawling with spies who pretend to be harmless butterfly fanciers.

Finally they allow him to make a phone call, and his graduate supervisor from the university has to come and bail him out. When he goes back to retrieve his bike, it's been pinched.

I get the bare bones of this from my parents over the beef stew. They don't know whether to be amused or alarmed. From my brother, however, I hear nothing of the sort. Instead I get a letter, written in pencil on a page torn from a loose-leaf notebook. His letters always begin without greeting and end without signature, as if they're part of one single letter, unrolling through time like an endless paper towel.

He's writing this letter, he says, from the top of a tree, where he's watching the football game over the stadium wall—cheaper than buying a ticket—and eating a peanut butter sandwich, cheaper than eating in a restaurant: he doesn't like monetary transactions. There are in fact several grease spots on the paper. He says he can see a bunch of pom-pom-covered capons jumping up and down. These must be the cheerleaders. He's living in a student dormitory with a lot of mucus membranes who do nothing but drool over girls and get pissed on American beer. In his opinion this takes some doing, as the stuff is weaker than shampoo and tastes like it into the bargain. In the mornings he eats prefrozen reheated fried eggs, which are square in shape and have ice crystals in the yolks. A triumph of modern technology, he says.

Apart from that he's enjoying himself, as he is hard at work on The Nature of the Universe. The burning question is: is the universe more like a giant ever-enlarging blimp, or does it pulsate, does it expand and contract? Probably the suspense is killing me, but I will just have to wait a few years till he works out the final answer. TUNE IN FOR THE NEXT THRILLING INSTALLMENT, he writes, in block letters.

I hear you've gone into the picture business, he continues in normal-sized writing. *I used to do that sort of thing when I was younger. I hope you're taking your cod liver oil pills and keeping out of trouble.* And that is the end of the letter.

I think of my brother sitting at the top of a tree, in California. He no longer knows who he's writing to, because I have surely changed beyond all recognition. And I no longer know who's writing. I think of him as staying always the same, but of course this can't be true. He must know things by now that he didn't know before, as I do.

Also: if he's eating a sandwich and writing a letter both at the same time, how is he holding on? He seems happy enough, up there in his perch of a sniper. But he should be more careful. What I have always assumed in him to be bravery may be merely an ignorance of consequences. He thinks he is safe, because he is what he says he is. But he's out in the open, and surrounded by strangers.

53

I sit in a French restaurant with Josef, drinking white wine and eating snails. They're the first snails I have ever eaten, this is the first French restaurant I have ever been in. It's the only French restaurant in Toronto, according to Josef. It's called La Chaumière, which Josef says means "thatched cottage." La Chaumière, is not however a thatched cottage, but a prosaic, dowdy building like other Toronto buildings. The snails themselves look like large dark pieces of snot; you eat them with a two-pronged fork. I think they are quite good, though rubbery.

Josef says they aren't fresh snails but have come out of a tin. He says this sadly, with resignation, as if it means the end, though the end of what is not clear; this is how he says many things.

It was the way he first said my name, for instance. That was back in May, in the last week of Life Drawing. Each of us was supposed to meet with Mr. Hrbik for an individual evaluation, to discuss our progress during the year. Marjorie and Babs were ahead of me, standing in the hall with take-out coffees. "Hi, kid," they said. Marjorie was telling a story about how a man exposed himself to her in Union Station, where she had gone to meet her daughter on the train from Kingston. Her daughter was my age, and going to Queen's.

"He had on a raincoat, would you believe," said Marjorie.

"Oh God," said Babs.

"So I looked him in the eye—the *eye*—and I said, 'Can't you do any better than that?' I mean, talk about weenies. No wonder the poor boob runs around in train stations trying to get somebody to look at it!"

"And?"

"Listen, what goes up must come down, eh?"

They snorted, spewing droplets of coffee, coughing out smoke. As usual I found them slightly disreputable: making jokes about things that were no joking matter.

Susie came out of Mr. Hrbik's office. "Hi, you guys," she said, trying for cheer. Her eyeshadow was smudged, her eyes pinkish. I'd been reading modern French novels, and William Faulkner as well. I knew what love was supposed to be: obsession, with undertones of nausea. Susie was the sort of girl who would go in for this kind of love. She would be abject, she would cling and grovel. She would lie on the floor, moaning, hanging onto Mr. Hrbik's legs, her hair falling like blond seaweed over the black leather of his shoes (he would have his shoes on, being about to stalk out of the door). From this angle, Mr. Hrbik was cut off at the knees and Susie's face was invisible. She would be squashed by passion, obliterated.

I was not sorry for her, however. I was a little envious.

"Poor bunny rabbit," Babs said behind her retreating back.

"Europeans," said Marjorie. "I don't believe for a minute he was ever divorced."

"Listen, maybe he was never even *married.* "

"What about those kids of his?"

"Most likely his nieces or something."

I scowled at them. Their voices were way too loud; Mr. Hrbik would hear them.

After they had gone it was my turn. I went in, and stood while Mr. Hrbik sat, going through my portfolio, which was spread out on his desk. I thought it was this that was making me nervous.

He flipped through the pages, hands, heads, bottoms, in silence, chewing his pencil. "This is nice," he said at last. "You have made progress. This is more relaxed, this line here."

"Where?" I said, leaning my hand on the desk, bending forward. He turned his head to the side, toward me, and there were his eyes. They were not purple after all but dark brown.

"Elaine, Elaine," he said sadly. He put his hand over

mine. Cold shot up my arm, into my stomach; I stood there frozen, revealed to myself. Is this what I'd been angling for, with my notions of rescue?

He shook his head, as if he'd given up or had no choice, then drew me down, between his knees. He didn't even stand up. So I was on the floor, on my knees, with my head tilted back, his hands caressing the back of my neck. I'd never been kissed that way before. It was like a perfume ad: foreign and dangerous and potentially degrading. I could get up and run for it, but if I stayed put, even for one more minute, there would be no more groping in car seats or movie theaters, no skirmishes over brassiere hooks. No nonsense, no fooling around.

We went to Josef's apartment in a taxi. In the taxi Josef sat quite far apart from me, although he kept his hand on my knee. I was not used to taxis then, and thought the driver was looking at us in the rearview mirror.

Josef's apartment was on Hazelton Avenue, which was not quite a slum although close to it. The houses there are old, close together, with frumpy little front gardens and pointed roofs and moldering wooden scrollwork around the porches. There were cars parked bumper to bumper along the sidewalk. Most of the houses were in pairs, attached together down one side. It was in one of these crumbling, pointy-roofed twin houses that Josef lived. He had the second floor.

A fat older man in shirtsleeves and suspenders was rocking on the porch of the house next to Josef's. He stared as Josef paid the taxi, then as we came up the front walk. "Nice day," he said.

"Isn't it?" I said. Josef paid no attention. He put his hand lightly on the back of my neck as we went up the narrow inner stairs. Everywhere he touched me felt heavy.

His apartment was three rooms: a front room, a middle room with a kitchenette, and a back room. The rooms were small, and there was little furniture. It was as if he'd just moved in, or was moving out. His bedroom was painted mauve. On the walls were several prints, which were of elongated figures, murkily colored. There was nothing else in this room but a mattress on the floor, covered with a Mexican blanket. I looked at it, and thought I was seeing adult life.

Josef kissed me, standing up this time, but I felt awkward. I was afraid someone would see in through the window. I was afraid he would ask me to take off my own clothes, that he would then turn me this way and that, looking at me from a distance. I didn't like being looked at from behind: it was a view over which I had no control. But if he asked this I would have to do it, because any hesitation on my part would place me beneath consideration.

He lay down on the mattress, and looked up as if waiting. After a moment I lay down beside him and he kissed me again, gently undoing my buttons. The buttons were on an outsize cotton shirt, which was what had replaced the turtlenecks now that it was warm. I put my arms around him, and thought: he was in the war.

"What about Susie?" I said. As soon as I said it I realized it was a high school question.

"Susie?" Josef asked, as if trying to remember her name. His mouth was against my ear; the name was like a regretful sigh.

The Mexican blanket was scratchy, which did not bother me: sex was supposed to be unpleasant the first time. I expected the smell of rubber too, and the pain; but there was not as much pain, and not nearly as much blood as everyone said.

Josef was not expecting the pain. "This is hurting you?" he said at one point. "No," I said, flinching, and he did not stop. He was not expecting the blood either. He would have to get his blanket cleaned, but he didn't mention this. He was considerate, and stroked my thigh.

Josef has gone on all summer. Sometimes he takes me to restaurants, with checked tablecloths and candles stuck in Chianti bottles; sometimes to foreign films about Swedes and Japanese, in small uncrowded theaters. But we always end up back at his apartment, under or on top of the Mexican blanket. His lovemaking is unpredictable; sometimes he is avid, sometimes routine, sometimes absent-minded, as if doodling. It's partly the unpredictability that keeps me hooked. This and his need, which seems to me at times helpless and beyond his control.

"Don't leave me," he says, running his hands over me; always before, not after. "I couldn't bear it." This is an old-fashioned thing to say, and in another man I would find it comical, but not in Josef. I am in love with his need. Only to think of it makes me feel suffused, inert, like the flesh of a watermelon. For this reason I've canceled my plans to return to the Muskoka resort, to work as I did last summer. Instead I've taken a job at the Swiss Chalet on Bloor Street. This is a place that serves nothing but chicken, "broasted," as it says on the sign. Chicken and dipping sauce, and cole-slaw and white buns, and one flavor of ice cream: Burgundy Cherry, which is a striking shade of purple. I wear a uniform with my name stitched on the pocket, as in high school gym class.

Josef sometimes picks me up there, after work. "You smell of chicken," he murmurs in the taxi, his face against my neck. I've lost all modesty in taxis; I lean against him, his hand around me, under my arm, on my breast, or I lie down along the seat, head in his lap.

Also I have moved out of home. On the nights when I'm with him, Josef wants me to stay all night. He wants to wake up with me asleep beside him, start to make love to me without waking me up. I've told my parents it's only for the summer, so I can be closer to the Swiss Chalet. They think it's a waste of money. They are racketing around up north somewhere and I would have the house to myself; but my idea of myself and my parents' idea of me no longer belong in the same place.

If I'd gone to Muskoka I wouldn't be living at home this summer either, but not living at home in the same city is different. Now I live with two of the other Swiss Chalet girls, student workers like myself, in a corridor-shaped apartment on Harbord Street. The bathroom is festooned with stockings and underpants; hair rollers perch on the kitchen counter like bristly caterpillars, dishes cake in the sink.

I see Josef twice a week, and know enough not to try calling him or seeing him at other times. Either he won't be there or he will be with Susie, because he hasn't stopped seeing her, not at all. But we are not to tell her about me; we are to keep it secret. "She would be so terribly hurt," he says. It's the last one in line who must bear the burden of knowing: if anyone is to be hurt it will have to be me. But I feel entrusted by him: we are in this together, this protecting of Susie. It's for her good. In this there is the satisfaction of all secrets: I know something she doesn't.

She's found out somehow that I'm working at the Swiss Chalet—probably it's Josef who's told her, casually, skirting discovery, probably he finds it exciting to think of us together—and once in a while she comes in for a cup of coffee, late in the afternoon when there's nobody much around. She's gained a little weight, and the flesh of her cheeks is puffy. I can see what she'll look like in fifteen years, if she isn't careful.

I am nicer to her than I ever have been. Also I'm a little wary of her. If she finds out, will she lose whatever grip on herself is left and go for me with a steak knife?

She wants to talk. She wants us to get together sometime. She still says "Josef and me." She looks forlorn.

Josef talks to me about Susie as if discussing a problem child. "She wants to get married," he says. He implies she is being unreasonable, but that to deny her this thing, this too-expensive toy, wounds him deeply just the same. I have no wish to put myself in the same category: irrational, petulant. I don't want to marry Josef, or anyone else. I have come to think of marriage as dishonorable, a crass trade-off rather than a free gift. And even the idea of marriage would diminish Josef, spoil him; this is not his place in the scheme of things. His place is to be a lover, with his secrecy and his almost-empty rooms, and his baleful memories and bad dreams. Anyway, I've put myself beyond marriage. I can see it back there, innocent and beribboned, like a child's doll: irretrievable. Instead of marriage I will be dedicated to my painting. I will end up with my hair dyed, wearing outlandish clothes and heavy, foreign silver jewellery. I will travel a lot. Possibly I will drink.

(There is of course the specter of pregnancy. You can't get a diaphragm unless you're married, rubbers are sold under the counter and only to men. There are those girls who went too far in back seats and got knocked up and dropped out of high school, or had strange, never-explained accidents. There are jocular terms for it: up the spout, bun in the oven. But such washroom notions have nothing to do with Josef and his experienced mauve bedroom. They also have nothing to do with me, wrapped as I am in dense minor-key enchantment. But I make little checkmarks on my pocket calendar, all the same.)

On the days when I have time off, when I'm not seeing Josef, I try to paint. Sometimes I draw with colored pencils. What I draw is the furniture in the apartment: the over-stuffed sofa from the Sally Ann strewn with shed clothing, the bulbous lamp lent by a roommate's mother, the kitchen stool. More often I don't have the energy, and end up reading murder mysteries in the bathtub.

Josef won't tell me about the war, or about how he got out of Hungary during the revolution. He says these things are too disturbing for him and he wants only to forget them. He says there are many ways to die and some are less pleasant than others. He says I am lucky I will never have to know things like this. "This country has no heroes," he says. "You should keep it that way." He tells me I am untouched. This is the way he wants me, he says. When he says these things he runs his hands over my skin as if he's erasing me, rubbing me smooth.

But he tells me his dreams. He's very interested in these dreams, and they are in fact like no dreams I can remember hearing about. There are red velvet curtains in them, red velvet sofas, red velvet rooms. There are white silk ropes in them, with tassels on the ends; there's a lot of attention to fabrics. There are decaying teacups.

He dreams of a woman wrapped up in cellophane, even over her face, and of another walking along the railing of a balcony dressed in a white shroud, and of another lying facedown in the bathtub. When he tells me these dreams, he doesn't look at me exactly; it's as if he's looking at a point several inches inside my head. I don't know how to respond, so I smile weakly. I'm a little jealous of these women in his dreams: none of them are me. Josef sighs and pats my hand. "You are so young," he says.

There is nothing that can be said in reply to this, although I don't feel young. Right now I feel ancient, and overworked and too warm. The constant odor of broasting chicken is taking away my appetite. It's late July, the Toronto humidity hangs like swamp gas over the city, and the air-conditioning at the Swiss Chalet broke down today. There were complaints. Someone upset a platter of quarters with buns and dipping sauce onto the kitchen floor, causing skids. The chef called me a stupid bitch.

"I have no country," says Josef mournfully. He touches my cheek tenderly, gazing into my eyes. "You are my country now."

I eat another tinned, inauthentic snail. It strikes me with no warning that I am miserable.

54

Cordelia has run away from home. This is not how she puts it.

She's tracked me down through my mother. I meet her for coffee, during my afternoon break, not at the Swiss Chalet. I could get the coffee free, but by now I want to be out of there as much as possible, away from the sickly back-room odor of raw poultry, the rows of naked chickens like dead babies, the mushed-up, lukewarm, dog dish debris of customers' meals. So instead we are in Murray's, down the street in the Park Plaza Hotel. It's medium clean, and although there's no air-conditioning there are ceiling fans. At least here I don't know what goes on in the kitchen.

Cordelia is thinner now, almost gaunt. The cheekbones of her long face are visible, her gray-green eyes are big in her face. Around each one of them is drawn a green line. She is tanned, her lips an understated orangey-pink. Her arms are angular, her neck elegant; her hair is pulled back like a ballerina's. She's wearing black stockings although it's summer, and sandals, although the sandals are not dainty women's summer sandals, but thick-soled and artistic, with primitive peasant buckles. Also a scoop-neck short-sleeved black jersey top that shows off her breasts, a full cotton skirt of a dull blue-green with abstract black swirls and squares on it, a wide black belt. She has on two heavy rings, one with a turquoise, and chunky square earrings, and a silver bracelet: Mexican silver. You would not say beautiful about her, but you would stare, as I am doing: for the first time in her life she looks distinguished.

We greeted each other on sight with the outstretched hands, the demi-hugs, the cries of surprise and delight that women are supposed to make who haven't seen each other for a while. Now I slump in Murray's, drinking wishy-washy coffee, while Cordelia talks and I wonder why I have agreed to this. I am at a disadvantage: I'm in my crumpled, gravy-spotted Swiss Chalet uniform, my armpits are sweaty, my feet hurt, my hair in this humidity is unruly and dank and curling like singed wool. There are dark circles under my eyes, because last night was one of Josef's nights.

Cordelia on the other hand is showing herself off to me. She wants me to see what has become of her, since her days of sloth and overeating and failure. She has reinvented herself. She's cool as a cucumber, and brimming with casual news.

What she is doing is working at the Stratford Shakespearean Festival. She is a bit-part player. "*Very* minor things," she says, waving her bracelet and rings dismissively, which means less minor than she says. "You know. Spear carrying, though of course I don't carry spears." She laughs, and lights a cigarette. I wonder if Cordelia has ever eaten snails, decide she is most likely on familiar terms with them; a depressing thought.

The Stratford Shakespearean Festival is quite famous by now. It was started several years ago in the town of Stratford, which has an Avon River running through it, and swans of both colors. I have read all this in magazines. People go there on the train, in buses, or in cars with picnic baskets; sometimes they stay all weekend and see three or four Shakespeare plays, one after the other. At first this festival was held in a big tent, like a circus. But now there is a real building, a strange, modern building, circular in shape. "So you have to project to three sides. It's such a strain on the voice," Cordelia says with a deprecating smile, as if she is projecting and straining her voice all in the line of work. She is like someone making herself up as she goes along. She's improvising.

"What do your parents think?" I say. This has been on my own mind lately: what parents think.

Her face closes down for a moment. "They're pleased I'm doing something," she says.

"What about Perdie and Mirrie?"

"You know Perdie," she says tightly. "Always the little

put-downs. But that's enough about me. What do *you* think of me?" This is an old joke of hers, and I laugh. "Seriously, what are you up to these days?" It's the tone I remember: polite but not too interested. "Since the last time I saw you."

I remember this last time with guilt. "Oh, nothing much," I say. "Going to school. You know." Right now it does look like nothing much. What have I really done all year? A smattering of art history, messing around with charcoal. There's nothing to show. There's Josef, but he's not exactly an accomplishment and I decide not to mention him.

"School!" says Cordelia. "Was I glad to see the end of *school*. God, what boredom." Stratford is only on in the summers, though. She will have to think of something else for the winter. Maybe the Earle Grey Players, going around to high schools. Maybe she will be ready for that.

She got the job at Stratford with the help of one of the Earle Grey cousins, who remembered her from her bedsheet days at Burnham. "People who know people," she says. She is one of Prospero's attendant spirits in *The Tempest*, and has to wear a body stocking, with a gauzy costume over top, sprinkled with dried leaves and spangles. "Obscene," she says. She's also a mariner in the first scene; she can get away with this because of her height. She's a court lady in *Richard III*, and she's the chief nun in *Measure for Measure*. In this one she actually speaks some lines. She recites for me, in a honey-colored Englishy voice:

> *Then, if you speak, you must not show your face,*
> *Or, if you show your face, you must not speak.*

"At rehearsal I kept getting mixed up," she says. She counts on her fingers. "Speak, hide face, show face, shut up." She puts her hands together in an attitude of prayer, bows forward, lowering her head. Then she gets up and does a full court curtsy out of *Richard III*, with the women shoppers having tea in Murray's gawking at her. "What I'd like to do next year is the First Witch in 'The Tartans.' 'When shall we three meet again, in thunder, lightning, or in rain?' The Old Man says I might be ready for it. He thinks it would be brilliant to have a *young* First Witch."

The Old Man, it turns out, is Tyrone Guthrie the director, from England and so famous I can't pretend not to have heard of him. "That's great," I say.

"Remember 'The Tartans' at Burnham? Remember that cabbage?" she says. "I was so humiliated."

I don't want to remember. The past has become discontinuous, like stones skipped across water, like postcards: I catch an image of myself, a dark blank, an image, a blank. Did I ever wear bat-wing sleeves and velveteen slippers, did I wear dresses like tinted marshmallows to formal dances, shuffle around the floor with some stranger's groin digging into mine? The dried corsages were thrown out long ago, the diplomas and class pins and photos must be down in my mother's cellar, in the steamer trunk along with the tarnishing silver. I glimpse those photos, rows and rows of lipsticked, spit-curled children. I would never smile, for those pictures. I would gaze stony-faced into the distance, beyond such adolescent diversions.

I remember my mean mouth, I remember how wise I thought I was. But I was not wise then. Now I am wise.

"Remember how we used to pinch things?" says Cordelia. "That was the only thing I really liked about that whole time."

"Why?" I say. I had not liked it much. I was always afraid of getting caught.

"It was something I could have," she says, and I'm not sure what she means.

Cordelia takes her sunglasses out of her shoulder bag and puts them on. There I am in her mirror eyes, in duplicate and monochrome, and a great deal smaller than life-size.

Cordelia gets me a free ticket to Stratford, so I can see her in action. I go on the bus. It's a matinee: I can get there, see the show, then take the bus back in time for my evening shift at the Swiss Chalet. The play is *The Tempest*. I watch for Cordelia, and when Prospero's attendants come on, with music and jittery lighting effects, I peer hard, trying to see which one she is, behind the disguise of costume. But I can't tell.

55

Josef is rearranging me. "You should wear your hair loose," he says, unpinning it from its ramshackle bun, running his hands through it to make it fluff out. "You look like a marvelous gypsy." He presses his mouth to my collarbone, untucks the bedsheet he's draped me in.

I stand still and let him do this. I let him do what he likes. It's August and too hot to move. Haze hangs over the city like wet smoke; it covers my skin with an oily film, seeps into my flesh. I move through the days like a zombie, going from one hour to the next without direction. I've stopped drawing the furniture at the apartment; I fill the bathtub with cool water and get into it, but I no longer read in there. Soon it will be time to go back to school. I can hardly think about it.

"You should wear purple dresses," says Josef. "It would be an improvement." He places me against the twilight of the window, turns me, stands back a little, running his hand up and down my side. I no longer care whether anyone can see in. I feel my knees begin to give, my mouth loosen. In our time together he does not pace or tug his hair, he moves slowly, gently, with great deliberation.

Josef takes me to the Park Plaza Hotel Roof Garden, in my new purple dress. It has a tight bodice, a low neck, a full skirt; it brushes against my bare legs as I walk. My hair is loose, and damp. I think it looks like a mop. But I catch a glimpse of myself, without expecting it, in the smoked-mirror wall of the elevator as we go up, and I see for an instant what Josef sees: a slim woman with cloudy hair, pensive eyes in a thin white face. I recognize the style: late nineteenth century. Pre-Raphaelite. I should be holding a poppy.

We sit on the outside patio, drinking Manhattans and looking over the stone balustrade. Josef has recently discovered a taste for Manhattans. This is one of the tallest buildings around. Below us Toronto festers in the evening heat, the trees spreading like worn moss, the lake zinc in the distance.

Josef tells me he once shot a man in the head; what disturbed him was how easy it was to do it. He says he hates the Life Drawing class, he will not go on with it forever, cooped up in this provincial deadwater teaching the rudiments to morons. "I come from a country that no longer exists," he says, "and you come from a country that does not yet exist." Once I would have found this profound. Now I wonder what he means.

As for Toronto, he says, it has no gaiety or soul. In any case, painting itself is a hangover from the European past. "It is no longer important," he says, waving it away with one hand. He wants to be in films, he wants to direct, in the United States. He will go there as soon as it can be arranged. He has good connections. There's a whole network of Hungarians, for instance. Hungarians, Poles, Czechoslovakians. There is more opportunity in films down there, to say the least, since the only films made in this country are short ones that come on before real movies, about leaves spiraling downward into pools or flowers opening in time-lapse photography, to flute music. The other people he knows are doing well in the United States. They will get him in.

I hold Josef's hand. His lovemaking these days is ruminative, as if he is thinking of something else. I discover I am somewhat drunk; also that I am afraid of heights. I have never been this high up in the air before. I think of standing close to the stone balustrade, toppling slowly over. From here you can see the United States, a thin fuzz on the horizon. Josef says nothing about me going there with him. I ask nothing.

Instead he says, "You are very silent." He touches my cheek. "Mysterious." I do not feel mysterious, but vacant.

"Would you do anything for me?" he says, gazing into my eyes. I sway toward him, far away from the earth. *Yes* would be so easy.

"No," I say. This is a surprise to me. I don't know where it has come from, this unexpected and stubborn truthfulness. It sounds rude.

"I did not think so," he says sadly.

Jon appears one afternoon in the Swiss Chalet. I don't recognize him at first because I don't look at him. I'm wiping off the table with a dishrag, every movement an effort, my arm heavy with lethargy. Last night I was with Josef, but tonight I won't be because it's not my night, it's Susie's night.

These days Josef rarely mentions Susie. When he does, it's with nostalgia, as if she's already a thing of the past, or beautifully dead, like someone in a poem. But this may be only his way of speaking. They may spend prosaic domestic evenings together, him reading the paper while she serves up a casserole. Despite his claim that I am a secret, they may discuss me the way Josef and I used to discuss Susie. This is not a comfortable thought.

I prefer to think of Susie as a woman shut inside a tower, up there in The Monte Carlo on Avenue Road, gazing out the window over the top of her painted sheet metal balcony, weeping feebly, waiting for Josef to appear. I can't imagine her having any other life apart from that. I can't see her washing out her underpants, for instance, and wringing them in a towel, hanging them on the bathroom towel rack, as I do. I can't imagine her eating. She is limp, without will, made spineless by love; as I am.

"Long time no see," says Jon. He leaps into focus beyond my wiping arm, grinning at me, his teeth white in a face more tanned than I remember. He's leaning on the table I'm wiping, wearing a gray T-shirt, old jeans cut off above the knees, running shoes with no socks. He looks healthier than he did in the winter. I've never seen him in the daytime before.

I'm conscious of my stained uniform: do I smell of underarm sweat, of chicken fat? "How did you get in here?" I say.

"Walked," he says. "How about a coffee?"

He has a summer job, with the Works Department, filling in potholes in the roads, tarring over the cracks made by frost heave; he does have a faint tarry smell about him. He's not what you would call clean. "How about a beer, later?" he says. This is a thing he's said often before: he wants a passport to the Ladies and Escorts, as usual. I'm not doing anything, so I say, "Why not? But I'll have to change."

After work I take the precaution of a shower and put on my purple dress. I meet him at the Maple Leaf and we go into the Ladies and Escorts. We sit there in the gloom, which is at least cool, and drink draft beer. It's awkward with just him: before there was always a group of them. Jon asks me what I've been up to and I say nothing much. He asks me if I've seen Uncle Joe around anywhere, and I say no.

"Probably he's disappeared into Susie's knickers," he says. "The lucky shit." He's still treating me like an honorary boy, still saying crude things about women. I'm surprised at the word *knickers*. He must have picked it up from Colin the Englishman. I wonder if he knows about me as well, whether he's making remarks about my knickers behind my back. But how could he?

He says the Works Department is good money, but he doesn't let on to the other guys that he's a painter, especially not to the old regulars. "They might think I'm a fruit or something," he says.

I drink more draft beer than I should, and then the lights flicker on and off and it's closing time. We walk out onto the hot night summer street, and I don't want to go home by myself.

"Can you get back all right?" says Jon. I say nothing. "Come on, I'll walk you," he says. He puts his hand on my shoulder and I smell his smell of tar and outdoors dust and sunny skin, and I begin to cry. I stand in the street, with the drunks staggering out of the Men Only, my hands pressed to my mouth, crying and feeling stupid.

Jon is startled. "Hey, pal," he says, patting me awkwardly. "What's wrong?"

"Nothing," I say. Being called *pal* makes me cry even more. I feel like a wet sock; I feel ugly. I hope he will think I've had too much to drink.

He puts an arm around me, gives a squeeze. "Come on," he says. "We'll go for coffee."

I stop crying as we walk along the street. We walk to a door beside a wholesale suitcase store, he takes out a key, and we go up the stairs in the dark. Inside the upstairs door he kisses me, with his tarry, beery mouth. There are no lights on. I put my arms around his waist and hold on as if I'm sinking into mud, and he lifts me like that and carries me through the dark room, bumping into the walls and furniture, and we fall together onto the floor.

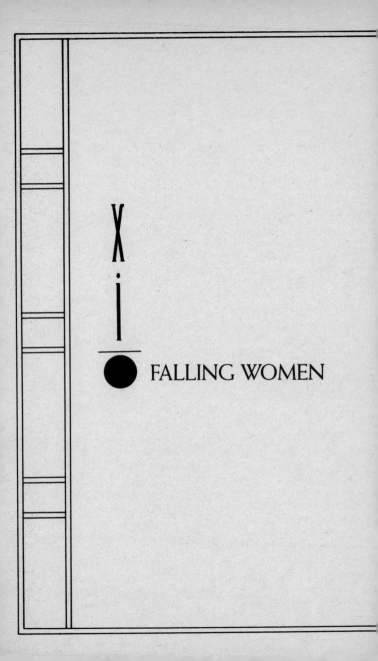

XI

FALLING WOMEN

56

I continue east along Queen Street, still a little dizzy from the wine at lunch. Tipsy was once the word. Alcohol's a depressant, it will let me down later, but right now I'm jaunty, I hum to myself, mouth slightly open.

Right here there's a group of statues, coppery-green, with black smears running down them like metal blood: a seated woman, holding a scepter, with three young soldiers marching forward grouped around her, their legs wound with bandagelike puttees, defending the Empire, their faces earnest, doomed, frozen into time. Above them on a stone tablet stands another woman, this time with angel wings: Victory or Death, or maybe both. This monument is in honor of the South African War, ninety years ago, more or less. I wonder if anyone remembers that war, or if anyone in all these cars barging forward ever even looks.

I head north on University Avenue, past the sterility of hospitals, along the old route of the Santa Claus Parade. The Zoology Building has been torn down, it must have been years ago. The window ledge where I once watched the soggy fairies and chilblained snowflakes, breathing in the smell of snakes and antiseptic and mice, is now empty air. Who else remembers where it used to be?

There are fountains up and down this roadway now, and squared-off beds of flowers, and new, peculiar statues. I follow the curve around the Parliament Building with its form of a squatting Victorian dowager, darkish pink, skirts huffed out, stolid. The flag I could never draw, demoted to the flag of a province, flies before it, bright scarlet, with the Union Jack in the top corner and all those impossible beavers and leaves encrested lower down. The new national flag flutters there as well, two red bands and a red maple leaf rampant on white, looking like a trademark for margarine of the cheaper variety, or an owl kill in snow. I still think of this flag as new, although they changed it long ago.

I cross the street, cut in behind a small church, left stranded here when they redeveloped. Sunday's sermon is announced on a billboard identical to the kind for supermarket specials: *Believing Is Seeing*. A vertical wave of plate glass breaks against it. Behind the polished façades, bouquets of teased cloth, buffed leather, cunning silver trinkets. Pasta to die for. Theology has changed, over the years: *just deserts* used to be what everyone could expect to get, in the end. Now it's a restaurant specializing in cakes. All they had to do was abolish guilt, and add an S.

I turn a corner, onto a side street, a double row of expensive boutiques: hand knits and French maternity outfits and ribbon-covered soaps, imported tobaccos, opulent restaurants where the wineglasses are thin-stemmed and they sell you location and overhead. The designer jeans emporium, the Venetian paper knickknack shop, the stocking boutique with its kicking neon leg.

These houses used to be semihovels; Josef's old territory, where beer-saturated fat men sat on the front porches, sweating in the August heat, while their children screamed and their dogs lay panting with frayed ropes tethering them to the fence, and paint peeled from their woodwork and the dispirited cat pee marigolds wilted along their cracked walkways. A few thousand dollars in the right place then and you'd be a millionaire today, but who could have guessed? Not me, going up the narrow stairs to Josef's second floor, with my breath quickening and his hand weighing on the small of my back, in the dying light of summer evenings: slow-paced, forbidden, sadly delicious.

I know more things about Josef now than I did then. I know them because I'm older. I know about his melancholy, his ambition, his desperation, the corners of emptiness in him that needed to be filled. I know the dangers.

What for instance was he doing with two women fifteen years younger than himself? If one of my daughters fell in love with such a man, I'd be frantic. It would be like the time Sarah and her best friend came rushing home from school, to tell me they'd seen their first flasher in the park. "Mummy, Mummy, a man had his pants down!"

To me it meant fear, and a ferocious anger. *Touch them and I'll kill you.* But to them it was merely noteworthy, and hilarious.

Or the first time I saw my own kitchen, after I had Sarah. I brought her home from the hospital and thought: *All those knives. All those sharp things and hot things.* All I could see was what might hurt her.

Maybe one of my daughters has a man like Josef, or a man like Jon, hidden away in her life, in secret. Who knows what grubby or elderly boys they are bending to their own uses, or to counterpoint me? All the while protecting me from themselves, because they know I would be horrified.

I see words on the front pages of newspapers that never used to be said out loud, much less printed—*sexual intercourse, abortion, incest*—and I want to hide their eyes, even though they are grown-up, or what passes for it. Because I am a mother, I am capable of being shocked; as I never was when I was not one.

I should get a little present for each of them, as I always did when they were younger and I went away. Once I knew by instinct what they would like. I don't any more. It's hard for me to remember exactly what age they've reached. I used to resent it when my mother would forget I was an adult, but I'm approaching the maundering phase myself, digging out the yellowing baby pictures, mooning over locks of hair.

I'm squinting into a window at some Italian silk scarves, wonderful indeterminate colors, gray-blue, sea-green, when I feel a touch on my arm, a chilly jump of the heart.

"Cordelia," I say, turning.

But it's not Cordelia. It's nobody I know. It's a woman, a girl really, Middle Eastern of some kind: a long full skirt to above the ankles, printed cotton, Canadian gum-soled boots incongruous beneath; a short jacket buttoned up, a kerchief folded straight across the forehead with a pleat at either side, like a wimple. The hand that touches me is lumpy in its northern mitten, the skin of the wrist between mitten and jacket cuff brownish, like coffee with double cream. The eyes are large, as in painted waifs.

"Please," she says. "They are killing many people." She doesn't say where. It could be a lot of places, or in between places; homelessness is a nationality now. Somehow the war never ended after all, it just broke up into pieces and got scattered, it gets in everywhere, you can't shut it out. Killing is endless now, it's an industry, there's money in it, and the good side and the bad side are pretty hard to tell apart.

"Yes," I say. This is the war that killed Stephen.

"Some are here. They have no, they have nothing. They would be killed . . ."

"Yes," I say. "I see." This is what I get for walking. In a car you're more insulated. And how do I know she is what she purports to be? She could be a dope addict. In the soft touch market, scams abound.

"I have with me a family of four. Two children. They are with me, it is my, it is my own responsibility." She stumbles a little on *responsibility*, but she gets it out. She's shy, she doesn't like what she's doing, this grabbing people on the street.

"Yes?"

"I am doing it." We look at each other. She is doing it. "Twenty-five dollars can feed a family of four for a month."

What can they be eating? Stale bread, cast-off doughnuts? Does she mean a week? If she can believe this, she deserves my money. I take off my glove, raid my purse, rustle bills, pink ones, blue ones, purple. It's obscene to have such power; also to feel so powerless. Probably she hates me.

"Here," I say.

She nods. She's not grateful, merely confirmed, in her opinion of me, or of herself. She takes off her bulky-knit mitten to receive the cash. I look at our hands, her smooth one, the nails pale moons, mine with its tattered cuticles, its skin of incipient toad. She tucks the bills in between the buttons of her jacket. She must have a purse in there, out of the reach of snatching. Then she slips on the mitt, dark red with a pink wool embroidered leaf.

"God will bless you," she says. She doesn't say Allah. Allah I might believe.

I walk away from her, pulling on my glove. Every day there's more of it, more of that silent wailing, those starving outstretched hands, *need need*, *help help*, there's no end.

57

In September I leave the Swiss Chalet and return to school. I also return to the cellar of my parents' house, because I can't afford not to. Both of these locations are hazardous: my life is now multiple, and I am in fragments. But I'm no longer lethargic. On the contrary I am alert, I crackle with adrenaline, despite the late-summer heat. It's treachery that does this for me, keeping on top of my own deceptions: I need to hide Josef from my parents, and Jon from all of them. I sneak around, heart in my mouth, dreading revelations; I avoid late nights, I evade and tiptoe. Strangely enough, this does not make me feel more insecure, but safer.

Two men are better than one, or at least they make me feel better. I am in love with both, I tell myself, and having two means that I don't have to make up my mind about either of them.

Josef offers me what he has always offered, plus fear. He tells me, casually, in the same way he told me about shooting a man in the head, that in most countries except this one a woman belongs to a man: if a man finds his woman with another man, he kills both of them and everyone excuses him. He says nothing about what a woman does, in the case of another woman. He tells me this while running his hand up my arm, over my shoulder, lightly across the neck, and I wonder what he suspects. He has taken to demanding speech from me; or else he puts his hand over my mouth. I close my eyes and feel him as a source of power, nebulous and shifting. I suspect there would be something silly about him, if I could see him objectively. But I can't.

• • •

As for Jon, I know what he offers. He offers escape, running away from the grown-ups. He offers fun, and mess. He offers mischief.

I consider telling him about Josef, to see what would happen. But the danger in this would be of a different order. He would laugh at me for sleeping with Josef, whom he considers ridiculous as well as old. He would not understand how I could take such a man seriously, he would not understand the compulsion. He would think less of me.

Jon's apartment over the luggage store is long and narrow and smells of acrylic and used socks, and has only two rooms plus the bathroom. The bathroom is purple, with red footprints painted up the wall, across the ceiling and down the opposite wall. The front room is painted stark white, the other one—the bedroom—is glossy black. Jon says this is to get back at the landlord, who is a prick. "When I move out, it'll take him fifteen coats to cover that up," he says.

Sometimes Jon lives in this apartment by himself; sometimes another person will be there, sometimes two, camping out on the floor in sleeping bags. These are other painters, on the lam from irate landlords or between odd jobs. When I ring the downstairs doorbell I never know who will open the door or what will be going on: the morning remains of an all-night party, a multiple argument, someone tossing their cookies in the toilet. "Tossing their cookies" is what Jon calls it. He thinks it's funny.

Different women pass me on the stairs, going up or down; or they are found hovering around the far end of the white room, where there's an improvised kitchen consisting of a hot plate and an electric kettle. It's never clear who these women are paired with; occasionally they are other art students, dropping in to talk. They don't talk much to one another though. They talk to the men, or are silent.

Jon's pictures hang in the white room or are stacked against the walls. They change almost weekly: Jon is productive. He paints very swiftly, in violent eye-burning acrylics, reds and pinks and purples, in frenzied loops and swirls. I feel I should admire these paintings, because I'm incapable of painting that way myself, and I do admire them, in monosyllables. But secretly I don't like them very much: I've seen things like this beside the highway, when something's been run over.

However, the pictures are not supposed to be pictures of anything you would recognize. They are a moment of process, trapped on the canvas. They are pure painting.

Jon is big on purity, but only in art: it doesn't apply to his housekeeping, which is an exuberant protest against all mothers and especially his own. He washes the dishes, when he washes them, in the bathtub, where scraps of crust and kernels of canned corn are to be seen caught in the drain. His living room floor is like a beach after the weekend. His bedsheets are a moment of process in themselves, but a moment that has gone on for some time. I prefer the top of his sleeping bag, which is less septic. The bathroom is like the bathrooms of service stations, on out-of-the-way roads, up north: a brown ring around the toilet bowl, which is likely to contain floating cigarette butts, handprints on the towels, if any, nondescript pieces of paper here and there on the floor.

At the moment I make no moves toward cleanliness. To do so would be to overstep the bounds, and to display a bourgeois lack of cool. "What are you, my mom?" I've heard him say, to one of the hovering women who was making feeble attempts to corral some of the moldier clutter. I don't want to be his mom, but rather a fellow conspirator.

Making love with Jon is not the leisurely, agonizing trance it is with Josef, but rambunctious, like puppies in mud. It's dirty, as in street fighting, as in jokes. Afterward we lie on top of his sleeping bag, eating potato chips out of the bag and giggling about nothing. Jon doesn't think women are helpless flowers, or shapes to be arranged and contemplated, as Josef does. He thinks they are smart or stupid. These are his categories. "Listen, pal," he says to me. "You've got more brains than most." This pleases me, but also dismisses me. I can take care of myself.

Josef begins asking me where I've been, what I've been doing. I am casual and sly. I hold Jon against him like an ace: if he can be duplicitous, then so can I. But he does not talk about Susie any more.

The last time I saw her was in late August, before I left the Swiss Chalet. She came in and had dinner by herself, a half chicken and some Burgundy Cherry ice cream. She'd been neglecting her hair, which was darker and straighter; her body had grown stubby, her face round. She ate in a mechanical way, as if eating was a chore, but she finished everything. It could be that she was eating for consolation, because of Josef: whatever else might happen, he would never marry her, and she must have known that. I assumed she was there to talk to me about him and I evaded her, brushing her away with a neutral smile. Her table wasn't one of mine.

But before she left she walked right up to me. "Have you seen Josef?" she asked. Her voice was plaintive, which annoyed me.

I lied, not well. "Josef?" I said, flushing. "No. Why would I?"

"I just thought you might know where he is," she said. She wasn't reproachful, but hopeless. She walked out, slumping like a middle-aged woman. With such an ass end, I thought, no wonder Josef's keeping away. He didn't like scrawny women but there was a limit in the other direction too. Susie was letting herself go.

Now, however, she calls me. It's late afternoon and I'm studying in the cellar when my mother summons me to the phone.

Susie's voice on the line is a soft, desperate wail. "Elaine," she says. "Please come over."

"What's the matter?" I say.

"I can't tell you. Just come over."

Sleeping pills, I think. That would be her style. And why me, why hasn't she phoned Josef? I feel like slapping her.

"Are you all right?" I say.

"No," she says, her voice rising. "I'm not all right. Something's gone wrong."

It doesn't occur to me to call a taxi. Taxis are for Josef; I'm used to going everywhere on buses and streetcars, and the subway. It takes me nearly an hour to get over to The

Monte Carlo. Susie didn't tell me her apartment number and I didn't think to ask, so I have to locate the superintendent. When I knock on the door, nobody answers, and I resort to the superintendent again.

"I know she's in there," I say, when he's reluctant to unlock the door for me. "She called me. It's an emergency."

When I finally get in, the apartment is dark; the drapes are drawn, the windows are closed, and there's an odd smell. Clothes are scattered here and there, jeans, winter boots, a black shawl I've seen Susie wearing. The furniture looks as if it's been picked out by her parents: a square-armed off-green sofa, a wheat-colored carpet, a coffee table, two lamps with the cellophane still on the shades. None of it goes with Susie as I've imagined her.

On the carpet there's a dark footprint.

Susie is behind the curtain that closes off the sleeping area. She's lying on the bed in her pink nylon shortie nightie, white as an uncooked chicken, eyes closed. The top covers of the bed and the pink tufted spread are on the floor. Underneath her, across the sheet, is a great splotch of fresh blood, spreading out like bright red wings to either side of her.

Desolation sweeps through me: I feel, for no good reason, that I have been abandoned.

Then I feel sick. I run into the bathroom and throw up. It's worse because the toilet bowl is dark red with blood. There are footprints of blood on the white and black tiled floor, fingerprints on the sink. The wastebasket is crammed with sopping sanitary pads.

I wipe my mouth on Susie's baby-blue towel, wash my hands in the blood-spattered sink. I don't know what to do next; whatever this is, I don't want to be involved. I have the fleeting, absurd idea that if she's dead I will be accused of murder. I think of sneaking out of the apartment, closing the door behind me, covering my tracks.

Instead I go back to the bed and feel Susie's pulse. I know that this is what you're supposed to do. Susie is still alive.

I find the superintendent, who calls an ambulance. I also call Josef, who is not there.

I ride to the hospital with Susie, in the back of the ambulance. She is now semiconscious, and I hold her hand, which is cold and small. "Don't tell Josef," she whispers to me. The pink nightie brings it home to me: she is none of the things I've thought about her, she never has been. She's just a nice girl playing dress-ups.

But what she's done has set her apart. It belongs to the submerged landscape of the things that are never said, which lies beneath ordinary speech like hills under water. Everyone my age knows about it. Nobody discusses it. Rumors are down there, kitchen tables, money exchanged in secret; evil old women, illegal doctors, disgrace and butchery. Down there is terror.

The two attendants are casual, and scornful. They have seen this before.

"What'd she use, a knitting needle?" one says. His tone is accusing: he may think I was helping her.

"I have no idea," I say. "I hardly even know her." I don't want to be implicated.

"That's what it usually is," he says. "Stupid kids. You'd think they'd have more sense."

I agree with him that she's been stupid. At the same time I know that in her place I would have been just as stupid. I would have done what she has done, moment by moment, step by step. Like her I would have panicked, like her I would not have told Josef, like her I would not have known where to go. Everything that's happened to her could well have happened to me.

But there is also another voice; a small, mean voice, ancient and smug, that comes from somewhere deep inside my head: *It serves her right.*

Josef, when he is finally located, is devastated. "The poor child, the poor child," he says. "Why didn't she tell me?"

"She thought you'd get mad at her," I say coldly. "Like her parents. She thought you'd kick her out, for getting pregnant."

Both of us know this is a possibility. "No, no," Josef says uncertainly. "I would have taken care of her." This could mean several things.

He calls the hospital, but Susie refuses to see him. Something has changed in her, hardened. She tells him she might never be able to have babies. She doesn't love him. She doesn't want to see him ever again.

Now Josef wallows. "What have I done to her?" he moans, tugging his hair.

He becomes more melancholy than ever; he doesn't want to go out for dinner, he doesn't want to make love. He stays in his apartment, which is no longer neat and empty but is filling up with disorganized parts of his life: take-out Chinese food containers, unwashed sheets.

He says he will never get over it, what he has done to Susie. This is how he thinks of it: something he's done, to Susie, to her inert and innocent flesh. At the same time he has been wounded by her: how can she treat him like this, cut him out of her life?

He expects me to console him, for his own guilt and the damage that's been done to him. But I am not good at this. I am beginning to dislike him.

"It was my child," he says.

"Would you have married her?" I ask. The spectacle of his suffering does not make me compassionate, but ruthless.

"You are cruel to me," says Josef. This was something he used to say before, in a sexual context, teasing. Now he means it. Now he is right.

Without Susie, whatever has been keeping us in equilibrium is gone. The full weight of Josef rests on me, and he is too heavy for me. I can't make him happy, and I resent my failure: I am not enough for him, I am inadequate. I see him as weak now, clinging, gutted like a fish. I can't respect a man who can allow himself to be reduced to such rubble by women. I look at his doleful eyes and feel contempt.

I make excuses, over the phone. I tell him I am very busy. One evening I stand him up. This is so deeply gratifying that I do it again. He tracks me down at the university, rumpled and unshaven and suddenly too old, and pleads with me as I walk between classes. I'm angered by this overlap of worlds.

"Who was that?" say the girls in the cashmere twin sets.
"Just someone I used to know," I say lightly.

Josef waylays me outside the museum and announces I have
driven him to despair: because of the way I've treated him,
he is leaving Toronto forever. He does not fool me: he was
planning to do this anyway. My mean mouth takes over.

"Good," I say.

He gives me a pained, reproachful stare, drawing himself
up into the proud, theatrical, poker-up-the-bum stance of
a matador.

I walk away from him. It's enormously pleasing to me,
this act of walking away. It's like being able to make people
appear and vanish, at will.

I do not dream about Josef. Instead I dream about Susie,
in her black turtleneck and jeans but shorter than she really
is, her hair cut into a pageboy. She's standing on a street I
know but do not recognize, among piles of smoldering autumn leaves, holding a coiled skipping rope, licking one half
of an orange Popsicle.

She is not drained and boneless, as I've last seen her.
Instead she is sly-eyed, calculating. "Don't you know what
a twin set is?" she says spitefully.

She continues to lick her Popsicle. I know I have done
something wrong.

58.

Time passes, and Susie fades. Josef does not reappear.

This leaves me with Jon. I have the sense that, like one of a pair of bookends, he is incomplete by himself. But I feel virtuous, because I'm no longer hiding anything from him. This makes no difference to him, however, since he didn't know I was hiding anything in the first place. He doesn't know why I am less casual about what he does with the rest of his time.

I decide I'm in love with him. Though I am too cagey to say it: he might object to the vocabulary, or think he's being pinned down.

I still go over to his long white and black apartment, still end up on top of his sleeping bag, although haphazardly: Jon isn't big on planning in advance, or on remembering. Sometimes when I arrive at his downstairs door there's no answer. Or else his phone gets cut off because he hasn't paid the bill. We are a couple, in a way, though nothing is explicit between us. When he's with me he's with me: that's about as far as he'll go in his definition of what is not yet called our relationship.

There are murky, smoky parties, with the lights turned out and candles flickering in bottles. The other painters are there, and assorted turtlenecked women, who have begun to appear in long, straight hair, parted in the middle. They sit in clumps, on the floor, in the dark, listening to folksongs about women being stabbed with daggers, and smoking marijuana cigarettes, which is what people do in New York. They refer to these as "dope" or "pot," and claim they loosen up your art.

Cigarettes of any kind make me choke, so I don't smoke them. Some nights I wind up in the back hall with one or another of the painters, because I would rather not see what Jon may be getting up to with the straight-haired girls. Whatever it is, I wish he would do it in secret. But he doesn't feel the need to hide anything: sexual possessiveness is bourgeois, and just a hangover from notions about the sanctity of private property. Nobody owns anybody.

He doesn't say all this. All he says is, "Hey, you don't own me."

Sometimes the other painters are merely stoned, or drunk, but sometimes they want to tell me their problems. They do this fumblingly, in starts and stops, in short words. Their problems are mostly about their girlfriends. Soon they will be bringing me their socks to darn, their buttons to sew on. They make me feel like an aunt. This is what I do instead of jealousy, in which there is no future. Or so I think.

Jon has given up on his paintings of swirls and innards. He says they are too romantic, too emotional, too sloppy, too sentimental. Now he's doing pictures in which all the shapes are either straight lines or perfect circles. He uses masking tape to get the lines straight. He works in blocks of flat color, no impasto showing.

He calls these paintings things like *Enigma: Blue and Red*, or *Variation: Black and White*, or *Opus 36*. They make your eyes hurt when you look at them. Jon says this is the point.

In the daytime I go to school.

Art and Archaeology is murkier and more velvety than last year, and filled with impasto and chiaroscuro. There are still Madonnas, but their bodies have lost their previous quality of suffused light and are more likely to be seen at night. There are still saints, though they no longer sit in quiet rooms or deserts, with their memento mori skulls and their doglike lions resting at their feet; instead they writhe in contorted poses, stuck full of arrows or tied to stakes. Biblical subjects tilt toward violence: Judith cutting off the head of Holofernes is now popular. There are a lot more classical gods and goddesses. There are wars, fights and slaughters, as before, but more confused and with intertwined arms and legs. There are still portraits of rich people, although in darker clothing.

As we run through the centuries, new things appear: ships by themselves, animals by themselves, such as dogs and horses. Peasants by themselves. Landscapes, with or without houses. Flowers by themselves, plates of fruit and cuts of meat, with or without lobsters. Lobsters are a favorite, because of the color.

Naked women.

There is considerable overlap: a naked goddess wreathed in flowers, with a couple of dogs standing by; biblical people with or without clothes, plus or minus animals, trees, and ships. Rich people pretending to be gods and goddesses. Fruit and slaughters are not usually combined, nor are gods and peasants. The naked women are presented in the same manner as the plates of meat and dead lobsters, with the same attention to the play of candlelight on skin, the same lusciousness, the same sensuous and richly rendered detail, the same painterly delight in tactility. (*Richly rendered*, I write. *Painterly delight in tactility*.) They appear served up.

I don't like these shadowy, viscous pictures. I prefer the earlier ones, with their daytime clarity, their calm arrested gestures. I have given up, too, on oil paints; I have come to dislike their thickness, their obliteration of line, their look of licked lips, the way they call attention to the brushstrokes of the painter. I can make nothing of them. What I want instead is pictures that seem to exist of their own accord. I want objects that breathe out light; a luminous flatness.

I draw with colored pencils. Or I paint in egg tempera, the technique of monks. Nobody teaches this any more, so I hunt through the library, searching for instructions. Egg tempera is difficult and messy, painstaking and, at first, heartbreaking. I muck up my mother's kitchen floor and pots, cooking the gesso, and ruin panel after panel before I can work out how to paint it on for a smooth working surface. Or I forget about my bottles of egg yolk and water, which go bad and stink up the cellar with a smell like sulfur. I use up a lot of egg yolks. The whites I separate carefully, and take upstairs to my mother, who makes them into meringue cookies.

I draw beside the picture window in the living room upstairs, when there is nobody home, or in the daylight from the cellar window. At night I use two gooseneck lamps, each of which takes three bulbs. None of this is adequate, but it's all I can manage. Later, I think, I will have a large studio, with skylights; though what I will paint in it is far from clear. Whatever it is will appear, even later, in colored plates, in books; like the work of Leonardo da Vinci, whose studies of hands and feet and hair and dead people I pore over.

I become fascinated with the effects of glass, and of other light-reflecting surfaces. I study paintings in which there are pearls, crystals, mirrors, shiny details of brass. I spend a long time over Van Eyck's *The Arnolfini Marriage*, going over the inadequate color print of it in my textbook with a magnifying glass; what fascinates me is not the two delicate, pallid, shoulderless hand-holding figures, but the pier glass on the wall behind them, which reflects in its convex surface not only their backs but two other people who aren't in the main picture at all. These figures reflected in the mirror are slightly askew, as if a different law of gravity, a different arrangement of space, exists inside, locked in, sealed up in the glass as if in a paperweight. This round mirror is like an eye, a single eye that sees more than anyone else looking: over this mirror is written, *Johannes de Eyck fuit hic. 1434.* It's disconcertingly like a washroom scribble, something you'd write with spray paint on a wall.

There is no pier glass in our house for me to practice on. So instead I paint ginger ale bottles, wineglasses, ice cubes from the refrigerator, the glazed teapot, my mother's fake pearl earrings. I paint polished wood, and metal: a copper-bottomed frying pan, as seen from the bottom, an aluminum double boiler. I fiddle over details, hunch over my pictures, dabbing at the highlights with tiny brushes.

I'm aware that my tastes are not fashionable, and so I pursue them in secret. Jon, for instance, would call this illustration. Any picture that's a picture of something recognizable is illustration, as far as he's concerned. There is no spontaneous energy in this kind of work, he would say. No process. I might as well be a photographer, or Norman

Rockwell. Some days I agree with him, because what have I done so far? Nothing that doesn't look like a random sampling from the Housewares Department of the *Eaton's Catalogue*. But I keep on.

On Wednesday evenings I take another night course: not Life Drawing, which is taught this year by an excitable Yugoslavian, but Advertising Art. The students are quite different from the Life Drawing bunch. They're mostly from the Commercial division of the Art College, not the Fine Arts one. Again they're mostly boys. Some of them have serious artistic ambitions, but they don't drink as much beer. They're cleaner and more earnest, and they want paying jobs when they graduate. So do I.

The teacher is an elderly man, thin and defeated-looking. He thinks he has failed in the real world, although he once created a famous illustration for canned pork and beans that I can remember from childhood. We ate a lot of canned pork and beans, during the war. His specialty is the rendering of smiles: the trick is to be able to do teeth, nice white even teeth, without putting in the separation between each tooth, which makes the smile appear too canine or too much like false teeth (which he himself has). He tells me I show ability in smiles, and that I could go far.

Jon teases me a little about this night course, but not as much as I thought he would. He refers to the teacher as Mr. Beanie Weenie, and lets it go at that.

59

I graduate from university, and discover that there's nothing much I can do with my degree. Or nothing I want to do at any rate. I don't want to go on to graduate work, I don't want to teach high school or be a curator's flunky in a museum.

By this time I've accumulated five night courses from the Art College, four of them in the Commercial area, and I trot them and my portfolio of smiles and dishes of caramel pudding and canned peach halves around to various ad agencies. For these purposes I buy a beige wool suit (on sale), medium-heel pumps to match, some pearl button earrings and a tasteful silk scarf (on sale) at Simpsons; this on the recommendation of my last night course instructor, in Layout and Design, who was a woman. She also recommended a haircut, but I would only go so far as a French roll, engineered with the help of some big rollers and hair-setting gel and a lot of bobby pins. Eventually I get a menial job doing mock-ups, and a small furnished two-bedroom apartment with kitchenette and separate entrance in a large crumbling house in the Annex, north of Bloor. I use the second bedroom for painting, and keep the door to it closed.

This place has a real bed, and a real kitchen sink. Jon comes for dinner and teases me about the towels I've bought (on sale), the ovenproof dishes I've acquired, my shower curtain. "*Better Homes and Gardens*, eh?" he says. He teases me about the bed, but he likes sleeping in it. He comes to my place, now, more often than I go to his.

My parents sell their house and move up north. My father has left the university and has gone back to research; he's now head of the Forest Insect Laboratory at Sault Ste.

Marie. He says Toronto is getting overpopulated, and also polluted. He says the lower Great Lakes are the world's largest sewer and that if we knew what was going into the drinking water we would all become alcoholics. As for the air, it's so full of chemicals we should be wearing gas masks. Up north you can still breathe.

My mother was not too happy to leave her garden, but made the best of it: "At least it's a chance to throw out a lot of that junk in the cellar," she said. They've started another garden in the Soo, although the growing season is shorter. In the summers though they're mostly on the road, driving from infestation to infestation. There is no shortage of insect life.

I don't miss my parents. Not yet. Or rather I don't want to be living with them. I am happy to be left to my own devices, my own messes. I can eat haphazardly now, snack on junk food and takeouts without worrying about balanced meals, go to bed when I like, let my dirty laundry rot, neglect the dishes.

I get a promotion. After a time I move to the art department of a publishing company, where I design book covers. At night, when Jon is not there, I paint. Sometimes I forget to go to bed, and find that it has become dawn and I have to change into my work clothes and go to work. I am groggy on those days, and have trouble hearing what is said to me; but nobody seems to notice.

I get postcards and the occasional short letter from my mother, sent from places like Duluth and Kapuskasing. She says the roads are getting too crowded. "Too many trailers," she says. I reply with news about my job, my apartment, and the weather. I don't mention Jon, because there is no news. News would be something definite and respectable, such as an engagement.

My brother Stephen is here and there. He has become more taciturn: he too now communicates by postcard. One comes from Germany, with a man in short leather pants on it and the message: *Great particle accelerator*; one from Nevada, with a cactus and the note, *Interesting life forms*. He goes to Bolivia on what I suppose to be a holiday, and sends a cigar-smoking woman in a high-crowned hat: *Excellent butterflies. Hope you are well.* At one point he gets married, which is announced by a postcard from San Francisco, with the Golden Gate Bridge and a sunset on it, and, *Got married. Annette sends regards.* This is all I hear about it until several years later, when he sends a postcard of the Statue of Liberty from New York, which says: *Got divorced.* I assume he has been puzzled by both events, as if they're not something he's done himself, on purpose, but things that have happened to him accidentally, like stubbing your toe. I think of him as walking into marriage as into a park, in a foreign country, at night, unaware of the possibilities for damage.

He turns up in Toronto to give a lecture, at a conference, notifying me in advance with a postcard of a statue of Paul Revere, from Boston: *Arrive Sun. 12th. My paper is on Mon. See you.*

I attend the lecture, not because I have high hopes for it on my own account—the title of it is "The First Pico-seconds and the Quest for a Unified Field Theory: Some Minor Speculations"—but because he is my brother. I sit nibbling my fingers as the university auditorium fills with the audience, which is composed largely of men. Most of them look like people I wouldn't have gone out with in high school.

Then my brother comes in, with the man who will introduce him. I haven't seen my brother for years; he's thinner, and his hair is beginning to recede. He needs glasses to read his text; I can see them poking out of his breast pocket. Someone has upgraded his wardrobe for him and he's wearing a suit and tie. These alterations don't make him appear more normal, however, but more anomalous, like a creature from an alien planet disguised in human clothing. He has a look of amazing brilliance, as if at any minute his head will light up and become transparent, disclosing a huge brightly colored brain inside. At the same time he looks rumpled and bewildered, as if he's just wakened from a pleasant dream to find himself surrounded by Munchkins.

The man introducing my brother says he needs no introduction, then goes on to list the papers he has written, the awards he has won, the contributions he has made. There is clapping, and my brother goes to the podium. He stands in front of a white projection screen, clears his throat, shifts from one foot to the other, puts on his glasses. Now he looks like someone who will turn up, later, on a stamp. He is ill at ease and I am nervous for him. I think he will mumble. But once he begins he is fine.

"When we gaze at the night sky," he says, "we are looking at fragments of the past. Not only in the sense that the stars as we see them are echoes of events that occurred light-years distant in time and space: everything up there and indeed everything down here is a fossil, a leftover from the first picoseconds of creation, when the universe crystallized out from the primal homogeneous plasma. In the first picosecond, conditions were scarcely imaginable. If we could travel in a time machine back toward this explosive moment, we would find ourselves in a universe replete with energies we do not understand and strangely behaving forces distorted beyond recognition. The farther back we probe, the more extreme these conditions become. Current experimental facilities can take us only a short way along this path. Beyond that point, theory is our only guide." After this he continues, in a language that sounds like English but is not, because I can't understand one word of it.

Luckily there is something to look at. The room darkens and the screen lights up, and there is the universe, or parts of it: the black void punctuated by galaxies and stars, white-hot, blue-hot, red. An arrow moves among them on the screen, searching and finding. Then there are diagrams and strings of numbers, and references to things that everyone here seems to recognize except me. There are, apparently, a great many more dimensions than four.

Murmurs of interest ripple through the room; there are whisperings, the rustling of paper. At the end, when the lights have come on again, my brother returns to language. "But what of the moment beyond the first moment?" he says. "Or does it even make sense to use the word *before*, since time cannot exist without space and space-time without events and events without matter-energy? But there is something that must have existed before. That something is the theoretical framework, the parameters within which the laws of energy must operate. Judging from the scanty but mounting evidence now available to us, if the universe was created with a *fiat lux*, that fiat must have been expressed, not in Latin, but in the one truly universal language: mathematics." This sounds a lot like metaphysics to me, but the men in the audience don't seem to take it amiss. There is applause.

I go to the reception afterward, which offers the usual university fare: bad sherry, thick tea, cookies out of a package. The numbers men murmur in groups, shake one another's hands. Among them I feel overly visible, and out of place.

I locate my brother. "That was great," I say to him.

"Glad you got something out of it," he says with irony.

"Well, math was never my totally strong point," I say. He smiles benignly.

We exchange news of our parents, who when last I heard from them were in Kenora, and heading west. "Still counting the old budworms, I guess," says my brother.

I remember how he used to throw up by the side of the road, and his smell of cedar pencils. I remember our life in tents and logging camps, the scent of cut lumber and gasoline and crushed grass and rancid cheese, the way we used to sneak around in the dark. I remember his wooden swords with the orange blood, his comic book collection. I see him crouching on the swampy ground, calling *Lie down, you're dead*. I see him dive-bombing the dishes with forks. All my early images of him are clear and sharp and Technicolor: his baggy-legged shorts, his striped T-shirt, his raggedy hair bleached by the sun, his winter breeches and leather helmet. Then there is a gap, and he appears again on the other side of it, unaccountably two years older.

"Remember that song you used to sing?" I say. "During the war. Sometimes you whistled it. 'Coming in on a Wing and a Prayer'?"

He looks perplexed, frowns a little. "I can't say I do," he says.

"You used to draw all those explosions. You borrowed my red pencil, because yours was used up."

He looks at me, not as if he doesn't remember these things himself, but as if he's puzzled that I do. "You can't have been very old then," he says.

I wonder what it was like for him, having a little sister tagging along. For me, he was a given: there was never a time when he didn't exist. But I was not a given, for him. Once he was singular, and I was an intrusion. I wonder if he resented me when I was born. Maybe he thought I was a pain in the bum; there's no doubt that he thought this sometimes. Considering everything and on the whole though, he made the best of me.

"Remember that jar of marbles you buried, under the bridge?" I say. "You would never tell me why you did it." The best ones, the red and blue puries, the waterbabies and cat's eyes, put into the ground, out of reach. He would have stamped the dirt down on top of the jar, and scattered leaves.

"I think I recall that," he says, as if not entirely willing to be reminded of his former, younger self. It disturbs me that he can remember some of these things about himself, but not others; that the things he's lost or misplaced exist now only for me. If he's forgotten so much, what have I forgotten?

"Maybe they're still down there," I say. "I wonder if anyone ever found them, when they built that new bridge. You buried the map, too."

"So I did," he says, smiling in his old, secret, maddening way. He still isn't telling, and I am reassured: despite his changed façade, his thinning hair and provisional suit, he is still the same person underneath.

After he has gone back, to wherever he's going next, I think of getting him a star named after himself, for his birthday. I have seen an advertisement for these: you send in your money, and you get a certificate with a star map, your own star marked on it. Possibly he would find this amusing. But I'm not sure that the word *birthday*, for him, would still have meaning.

60
.

 Jon has given up his eye-damaging geometrical shapes and is painting pictures that look like commercial illustrations: huge Popsicles, giant salt and pepper shakers, peach halves in syrup, paper dishes overflowing with french fries. He does not talk about purity any more but of the necessity of using common cultural sign systems to reflect the iconic banality of our times. I think I could give him a few tips from my own professional experience: his peach halves could be glossier, for instance. But I don't say this.

Increasingly, Jon paints these things in my living room. He's been gradually moving in his things, beginning with the paints and canvas. He says he can't paint at his place because there are too many people in it, which is true: the front room is silting up with American draft dodgers, a shifting population, all of whom seem to be friends of friends. Jon has to step over them to get to the walls, because they lie around on their sleeping bags, forlorn and smoking dope, wondering what to do next. They are depressed because Toronto isn't the United States without a war on, as they thought it would be, but some limbo they have strayed into by accident and can't get out of. Toronto is nowhere, and nothing happens in it.

Jon stays over three or four nights a week. I don't ask what he does on the other nights.

He thinks he is making a large concession, to something he assumes I want. And maybe I do want it. When I'm alone, I let the dishes accumulate in the sink, I allow colored fur to grow in jars of leftovers, I use up all my underpants before washing any of them. But Jon turns into a model of tidiness and efficiency. I get up in the morning and make coffee for him, I set two places at the table, with my newly acquired ovenproof earthenware in off-white, with speckles. I don't even mind doing his laundry at the Laundromat, along with my own.

Jon is not used to having all these clean clothes. "You're the sort of girl who should get married," he says one day, when I appear with a pile of folded shirts and jeans. I think this may be an insult, but I'm not sure.

"Do your own laundry then," I say.

"Hey," he says, "don't be like that."

On Sundays we sleep late, make love, go for walks, holding hands.

One day, when nothing has changed, nothing has been done or happened that is any different from usual, I discover I am pregnant. My first reaction is unbelief. I count and recount, wait another day, then another, listening to the inside of my body as if for a footfall. Finally I slink off to the drugstore with some pee in a bottle, feeling like a criminal. Married women go to their doctors. Unmarried women do this.

The man in the drugstore tells me the results are positive. "Congratulations," he says, with disapproving irony. He can see right through me.

I'm afraid to tell Jon. He will expect me to go and have it out, like a tooth. He will say "it." Or he will want me to sit in the bathtub while he pours boiling water into it; he will want me to drink gin. Or else he will vanish. He's said, often enough, that artists can't live like other people, tied down to demanding families and expensive material possessions.

I think about things I've heard: drinking a lot of gin, knitting needles, coat hangers; but what do you do with them? I think about Susie and her wings of red blood. Whatever it was she did, I will not do it. I am too frightened. I refuse to end up like her.

I go back to my apartment, lie down on the floor. My body is numb, inert, without sensation. I can hardly move, I can hardly breathe. I feel as if I'm at the center of nothingness, of a black square that is totally empty; that I'm exploding slowly outward, into the cold burning void of space.

When I wake up it's the middle of the night. I don't know where I am. I think I'm back in my old room with the cloudy light fixture, in my parents' house, lying on the floor because I've fallen out of bed, as I used to do when we had the army cots. But I know that the house has been sold, that my parents are no longer there. I have somehow been overlooked, left behind.

This is only the end of a dream. I get up, turn on the lights, make myself some hot milk, sit at the kitchen table, shivering with cold.

Until now I've always painted things that were actually there, in front of me. Now I begin to paint things that aren't there.

I paint a silver toaster, the old kind, with knobs and doors. One of the doors is partly open, revealing the red-hot grill within. I paint a glass coffee percolator, with bubbles gathering in the clear water; one drop of dark coffee has fallen, and is beginning to spread.

I paint a wringer washing machine. The washing machine is a squat cylinder of white enamel. The wringer itself is a disturbing flesh-tone pink.

I know that these things must be memories, but they do not have the quality of memories. They are not hazy around the edges, but sharp and clear. They arrive detached from any context; they are simply there, in isolation, as an object glimpsed on the street is there.

I have no image of myself in relation to them. They are suffused with anxiety, but it's not my own anxiety. The anxiety is in the things themselves.

I paint three sofas. One of them is chintz, in dirty rose; one is maroon velvet, with doilies. The one in the middle is apple-green. On the middle cushion of the middle sofa is an egg cup, five times life-size, with a broken eggshell in it.

I paint a glass jar, with a bouquet of nightshade rising out of it like smoke, like the darkness from a genie's bottle.

The stems twist and intertwine, the branches cluster with red berries, purple flowers. Scarcely visible, far back in the dense tangle of the glossy leaves, are the eyes of cats.

In the daytime I go to work, come back, talk, and eat. Jon comes over, eats, sleeps, and goes away. I watch him with detachment; he notices nothing. Every move I make is sodden with unreality. When no one is around, I bite my fingers. I need to feel physical pain, to attach myself to daily life. My body is a separate thing. It ticks like a clock; time is inside it. It has betrayed me, and I am disgusted with it.

I paint Mrs. Smeath. She floats up without warning, like a dead fish, materializing on a sofa I am drawing: first her white, sparsely haired legs without ankles, then her thick waist and potato face, her eyes in their steel rims. The afghan is draped across her thighs, the rubber plant rises behind her like a fan. On her head is the felt hat like a badly done-up package that she used to wear on Sundays.

She looks out at me from the flat surface of paint, three-dimensional now, smiling her closed half-smile, smug and accusing. Whatever has happened to me is my own fault, the fault of what is wrong with me.

Mrs. Smeath knows what it is. She isn't telling.

One picture of Mrs. Smeath leads to another. She multiplies on the walls like bacteria, standing, sitting, flying, with clothes, without clothes, following me around with her many eyes like those 3-D postcards of Jesus you can get in the cheesier corner stores. Sometimes I turn her faces to the wall.

61

I wheel Sarah along the street in her stroller, avoiding the mounds of melting slush. Although she is over two, she still can't walk fast enough in her red rubber boots to keep up when we go shopping. Also this way I can hang the grocery bags from the stroller handle, or tuck them in around her. I know a great many such minor tricks now, involving objects and gadgets and the rearrangement of space, that I didn't need to know before.

We're living in a larger place now, the three of us: the upper two stories of a red brick semidetached house with a sagging wooden square-pillared porch, on a side street west along Bloor. There are a lot of Italians around here. The older women, the married ones and the widows, wear black clothes and no makeup, as I used to do. When I was in the later months of pregnancy, they would smile at me, as if I was almost one of them. Now they smile at Sarah first.

I myself wear miniskirts in primary colors, with tights underneath and boots, and an ankle-length coat over top. I am not entirely satisfied with this clothing. It's hard to sit down in. Also I've put on some weight, since having Sarah. These skimpy skirts and tiny bodices were designed for women a lot skinnier than I am, of which there now seem to be dozens, hundreds: weasel-faced girls with long hair hanging to the place where their bums ought to be, their chests flat as plywood, making me feel bulbous by comparison.

A new vocabulary has come with them. *Far out*, they say. *Cosmic. Blew my mind. Uptight. Let it all hang out.* I consider myself too old for such words: they are for young people, and I am no longer young. I have found a gray hair behind my left ear. In a couple of years I will be thirty. Over the hill.

I wheel Sarah up the walk, unbuckle her, set her at the foot of the porch steps, unhook and lift out the grocery bags, fold up the stroller. I walk Sarah up the steps to the front door: these steps can be slippery. I go back for the bags and the stroller, lug them up the steps, fumble in my purse for the key, open the door, lift Sarah inside, then the bags and stroller, close and lock the door. I walk Sarah up the inside stairs, open the inside door, put her inside, close the baby gate, go back down for the bags, carry them up, open the gate, go in, close the gate, go into the kitchen, set the bags on the table, and begin to unpack: eggs, toilet paper, cheese, apples, bananas, carrots, hot dogs, and buns. I worry about serving too many hot dogs: when I was young they were carnival food, and supposed to be bad for you. You might get polio from them.

Sarah is hungry, so I stop unpacking the groceries to get her a glass of milk. I love her ferociously, and am frequently irritated by her.

For the first year I was tired all the time, and fogged by hormones. But I'm coming out of it now. I'm looking around me.

Jon comes in, scoops Sarah up, gives her a kiss, tickles her face with his beard, carries her squealing off into the living room. "Let's hide on Mummy," he says. He has a way of putting the two of them into the same camp, in pretended league against me, that annoys me more than it should. Also I don't like it when he calls me Mummy. I am not his mummy, but hers. But he too loves her. This was a surprise, and I'm not finished being grateful for it. I don't yet see Sarah as a gift I have given him, but one he has allowed me. It's because of her that we got married, at City Hall, for the oldest of reasons. One that was nearly obsolete. But we didn't know that.

Jon, who is a lapsed Lutheran from Niagara Falls, thought we should go there for our honeymoon. He broke up over the word *honeymoon*. He thought it would be a sort of joke: self-conscious corniness, like a painting of a giant

Coke bottle. "Amazing visuals," he said. He wanted to take me to the waxworks, the flower clock, the *Maid of the Mist*. He wanted us to get satin shirts with our names embroidered on the pockets and NIAGARA FALLS across the back. But I was silently offended by this approach to our marriage. Whatever else we were getting into as the weeks passed one another and my body swelled like a slow flesh balloon, it was not a joke. So we ended by not going.

Right after we were married, I lapsed into a voluptuous sloth. My body was like a feather bed, warm, boneless, deeply comforting, in which I lay cocooned. It may have been the pregnancy, sponging up my adrenaline. Or it may have been relief. Jon glowed for me then like a plum in sunlight, richly colored, perfect in form. I would lie in bed beside him or sit at the kitchen table, running my eyes over him like hands. My adoration was physical, and wordless. I would think *Ah*, nothing more. Like a breath breathed out. Or I would think, like a child, *Mine*. Knowing it wasn't true. *Stay that way*, I would think. But he could not.

Jon and I have begun to have fights. Our fights are secret fights, conducted at night, when Sarah is asleep: a squabbling in undertones. We keep them from her, because if they are frightening to us—as they are—how much more frightening will they be to her?

We thought we were running away from the grown-ups, and now we are the grown-ups: this is the crux of it. Neither of us wants to take it on, not the whole thing. We compete, for instance, over which of us is in worse shape. If I get a headache, he gets a migraine. If his back hurts, my neck is killing me. Neither one of us wants to be in charge of the Band-Aids. We fight over our right to remain children.

At first I do not win these fights, because of love. Or so I say to myself. If I were to win them, the order of the world would be changed, and I am not ready for that. So instead I lose the fights, and master different arts. I shrug, tighten my mouth in silent rebuke, turn my back in bed, leave questions unanswered. I say, "Do it however you like," provoking sullen fury from Jon. He does not want just capitulation, but admiration, enthusiasm, for himself and his ideas, and when he doesn't get it he feels cheated.

Jon has a job now, supervising part-time at a co-op graphics studio. I am part-time as well. Between the two of us, we can manage to cover the rent.

Jon is no longer painting on canvas, or on anything flat. In fact he is no longer painting. Flat surfaces with paint on them he calls "art-on-the-wall." There is no reason for art to be on the wall, there's no reason for it to have a frame around it or paint on it. Instead he is making constructions, out of things he gathers from junk heaps or finds here and there. He makes wooden boxes with compartments, each containing a different item: three pairs of outsize ladies' panties in fluorescent colors, a plaster hand with long false nails glued onto it, an enema bag, a toupee. He makes a motorized furry bedroom slipper that runs around on the floor by itself, and a family of diaphragms fitted up with monster movie eyes and mouths and jumping legs underneath that hop around on the table like radiation-damaged oysters. He's decorated our bathroom in red and orange, with purple mermaids swimming on the walls, and hooked up the toilet seat so that it plays "Jingle Bells" when it's raised. This is for Sarah's benefit. He makes toys for her as well, and lets her play with ends of wood and leftover pieces of cloth and some of his less dangerous tools, while he's working.

That's when he's here. Which is by no means most of the time.

For the first year after Sarah was born I didn't paint at all. I was freelancing then, working at home, and just keeping up with the few book cover assignments I'd taken on was a major effort. I felt clogged, as if swimming with my clothes on. Now that I'm half a day at work, it's better.

I've done some of what I call my own work as well, although hesitantly: my hands are out of practice, my eyes disused. Most of what I do is drawing, because the preparation of the surface, the laborious underpainting and detailed concentration of egg tempera are too much for me. I have lost confidence: perhaps all I will ever be is what I am now.

I'm sitting on a wooden folding chair, on a stage. The curtains are open and I can see the auditorium, which is small, battered, and empty. Also on the stage is a stage set, not yet dismantled, for a play which has just closed. The set consists of the future, which will be sparsely furnished, but will contain a good many cylindrical black columns and several austere flights of stairs.

Arranged around the columns on other wooden chairs, and sitting here and there on the stairs, are seventeen women. Every one of them is an artist, or something like it. There are several actresses, two dancers, three painters besides me. There's one magazine writer, and an editor from my own publishing company. One woman is a radio announcer (daytime classical music), one does puppet shows for children, one is a professional clown. One is a set designer, which is why we're here: she got us the space for this meeting. The reason I know all of this is that we had to say our names, going around the circle, and what we do. Not for a living: for a living is different, especially for the actresses. Also for me.

This is a meeting. It's not the first such meeting I've been to, but I still find it startling. For one thing, it's all women. That in itself is unusual, and has an air of secrecy about it, and an unfocused, attractive dirtiness: the last all-women gathering I was at was Health Class in high school, where the girls were separated off from the boys so they could be told about the curse. Not that the word was used. "Those days" was the accepted, official phrase. It was explained that tampons, although not recommended for young girls, which we knew meant virgins, could not get lost inside you and end up in your lung. There was considerable giggling, and when the teacher spelled *blood*—"B-L-O-O-D"—one girl fainted.

Today there is no giggling or fainting. This meeting is about anger.

Things are being said that I have never consciously thought about before. Things are being overthrown. Why, for instance, do we shave our legs? Wear lipstick? Dress up in slinky clothing? Alter our shapes? What is wrong with us the way we are?

It's Jody asking these questions, one of the other painters. She does not dress up or alter her shape. She wears workboots, and striped coveralls, one leg of which she hauls up to show us the real leg underneath, which is defiantly, resplendently hairy. I think of my own cowardly, naked legs, and feel brainwashed, because I know I cannot go all the way. I draw the line at armpits.

What is wrong with us the way we are is men.

Many things are said about men. Two of these women have been raped, for instance. One has been beaten up. Others have been discriminated against at work, passed over or ignored; or their art has been ridiculed, dismissed as too feminine. Others have begun to compare their salaries with those of men, and have found them to be less.

I have no doubt that all of these things are true. Rapists exist, and those who molest children and strangle girls. They exist in the shadows, like the sinister men who lurk in ravines, not one of whom I have ever seen. They are violent, wage wars, commit murders. They do less work and make more money. They shove the housework off on women.

They are insensitive and refuse to confront their own emotions. They are easily fooled, and wish to be: for instance, with a few gasps and wheezes they can be conned into thinking they are sexual supermen. There are giggles of recognition over this. I begin to wonder if I've been faking orgasm without knowing it.

But I am on shaky ground, in this testifying against men, because I live with one. Women like me, with a husband, a child, have been referred to with some scorn as *nukes*, for *nuclear family*. *Pronatalist* is suddenly a bad word. There are some other nukes in this group, but they are not in the majority and say nothing in their own defence. It seems to be worthier to be a woman with a child but no man. That way you've paid your dues. If you stay with the man, whatever problems you are having are your own fault.

None of this is actually said.

These meetings are supposed to make me feel more powerful, and in some ways they do. Rage can move mountains. In addition, they amaze me: it's shocking, and exciting, to hear such things emerging from the mouths of women. I begin to think that women I have thought were stupid, or wimps, may simply have been hiding things, as I was.

But these meetings also make me nervous, and I don't understand why. I don't say much, I am awkward and uncertain, because whatever I do say might be the wrong thing. I have not suffered enough, I haven't paid my dues, I have no right to speak. I feel as if I'm standing outside a closed door while decisions are being made, disapproving judgments are being pronounced, inside, about me. At the same time I want to please.

Sisterhood is a difficult concept for me, I tell myself, because I never had a sister. Brotherhood is not.

I work at night, when Sarah is asleep, or in the early morning. Right now I am painting the Virgin Mary. I paint her in blue, with the usual white veil, but with the head of a lioness. Christ lies in her lap in the form of a cub. If Christ is a lion, as he is in traditional iconography, why wouldn't the Virgin Mary be a lioness? Anyway it seems to me more accurate about motherhood than the old bloodless milk-and-water Virgins of art history. My Virgin Mary is fierce, alert to danger, wild. She stares levely out at the viewer with her yellow lion's eyes. A gnawed bone lies at her feet.

I paint the Virgin Mary descending to the earth, which is covered with snow and slush. She is wearing a winter coat over her blue robe, and has a purse slung over her shoulder. She's carrying two brown paper bags full of groceries. Several things have fallen from the bags: an egg, an onion, an apple. She looks tired.

Our Lady of Perpetual Help, I call her.

Jon does not like me painting at night. "When else can I do it?" I say. "You tell me." There is only one answer, one that would not involve the loss of his own time: *Don't do it at all*. But he doesn't say this.

He doesn't say what he thinks of my paintings, but I know anyway. He thinks they are irrelevant. In his mind, what I paint is lumped in with the women who paint flowers. *Lumped* is the word. The present tense is moving forward, discarding concept after concept, and I am off to the side somewhere, fiddling with egg tempera and flat surfaces, as if the twentieth century has never happened.

There is freedom in this: because it doesn't matter what I do, I can do what I like.

We have begun to slam doors, and to throw things. I throw my purse, an ashtray, a package of chocolate chips, which breaks on impact. We are picking up chocolate chips for days. Jon throws a glass of milk, the milk, not the glass: he knows his own strength, as I do not. He throws a box of Cheerios, unopened.

The things I throw miss, although they are worse things. The things he throws hit, but are harmless.

I begin to see how the line is crossed, between histrionics and murder.

Jon smashes things, and glues the shards into place in the pattern of breakage. I can see the appeal.

Jon sits in the living room, having a beer with one of the painters. I am in the kitchen, slamming around the pots.

"What's with her?" says the painter.

"She's mad because she's a woman," Jon says. This is something I haven't heard for years, not since high school. Once it was a shaming thing to say, and crushing to have it said about you, by a man. It implied oddness, deformity, sexual malfunction.

I go to the living room doorway. "I'm not mad because I'm a woman," I say. "I'm mad because you're an asshole."

62

Some of us from the meetings are having a group show, of women only. This is risky business, and we know it. Jody says we could get trashed, by the male art establishment. Their line these days is that great art transcends gender. Jody's line is that art so far has been mostly men admiring one another. A woman artist can get admired by them only as a sideline, a sort of freaky exception. "Titless wonders," says Jody.

We could get trashed by women as well, for singling ourselves out, putting ourselves forward. We could be called elitist. There are many pitfalls.

There are four of us in on the show. Carolyn, who has an angelic moon face framed in a Dutch cut with dark bangs, calls herself a fabric artist. Some of her pieces are patchwork quilts, in inventive designs. One has condoms stuffed with tampons (unused), glued onto it in the shapes of letters, spelling out WHAT IS LUV? Another is done in florals, with an appliquéd message:

UP YOUR

MAN

IFESTO!

Or else she makes wall hangings out of toilet paper twisted like rope, braided and woven with reels of outdated girlie movies, the kind that used to be called "art films." "Used porn," she says cheerfully. "Why not recycle it, eh?"

Jody does store mannequins, sawn apart, the pieces glued back together in disturbing poses. She fixes them up with paint and collage and steel wool stuck on at appropriate places. One hangs from a meat hook, stuck through the solar plexus, another has trees and flowers painted all over her face like fine tattooing, with a delicacy I wouldn't have suspected from Jody. Another has the heads of six or seven old dolls attached to her stomach. I recognize some of them: Sparkle Plenty, Betsy Wetsy, Barbara Ann Scott.

Zillah is blond and skimpy, like the frail flower girls of a few years back. She calls her pieces *Lintscapes*. They are made from the wads of feltlike fuzz that accumulate on drier filters and can be peeled off in sheets. I have admired these myself as I stuffed them into the wastebasket: their texture, their soft colors. Zillah has bought a number of towels in different shades and run them repeatedly through the dryer, to get shades of pink, of gray-green, of off-white, as well as the standard underneath-the-bed gray. These she has cut and shaped and glued carefully to a backing, to form multilayered compositions that resemble cloudscapes. I am entranced by them, and wish I had thought of this first. "It's like making a soufflé," Zillah says. "One breath of cold air and you're dead in the water."

Jody, who is more in charge than anyone, has gone through my paintings and chosen the ones for the show. She's taken some of the still lifes, *Wringer*, *Toaster*, *Deadly Nightshade*, and *Three Witches*. *Three Witches* is the one of the three different sofas.

Apart from the still lifes, what I'm showing is mostly figurative, although there are a couple of constructions made from drinking straws and uncooked macaroni, and one called *Silver Paper*. I didn't want to include these, but Jody liked them. "Domestic materials," she said.

The Virgin Mary pieces are in the show, and all of the Mrs. Smeaths. I thought there were too many of her, but Jody wanted them. "It's woman as anticheesecake," she said. "Why should it always be young, beautiful women? It's good to see the aging female body treated with compassion, for a change." This, only in more high-flown language, is what she's written in the catalogue.

The show is held in a small defunct supermarket, west on Bloor Street. It is to be converted to a hamburger heaven, shortly; but meanwhile it's empty, and one of the women who knows a cousin of the wife of the developer who owns it has managed to persuade him to let us use it for two

weeks. She told him that in the Renaissance the most famous
dukes were known for their aesthetic taste and patronage of
the arts, and this idea appealed to him. He doesn't know
it's an all-woman show; just some artists, is what she told
him. He says it's okay with him as long as we don't get the
place dirty.

"What's to get dirty?" says Carolyn, as we look around.
She's right, it's dirty enough already. The produce counters
and shelves have been torn up, there are patches ripped off
the erstwhile linoleum tile flooring where the wide bare
boards show through, lights dangle in wire cages; only some
of them work. The checkout counters are still in place,
though, and there are a few tattered signs drooping on the
walls: SPECIAL 3/95¢. FRESH FROM CALIFORNIA. MEAT
LIKE YOU LIKE IT.

"We can make this space work for us," says Jody, strid-
ing around with her hands in her coverall pockets.

"How?" says Zillah.

"I didn't take judo for nothing," says Jody. "Let the
momentum of the enemy carry him off balance."

In practice this means that she appropriates the MEAT LIKE
YOU LIKE IT sign and incorporates it into one of her constructions, an especially violent dismemberment in which
the mannequin, dressed only in ropes and leather straps,
has ended up with her head tucked upside-down under her
arm.

"If you were a man you'd get stomped for that," Carolyn
tells her.

Jody smiles sweetly. "But I'm not one."

We work for three days, arranging and rearranging. After we have the stuff in place, there are the rented trestle
tables to be assembled for the bar, the hooch and eats to be
bought. *Hooch* and *eats* are Jody's words. We get Canadian
wine in gallon jugs, Styrofoam cups to serve it in, pretzels
and potato chips, hunks of cheddar cheese wrapped in plastic film, Ritz crackers. This is what we can afford; but also
there's an unspoken rule that the food has to be unwaver-
ingly plebeian.

Our catalogue is a couple of mimeographed sheets stapled together at the top corner. This catalogue is supposed to be a collective effort, but in fact Jody has written most of it, because she has the knack. Carolyn makes a banner, out of bedsheets dyed to look as if someone's bled on them, to hang above the outside door:

F(OUR) FOR ALL.

"What's that supposed to mean?" says Jon, who has dropped by, supposedly to pick me up, really to see. He is suspicious of my doings with women, although he will not demean himself by saying so. He does however refer to them as "the girls."

"It's a pun on *free for all*," I tell him, although I know he knows this. "Plus it encapsulates the word *our*." *Encapsulate* is also one of Jody's words.

He does not comment.

It's the banner that attracts the newspapers: this kind of thing is new, it's an event, and it promises disruption. One newspaper sends a photographer, in advance, who says, jokingly, "Come on, girls, burn a few bras for me," while he's taking our pictures.

"Pig," says Carolyn in a low voice.

"Cool it," says Jody. "They love it when you freak."

Before the opening, I come to the gallery early. I pace around the show, up and down the former aisles, around the checkout counters where Jody's sculptures pose like models on a runway, past the wall where Carolyn's quilts yell defiance. This is strong work, I think. Stronger than mine. Even Zillah's gauzy constructions appear to me to have a confidence and subtlety, an assurance, that my own paintings lack: in this context my pictures are too highly finished, too decorative, too merely pretty.

I have strayed off course, I have failed to make a statement. I am peripheral.

I drink some of the awful wine and then some more, and feel better; although I know that later I will feel worse.

The stuff tastes like something you'd use to tenderize pot roast.

I stand against the wall, beside the door, hanging onto my Styrofoam cup. I'm standing here because it's the exit. Also the entrance: people arrive, and then more people.

Many, most of these people are women. There are all kinds of them. They have long hair, long skirts, jeans and overalls, earrings, caps like construction workers', lavender shawls. Some of them are other painters, some just look like it. Carolyn and Jody and Zillah are here by now, and there are greetings called, squeezes of the arm, kisses on cheeks, shrieks of delight. They all seem to have more friends than I do, more close women friends. I've never really considered it before, this absence; I've assumed that other women were like me. They were, once. And now they are not.

There is Cordelia, of course. But I haven't seen her for years.

Jon is not here yet, although he said he would come. We even got a baby-sitter so he could. I think maybe I will flirt with someone, someone inappropriate, just to see what could happen; but there aren't many possibilities, because there aren't many men. I make my way through the crowd with another Styrofoam cup of the dreadful red marinade, trying not to feel left out.

Right behind me a woman's voice says, "Well, they certainly are *different*." It's the quintessential Toronto middle-class-matron put-down, the ultimate disapproval. It's what they say about slums. It would not look good over the sofa, is what she means. I turn and look at her: a well-cut silver-gray suit, pearls, a suave scarf, expensive suede shoes. She's convinced of her own legitimacy, her right to pronounce: I and my kind are here on sufferance.

"Elaine, I'd like you to meet my mother," says Jody. The idea of this woman being Jody's mother is breathtaking. "Mum, Elaine did the flower painting. The one you like?"

She means *Deadly Nightshade*. "Oh yes," says Jody's mother, smiling warmly. "You girls are all so gifted. I did like that one, the colors are lovely. But what are all those eyes doing in it?"

This is so much what my own mother would say that I am swept with longing. I want my mother to be here. She would dislike most of this, the cut-up mannequins especially; she wouldn't understand it at all. But she would smile, and dredge up something nice to say. Very recently I would have derided such talents. Now I have need of them.

I get myself another cup of wine and a Ritz cracker with some cheese on it, and peer through the crowd for Jon, for anyone. What I see, over the heads, is Mrs. Smeath.

Mrs. Smeath is watching me. She lies on the sofa with her turbanlike Sunday hat on, the afghan wrapped around her. I have named this one *Torontodalisque: Homage to Ingres*, because of the pose, and the rubber plant like a fan behind her. She sits in front of a mirror with half of her face peeling off, like the villain in a horror comic I once read; this one is called *Leprosy*. She stands in front of her sink, her wicked paring knife in one hand, a half-peeled potato in the other. This one is called AN·EYE·FOR·AN·EYE.

Next to this is *White Gift*, which is in four panels. In the first one, Mrs. Smeath is wrapped up in white tissue paper like a can of Spam or a mummy, with just her head sticking out, her face wearing its closed half-smile. In the next three she's progressively unwrapped: in her print dress and bib apron, in her back-of-the-catalogue *Eaton's* flesh-colored foundation garment—although I don't expect she possessed one—and finally in her saggy-legged cotton underpants, her one large breast sectioned to show her heart. Her heart is the heart of a dying turtle: reptilian, dark-red, diseased. Across the bottom of this panel is stenciled: THE·KINGDOM·OF·GOD·IS·WITHIN·YOU.

It's still a mystery to me, why I hate her so much.

I look away from Mrs. Smeath, and there is another Mrs. Smeath, only this one is moving. She's just inside the door and heading toward me. She's the same age as she was. It's as if she's stepped down off the wall, the walls: the same round raw potato face, the hulky big-boned frame, the glittering spectacles and hairpin crown. My gut clenches in fear; then there's that rancid hate, flashing up in an instant.

But of course this can't be Mrs. Smeath, who must be much older by now. And it isn't. The hairpin crown was an optical illusion: it's just hair, graying and cropped short. It's Grace Smeath, charmless and righteous, in shapeless, ageless clothing, dun in color; she is ringless and without ornament. By the way she stalks, rigid and quivering, lips pinched, the freckles standing out on her root-white skin like bug bites, I can see that this will not be transformed into a light social occasion by any weak-chinned smiling of mine.

I try anyway. "Is it Grace?" I say. Several nearby people have stopped in mid-word. This is not the sort of woman who usually frequents gallery openings, of any kind.

Grace clumps relentlessly forward. Her face is fatter than it used to be. I think of orthopedic shoes, lisle stockings, underwear laundered thin and gray, coal cellars. I am afraid of her. Not of anything she could do to me, but of her judgment. And here it comes.

"You are disgusting," she says. "You are taking the Lord's name in vain. Why do you want to hurt people?"

What is there to be said? I could claim that Mrs. Smeath is not Grace's mother but a composition. I could mention the formal values, the careful use of color. But *White Gift* is not a composition, it's pictures of Mrs. Smeath, and indecent pictures at that. It's washroom graffiti raised to a higher order.

Grace is staring past me at the wall: there are not just one or two foul pictures to be appalled by, there are many. Mrs. Smeath in metamorphosis, from frame to frame, naked, exposed and desecrated, along with the maroon velvet chesterfield, the sacred rubber plant, the angels of God. I have gone way too far.

Grace's hands are fists, her fatted chin is trembling, her eyes are pink and watery, like a laboratory rabbit's. Is that a tear? I am aghast, and deeply satisfied. She is making a spectacle of herself, at last, and I am in control.

But I look again, more closely: this woman is not Grace. She doesn't even look like Grace. Grace is my age, she would not be this old. There's a generic resemblance, that's all. This woman is a stranger.

"You ought to be ashamed of yourself," says the woman who is not Grace. Her eyes narrow behind her glasses. She raises her fist, and I drop my glass of wine. Red splashes the wall and floor.

What she has in her clenched hand is a bottle of ink. With a shaky twist she unscrews the top, and I hold my breath, with fright but also curiosity: is it me she'll throw it at? For throwing is clearly her intention. There are gasps around us, this is happening fast, Carolyn and Jody are pushing forward.

The woman who is not Grace hurls the ink, bottle and all, straight at *White Gift*. The bottle careens and thuds to the carpet, ink pours down over the skyscape, veiling Mrs. Smeath in Parker's Washable Blue. The woman gives me a triumphant smile and turns, not stalking now but scurrying, heading for the door.

I have my hands over my mouth, as if to scream. Carolyn envelops me, hugging. She smells like a mother. "I'll call the police," she says.

"No," I say. "It will come off." And it probably will, because *White Gift* is varnished, and painted on wood. Maybe there won't even be a dent.

There are women gathering around me, the rustle of their feathers, a cooing. I am soothed and consoled, patted, cherished as if in shock. Maybe they mean it, maybe they like me after all. It's so hard for me to tell, with women.

"Who was that?" they ask.

"Some religious nut case," says Jody. "Some reactionary."

I will be looked at, now, with respect: paintings that can get bottles of ink thrown at them, that can inspire such outraged violence, such uproar and display, must have an odd revolutionary power. I will seem audacious, and brave. Some dimension of heroism has been added to me.

FEATHERS FLY AT FEMINIST FRACAS, says the paper. The picture is of me cringing, hands over my mouth, Mrs.

Smeath bare-naked and dripping with ink in the background. This is how I learn that women fighting is news. There's something titillating about it, upended and comic, like men in evening gowns and high heels. *Hen fighting*, it's called.

The show itself attracts bad adjectives: "abrasive," "aggressive" and "shrill." It's mostly Jody's statues and Carolyn's quilts that are called these things. Zillah's lintscapes are termed "subjective," "introverted" and "flimsy." Compared with the rest of them, I get off easy: "naïve surrealism with a twist of feminist lemon."

Carolyn makes a bright yellow banner with the words "abrasive," "aggressive" and "shrill" on it in red, and hangs it outside the door. A great many people come.

63

I'm waiting, in a waiting room. The waiting room has several nondescript blondwood chairs in it, with seats upholstered in olive green, and three end tables. This furniture is a clunky imitation of the early Scandinavian furniture of ten or fifteen years ago, now drastically out of style. On one of the tables there are some thumbed *Reader's Digest* and *Maclean's* magazines, and on another an ashtray, white with a rosebud trim. The carpet is an orangey-green, the walls an off-yellow. There is one picture, a lithoprint of two coy, grisly children in pseudo-peasant costume, vaguely Austrian, using a mushroom for an umbrella.

The room smells of old cigarette smoke, old rubber, the worn intimacy of cloth too long against flesh. On top of that, an overlay of floor wash antiseptic, seeping in from the corridors beyond. There are no windows. This room sets me on edge, like fingernails on a blackboard. Or like a dentist's waiting room, or the room where you'd wait before a job interview, for a job you didn't want to get.

This is a discreet private loony bin. A rest home, it's called: The Dorothy Lyndwick Rest Home. The sort of place well-off people use for stowing away those members of their families who are not considered fit to run around in public, in order to keep them from being carted off to 999 Queen, which is neither discreet nor private.

999 Queen is both a real place and high school shorthand for all funny farms, booby hatches, and nuthouses that could possibly be imagined. We had to imagine them, then, never having seen one. "999 Queen," we would say, sticking our tongues out the sides of our mouths, crossing our eyes, making circles near our ears with our forefingers. Craziness

was considered funny, like all other things that were in reality frightening and profoundly shameful.

I am waiting for Cordelia. Or I think it will be Cordelia: her voice on the phone did not sound like her, but slower and somehow damaged. "I saw you," is what she said, as if we had been talking together only five minutes before. But in fact it had been seven years, or eight, or nine: the summer she worked at the Stratford Shakespearean Festival, the summer of Josef. "In the paper," she added. And then a pause, as if this was a question.

"Right," I said. Then, because I knew I should, "Why don't we get together?"

"I can't go out," Cordelia said, in the same slowed-down voice. "You'll have to come here."

And so I am here.

Cordelia comes through a door at the far end of the room, walking carefully, as if balancing, or lame. But she is not lame. Behind her is another woman, with the optimistic, false, toothy smile of a paid attendant.

It takes me a moment to recognize Cordelia, because she doesn't look at all the same. Or rather she doesn't look the way she did when I last saw her, in her wide cotton skirt and barbaric bracelet, elegant and confident. She is in an earlier phase, or a later one: the soft green tweeds and tailored blouses of her good-taste background, which now appear matronly on her, because she has put on weight. Or has she? Flesh has been added, but it has slid down, toward the middle of her body, like mud sliding down a hill. The long bones have risen to the surface of her face, the skin tugged downward on them as if by irresistible gravitational pull. I can see how she'll be when she's old.

Someone has done her hair. Not her. She would never make it in tight little waves like that.

Cordelia stands uncertainly, squinting a little, head poking forward and swinging imperceptibly from side to side, the way an elephant's does, or some slow, bewildered animal. "Cordelia," I say, standing up.

"There's your friend," says the woman, smiling relentlessly. She takes Cordelia by the arm and gives a small tug, to start her in the right direction. "There you are," I say, falling already into the trap of addressing her like a child. I come forward, give her an awkward kiss. I find to my surprise that I'm glad to see her.

"Better late than never," Cordelia says, with the same hesitation, the thickness in her voice I've heard over the phone. The woman steers her to the chair across from mine, settles her down into it with a little push, as if she's elderly, and stubborn.

Suddenly I'm outraged. No one has a right to treat Cordelia this way. I scowl at the woman, who says, "How nice of you to come! Cordelia enjoys a visit, don't you, Cordelia?"

"You can take me out," Cordelia says. She looks up at the woman, for approval.

"Yes, that's right," says the woman. "For tea or something. If you promise to bring her back, that is!" She gives a cheery laugh, as if this is a joke.

I take Cordelia out. The Dorothy Lyndwick Rest Home is in High Park, a suburb where I've never been before and don't know my way around, but there's a corner café a few blocks along. Cordelia knows it, and her way there. I don't know whether I should take her arm or not, and so I don't; I walk along beside her, watchful at crossings as if she's blind, slowing my pace to hers.

"I don't have any money," says Cordelia. "They won't let me have any. They even get my cigarettes for me."

"That's all right," I say.

We ease into a booth, order coffee and two toasted Danishes. I give the order: I don't want the waitress staring. Cordelia fumbles, produces a cigarette. Her hand, lighting it, is shaky. "Great flaming blue-headed balls of Jesus," she says, making an effort with the syllables. "It's good to be out of there." She tries a laugh, and I laugh with her, feeling culpable and accused.

I should ask her things: what has she been doing, for these years we've skipped? What about her acting, what became of that? Did she get married, have children? What exactly has been going on, to bring her where she is? But all of this is beside the point. It's detachable, it's been added on. The main thing is Cordelia, the fact of her now.

"What the shit have they got you on?" I say.

"Some sort of tranquilizers," she says. "I hate them. They make me drool."

"What for?" I say. "How did you end up in that nuthatch anyway? You aren't any crazier than I am."

Cordelia looks at me, blowing out smoke. "Things weren't working out very well," she says after a while.

"So?" I say.

"So. I tried pills."

"Oh, Cordelia." Something goes through me with a slice, like watching a child fall, mouth down on rock. "Why?"

"I don't know. It just came over me. I was tired," she says.

There is no point telling her she shouldn't have done such a thing. I do what I'd do in high school: I ask for the details. "So did you conk out?"

"Yes," she said. "I checked into a hotel, to do it. But they figured it out—the manager or someone. I had to get my stomach pumped. That was revolting. Vomit-making, you could say."

She does what would be a laugh, except that her face is so rigid. I think I may cry. At the same time I'm angry with her, though I don't know why. It's as if Cordelia has placed herself beyond me, out of my reach, where I can't get at her. She has let go of her idea of herself. She is lost.

"Elaine," she says, "get me out."

"What?" I say, brought up short.

"Help me get out of there. You don't know what it's like. You have no privacy." This is the closest to pleading she's ever come.

A phrase comes to me, a remnant left over from boys, from Saturday afternoons, reading the comics: *Pick on somebody your own size*. "How could I do that?" I say.

"Visit me tomorrow and we'll go in a taxi." She sees me hesitate. "Or just lend me the money. That's all you have to do. I can hide the pills in the morning, I won't take them. Then I'll be all right. I know it's those pills that're keeping me like this. Just twenty-five dollars is all I need."

"I don't have a lot of money with me," I say, which is true enough, but an evasion. "They'd catch you. They'd know you were off the pills. They could tell."

"I can fool them any day," Cordelia says, with a flicker of her old cunning. Of course, I think, she's an actress. Or was. She can counterfeit anything. "Anyway, those doctors are so dumb. They ask all these questions, they believe anything I tell them, they write it all down."

There are doctors, then. More than one. "Cordelia, how can I take the responsibility? I haven't even talked, I haven't talked to anyone."

"They're all assholes," she says. "There's nothing wrong with me. You know, you said yourself." There's a frantic child in there, behind that locked, sagging face.

I have an image of spiriting Cordelia away, rescuing her. I could do it, or something like it; but then where would she end up? Hiding out in our apartment, sleeping on an improvised bed like the draft dodgers, a refugee, a displaced person, smoking up the kitchen with Jon wondering who the hell she is and why she's there. Things are uneven between us as it is; I'm not sure I can afford Cordelia. She'd be one more sin of mine, to be chalked up to the account he's keeping in his head. Also I am not feeling totally glued together myself.

And there's Sarah to think of. Would she take to this Auntie Cordelia? How is Cordelia with small children? And exactly how sick in the head is she, anyway? How long before I'd come back and find her out cold on the bathroom floor, or worse? In the middle of a bright red sunset. Jon's work table is an arsenal, there are little saws lying around, little chisels. Maybe it would just be melodrama, a skin-deep slash or two, her old theatricality; though perhaps theatrical people are not less risky, but more. In the interests of the role they'll sacrifice anything.

"I can't, Cordelia," I say gently. But I don't feel gentle toward her. I am seething, with a fury I can neither explain nor express. *How dare you ask me?* I want to twist her arm, rub her face in the snow.

The waitress brings the bill. "Are you sufficiently so-phonsified?" I say to Cordelia, trying for lightness, and a change of subject. But Cordelia has never been stupid.

"So you won't," she says. And then, forlornly: "I guess you've always hated me."

"No," I say. "Why would I? No!" I am shocked. Why would she say such a thing? I can't remember ever hating Cordelia.

"I'll get out anyway," she says. Her voice is not thick now, or hesitant. She has that stubborn, defiant look, the one I remember from years ago. *So?*

I walk her back, deposit her. "I'll come to visit you," I say. I intend to, but know at the same time that the chances are slim. She'll be all right, I tell myself. She was like this at the end of high school, and then things got better. They could again.

On the streetcar going back, I read the advertisements: a beer, a chocolate bar, a brassiere turning into a bird. I imitate relief. I feel free, and weightless.

But I am not free, of Cordelia.

I dream Cordelia falling, from a cliff or bridge, against a background of twilight, her arms outspread, her skirt open like a bell, making a snow angel in the empty air. She never hits or lands; she falls and falls, and I wake with my heart pounding and gravity cut from under me, as in an elevator plummeting out of control.

I dream her standing in the old Queen Mary school-yard. The school is gone, there is nothing but a field, and the hill behind with the scrawny evergreen trees. She is wearing her snowsuit jacket, but she is not a child, she's the age she is now. She knows I have deserted her, and she is angry.

After a month, two months, three, I write Cordelia a note, on flowered notepaper of the sort that doesn't leave much space for words. I purchase this notepaper specially. My note is written with such false cheerfulness I can barely stand to lick the flap of the envelope. In it I propose another visit.

But my note comes back in the mail, with *address unknown* scrawled across it. I examine this writing from every angle, trying to figure out if it could be Cordelia's, disguised. If it isn't, if she's no longer at the rest home, where has she gone? She could ring the doorbell at any minute, call on the phone. She could be anywhere.

I dream a mannequin statue, like one of Jody's in the show, hacked apart and glued back together. It's wearing nothing but a gauze costume, covered with spangles. It ends at the neck. Underneath its arm, wrapped in a white cloth, is Cordelia's head.

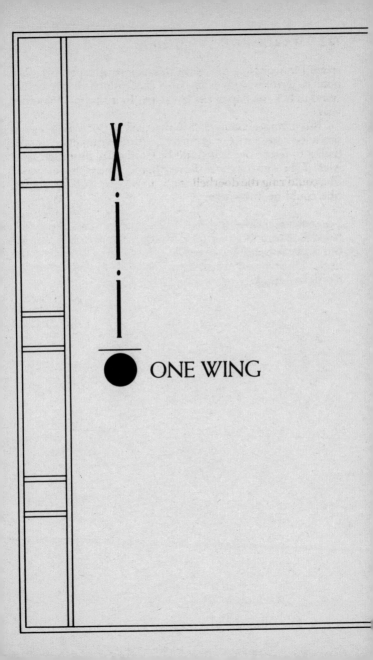

ONE WING

64
•

*I*n the corner of a parking lot, among the sumptuous boutiques, they've reconstructed a forties diner. 4-D's Diner, it's called. Not a renovation, brand-new.

They couldn't tear this stuff down fast enough, once.

Inside it's pretty authentic, except that it looks too clean; and it's less forties than early fifties. They have a soda fountain countertop, with stools along it topped in acid lime-green, and vinyl-padded booths in a shade of shiny purple that looks like the skin of an early shark-finned convertible. A jukebox, chrome coat trees, grainy black-and-white photos on the walls, of real forties diners. The waitresses have white uniforms with black tab trim, although the shade of their red lipstick isn't quite right and they should have run it around the edges of their mouths. The waiters have those soda jerk caps set at an angle, and the right haircuts, a close shave up the back of the neck. They're doing a roaring business. Kids in their twenties, mostly.

Really it's like Sunnysides, done over as a museum. They could have Cordelia and me in here, in our bat wing sleeves and cinch belts, stuffed and mounted or made of wax, drinking our milkshakes, looking as bored as we could.

The last time I saw Cordelia, she was going through the door of the rest home. That was the last time I talked to her. Although it wasn't the last time she talked to me.

There are no avocado and sprout sandwiches, the coffee is not espresso, the pie is coconut cream and no worse than it was then. This is what I have, coffee and pie, sitting in one of the purple booths, watching young people exclaim over what they think is the quaintness of the past.

The past isn't quaint while you're in it. Only at a safe

distance, later, when you can see it as décor, not as the shape your life's been squeezed into.

They have Elvis Presley zucchini molds now: you clamp them around your zucchini while it's young, and as it grows it's deformed into the shape of Elvis Presley's head. Is this why he sang? To become a zucchini? Vegetarianism and reincarnation are in the air, but that's taking it too far. I'd rather come back as a sow bug, myself; or a stir-fried shrimp. Though I suppose the whole idea's more lenient than Hell.

"You've done it well," I say to the waitress. "Of course the prices are wrong. It was ten cents for a coffee, then."

"Really," she says, not as a question. She gives me a dutiful smile: *Boring old frump*. She is half my age, living, already, a life I can't imagine. Whatever her guilts are, her hates and terrors, they are not the same. What do they do about AIDS, these girls? They can't just roll around in the hay, the way we did. Is there a courtship ritual that involves, perhaps, an exchange of doctors' telephone numbers? For us it was pregnancy that was the scary item, the sexual booby trap, the thing that could finish you off. Not any more.

I pay the bill, overtip, gather up my packages, an Italian scarf for each of my daughters, a fountain pen for Ben. Fountain-pens are coming back. Somewhere in Limbo, all the old devices and appliances and costumes are lined up, waiting their turn for re-entry.

I walk up the street, along to the corner. The next street is Josef's. I count houses: this one must be his. The front's been ripped out and glassed over, the lawn is paving stone. There's an antique child's rocking horse in the window, a threadbare quilt, a wooden-headed doll with a battered face. Onetime throw-outs, recycled as money. Nothing so indiscreet as a price tag, which means outrageous.

I wonder what became of Josef, eventually. If he's still alive he must be sixty-five, or more. If he was a dirty old man then, how dirty is he now?

He did make a film. I think it was him; in any case, the director's name was the same. I saw it by accident, at a film festival. This was a lot later, when I was already living in Vancouver.

It was about two women with nebulous personalities and cloudy hair. They wandered through fields with the wind blowing their thin dresses against their thighs, and gazed inscrutably. One of them took apart a radio and dropped the pieces into a stream, ate a butterfly, and cut the throat of a cat, because she was deranged. These things wouldn't have been as appealing if she had been ugly, instead of blond and ethereal. The other one made little slashes on the skin of her thigh, using an old-fashioned straight razor that had belonged to her grandfather. Toward the end she jumped off a railway overpass, into a river, her dress fluttering like a window curtain. Except for the colors of their hair, it was hard to tell the two of them apart.

· The man in this film was in love with both of them and couldn't make up his mind. Hence their craziness. This is what convinced me that it must have been Josef: it wouldn't have occurred to him that they might have had reasons of their own for being crazy, apart from men.

None of the blood in this film was real blood. Women were not real to Josef, any more than he was real to me. This was why I could treat his sufferings with such scorn and unconcern: he wasn't real. The reason I've never dreamed about him was that he belonged already to the world of dreams: discontinuous, irrational, obsessive.

I was unfair to him, of course, but where would I have been without unfairness? In thrall, in harness. Young women need unfairness, it's one of their few defenses. They need their callousness, they need their ignorance. They walk in the dark, along the edges of high cliffs, humming to themselves, thinking themselves invulnerable.

I can't blame Josef for his film. He was entitled to his own versions, his own conjurings; as I am. I may have served his ends, but he served mine as well.

There is *Life Drawing*, for instance, hanging right now on the gallery wall, Josef preserved in aspic and good enough to eat. He is on the left side of the picture, stark-naked but turned with a twist half away from the viewer, so what you get is the ass end, then the torso in profile. On the right side is Jon, in the same position. Their bodies are somewhat idealized: less hairy than they really were, the muscle groups in higher definition, the skin luminous. I thought about putting Jockey shorts on them, in deference to Toronto, but decided against it. Both of them have wonderful bums.

Each of them is painting a picture, each picture is on an easel. Josef's is of a voluptuous but not overweight woman, sitting on a stool with a sheet draped between her legs, her breasts exposed; her face is Pre-Raphaelite, brooding, consciously mysterious. Jon's painting is a series of intestinal swirls, in hot pink, raspberry ripple red and Burgundy Cherry purple.

The model is seated on a chair between them, face front, bare feet flat on the floor. She's clothed in a white bedsheet, wrapped around her below the breasts. Her hands are folded neatly in her lap. Her head is a sphere of bluish glass.

I sit with Jon at a table in the roof bar of the Park Plaza Hotel, drinking white wine spritzers. My suggestion: I wanted to see it again. Outside, the skyline has changed: the Park Plaza is no longer the tallest building around, but a squat leftover, dwarfed by the svelte glassy towers that rise around it. Due south is the CN Tower, lifting up like a huge inverted icicle. This is the sort of architecture you used to see only in science fiction comic books, and seeing it pasted flat against the monotone lake-sky I feel I've stepped not forward in time but sideways, into a universe of two dimensions.

But inside the bar not a lot has changed. The place still looks like a high-class Regency bordello. Even the waiters, with their good-grooming hair and air of harried discretion, look the same, and probably are. The management used to keep ties in the coat check, for gentlemen who'd forgotten them. *Forgotten* was the word, because surely no gentleman would deliberately choose to go tieless. It was a big thing when this place was cracked by women in pant suits. A chic black model did it: they couldn't refuse to let her in, she could have hit them with racism. Even this memory dates me, and the little thrill of triumph that goes with it: what woman, now, would think of a pant suit as liberation?

I didn't used to come here with Jon. He would have sneered, then, at the upholstered period chairs, the looped drapes, the men and women cut from a glossy whisky ad. It was Josef I came with, Josef whose hand I touched, across the surface of the table. Not Jon's, as now.

It's only the ends of the fingers, only lightly. This time we don't say much: there's none of the verbal prodding there was at lunch. There's a shared vocabulary, of monosyllable and silence; we know why we're here. Going down in the elevator, I look into the smoked-mirror wall and see my face in the dark glass obscured by time, as a stone overgrown. I could be any age.

We take a taxi back to the warehouse, our hands resting side by side on the seat. We go up the stairs to the studio, slowly, so we won't get out of breath: neither one of us wants to be caught out by the other in a middle-aged wheeze. Jon's hand is on my waist. It's familiar there; it's like knowing where the light switch is, in a house you once lived in but haven't been back to for years. When we reach the door, before we go in, he pats me on the shoulder, a gesture of encouragement, and of wistful resignation.

"Don't turn on the light," I say.

Jon puts his arms around me, his face in the angle of my neck. It's a gesture less of desire than of fatigue.

The studio is the purplish gray of autumn twilight. The plaster casts of arms and legs glimmer whitely, like broken statues in a ruin. There's a scatter of my clothes in the corner, empty cups dotted here and there, on the work counter, by the window, marking my daily trails, claiming space. This room seems like mine now, as if I've been living here all along, no matter where else I've been or what else I've been doing. It's Jon who has been away, and has returned at last.

We undress each other, as we used to do at first; but more shyly. I don't want to be awkward. I'm glad it's dusk; I'm nervous about the backs of my thighs, the wrinkling above my knees, the soft fold across my stomach, not fatness exactly but a pleat. The hair on his chest is gray, a shock.

I avoid looking at the small beer belly that's grown on him, though I'm aware of it, of the changes in his body, as he must be of mine.

When we kiss, it's with a gravity we lacked before. Before we were avid, and selfish.

We make love for the comfort of it. I recognize him, I could recognize him in total darkness. Every man has his own rhythm, which remains the same. In this there is the relief of greeting.

I don't feel I'm being disloyal to Ben, only loyal to something else; which predates him, which has nothing to do with him. An old score.

Also I know it's something I'll never do again. It's the last look, before turning away, at some once-visited, once-extravagant place you know you won't go back to. An evening view, of Niagara Falls.

We lie together under the duvet, arms around each other. It's hard to remember what we used to fight about. The former anger is gone, and with it that edgy, jealous lust we used to have for each other. What's left is fondness, and regret. A diminuendo.

"Come to the opening?" I say. "I'd like you to."

"No," he says. "I don't want to."

"Why not?"

"I'd feel bad," he says. "I wouldn't want to see you that way."

"What way?" I say.

"With all those people, slobbering over you."

What he means is that he doesn't want to be merely an onlooker, that there's no room for him in all that, and he's right. He doesn't want to be just my ex-husband. He would be dispossessed, of me and of himself. I realize I don't want it either, I don't really want him to be there. I need him to be, but I don't want it.

I turn, lean on my elbow, kiss him again, on the cheek this time. The hair down low, behind his ears, is already turning white. I think, we did that just in time. It was almost too late.

65

*W*ith Jon it's like falling downstairs. Up until now there have been preliminary stumblings, recoveries, a clutching for handholds. But now all balance is lost and we plunge down headlong, both of us, noisily and without grace, gathering momentum and abrasions as we go.

I enter sleep angry and dread waking up, and when I do wake I lie beside the sleeping body of Jon, in our bed, listening to the rhythm of his breathing and resenting him for the oblivion he still controls.

For weeks he has been more silent than usual, and home less. Home less, that is, when I am home. When I'm away at work he is there all right, even when Sarah's in preschool. I've begun to find signs, tiny clues left in my way like breadcrumbs dropped on a trail: a cigarette with a pink mouthmark on it, two used glasses in the sink, a hairpin that is not mine, beneath a pillow that is. I clean up and say nothing, hoarding these things for times of greater need.

"Someone named Monica called you," I tell him.

It's morning, and there's a whole day to get through. A day of evasion, suppressed anger, false calm. We are well beyond throwing things, by now.

He's reading the paper. "Oh?" he says. "What did she want?"

"She said to tell you Monica called," I say.

He comes back late at night and I'm in bed, feigning sleep, my head churning. I think of subterfuges: examining his shirts for perfume, tailing him along the street, hiding in the closet and jumping out, red-hot with discovery. I think of other things I could do. I could leave, go somewhere unspecified, with Sarah. Or I could demand that we talk things through. Or I could pretend nothing is happening, continue on with our lives as usual. This would have been the advice offered in women's magazines, of a decade ago: wait it out.

I see these things as scenarios, to be played through and discarded, perhaps simultaneously. None of them precludes the others.

In real life, the days go on as usual, darkening to winter and heavy with the unspoken.

"You had a thing with Uncle Joe, didn't you?" Jon says casually. It's a Saturday, and we're making a stab at normality by taking Sarah to Grange Park, to play in the snow.

"Who?" I say.

"You know. Josef what's-his-name. The old stick man."

"Oh, him," I say. Sarah is over by the swings with some other kids. We're sitting on a bench, having cleared the snow. I think I should be making a snowman, or doing some other thing good mothers are supposed to do. But I'm too tired.

"But you did, didn't you?" Jon says. "At the same time as me."

"Where did you get that idea?" I say. I know when I'm being accused. I run over my own ammunition: the hairpins, the lipstick, the phone calls, the glasses in the sink.

"I'm not a moron, you know. I figured it out."

He has jealousies of his own then, wounds of his own to lick. Things I have inflicted. I should lie, deny everything. But I don't want to. Josef, at the moment, gives me a little pride.

"That was years ago," I say. "Thousands of years ago. It wasn't important."

"Like shit," he says. I once thought he would ridicule me, if he found out about Josef. The surprise is that he takes him seriously.

That night we make love, if that is any longer the term for it. It's not shaped like love, not colored like it, but harsh, war-colored, metallic. Things are being proved. Or repudiated.

In the morning he says, "Who else has there been?" Out of nowhere. "How do I know you weren't hopping into the sack with every old fart around?"

I sigh. "Jon," I say. "Grow up."

"How about Mr. Beanie Weenie?" he keeps on.

"Oh, come on," I say. "You were hardly the angel. Your place was crawling with all those skinny girls. You didn't want strings, remember?"

Sarah is still in her crib asleep. We are safe, we can get down to it, this telling of bad truths which are not entirely true. Once you start, it's difficult to stop. There is even a certain relish in it.

"At least I was open about it," he says. "I didn't sneak around. I didn't pretend to be so goddamn pure and faithful, the way you did."

"Maybe I loved you," I say. I notice the past tense. So does he.

"You wouldn't know love if you fell over it," he says.

"Not like Monica?" I say. "You're not being very open right now. I've found those hairpins, in my own bed. You could at least have the decency to do it somewhere else."

"How about you?" he says. "You're always going out, you get around."

"Me?" I say. "I don't have the time. I don't have time to think, I don't have time to paint, I barely have time to shit. I'm too busy paying the goddamn rent."

I've said the worst thing, I've gone too far. "That's it," says Jon. "It's always you, what you contribute, what you put up with. It's never me." He hunts for his jacket, heads for the door.

"Going to see Monica?" I say, with as much venom as I can dredge up. I hate it, this schoolyard bickering. I want embraces, tears, forgiveness. I want them to arrive by themselves, with no effort on my part, like rainbows.

"Trisha," he says. "Monica is just a friend."

It's winter. The heat goes off, comes on again, goes off, at random. Sarah has a cold. She coughs at night and I get up for her, feeding her spoonfuls of cough syrup, bringing her drinks of water. In the daytime we are both exhausted.

I am sick a lot myself this winter. I get her colds. I lie in bed on weekend mornings, looking up at the ceiling, my head clogged and cottony. I want glasses of ginger ale, squeezed orange juice, the sound of distant radios. But these things are gone forever, nothing arrives on a tray. If I want ginger ale I'll have to go to the store or the kitchen, buy it or pour it myself. In the main room Sarah watches cartoons.

I don't paint at all any more. I can't think about painting. Although I've received a junior grant from a government arts program, I can't organize myself enough to lift a brush. I push myself through time, to work, to the bank to get money, to the supermarket to buy food. Sometimes I watch daytime soaps on television, where there are more crises and better clothes than in real life. I tend to Sarah.

I don't do anything else. I no longer go to the meetings of women, because they make me feel worse. Jody phones and says we should get together, but I put her off. She would jolly me along, make bracing and positive suggestions I know I can't live up to. Then I would only feel more like a failure.

I don't want to see anyone. I lie in the bedroom with the curtains drawn and nothingness washing over me like a sluggish wave. Whatever is happening to me is my own fault. I have done something wrong, something so huge I can't even see it, something that's drowning me. I am inadequate and stupid, without worth. I might as well be dead.

One night Jon does not come back. This is not usual, it isn't our silent agreement: even when he stays out late he is always in by midnight. We haven't had a fight this day; we've hardly spoken. He hasn't phoned to say where he is. His intention is clear: he has left me behind, in the cold.

I crouch in the bedroom, in the dark, wrapped in Jon's old sleeping bag, listening to the wheezing sound of Sarah breathing and the whisper of sleet against the window. Love blurs your vision; but after it recedes, you can see more clearly than ever. It's like the tide going out, revealing whatever's been thrown away and sunk: broken bottles, old gloves, rusting pop cans, nibbled fishbodies, bones. This is the kind of thing you see if you sit in the darkness with open eyes, not knowing the future. The ruin you've made.

My body is inert, without will. I think I should keep moving, to circulate my blood, as you are supposed to do in a snowstorm so you won't freeze to death. I force myself to stand up. I will go to the kitchen and make tea.

Outside the house a car slides by, through the mushy snow, a muffled rushing. The main room is dark, except for the light coming in from the lampposts on the street, through the window. The things on Jon's work table glint in this half-light: the flat blade of a chisel, the head of a hammer. I can feel the pull of the earth on me, the dragging of its dark curve of gravity, the spaces between the atoms you could fall so easily through.

This is when I hear the voice, not inside my head at all but in the room, clearly: *Do it. Come on. Do it.* This voice doesn't offer a choice; it has the force of an order. It's the difference between jumping and being pushed.

The Exacto knife is what I use, to make a slash. It doesn't even hurt, because right after that there's a whispering sound and space closes in and I'm on the floor. This is how Jon finds me. Blood is black in the darkness, it does not reflect, so he doesn't see until he turns on the light.

I tell the people at Emergency that it was an accident. I am a painter, I say. I was cutting canvas and my hand slipped. It's my left wrist, so this is plausible. I'm frightened, I want to hide the truth: I have no intention of being stuffed into 999 Queen Street, now or ever.

"In the middle of the night?" the doctor says.

"I often work at night," I say.

Jon backs me up. He's just as scared as I am. He tied my wrist up in a tea towel and drove me to the hospital. I leaked through the towel, onto the front seat.

"Sarah," I said, remembering her.

"She's downstairs," Jon said. Downstairs is the landlady, a middle-aged Italian widow.

"What did you tell her?" I asked.

"I said it was your appendix," Jon said. I laughed, a little. "What the hell got into you?"

"I don't know," I said. "You'll have to get this car cleaned." I felt white, drained of blood, cared for, purified. Peaceful.

"Are you sure you don't want to talk to someone?" the doctor in Emergency says.

"I'm fine now," I say. The last thing I want to do is talking. I know what he means by *someone*: a shrink. Someone who will tell me I'm nuts. I know what kind of people hear voices: people who drink too much, who fry their brains with drugs, who slip off the rails. I feel entirely steady, I'm not even anxious any more. I've already decided what I will do, afterward, tomorrow. I'll wear my arm in a sling and say I broke my wrist. So I don't have to tell him, or Jon, or anyone else, about the voice.

I know it wasn't really there. Also I know I heard it.

It wasn't a frightening voice, in itself. Not menacing but excited, as if proposing an escapade, a prank, a treat. Something treasured, and secret. The voice of a nine-year-old child.

66
.

*T*he snow has melted, leaving a dirty filigree, the wind is blowing around the grit left over from winter, the crocuses are pushing up through the mud of the desolate smashed-down lawns. If I stay here I will die.

It's the city I need to leave as much as Jon, I think. It's the city that's killing me.

It will kill me suddenly. I'll be walking along the street, thinking of nothing in particular, and all at once I will turn sideways and dive off the curb, to be smashed by a speeding car. I will topple in front of a subway train without warning, I'll plunge from a bridge without intention. All I will hear will be that small voice, inviting and conspiratorial, gleeful, urging me over. I know I'm capable of such a thing.

(Worse: although I'm afraid of this idea and ashamed of it, and although in the daytime I find it melodramatic and ludicrous and refuse to believe in it, I also cherish it. It's like the secret bottle stashed away by alcoholics: I may have no desire to use it, right now, but I feel more secure knowing it's there. It's a fallback, it's a vice, it's an exit. It's a weapon.)

At night I sit beside Sarah's crib, watching the flutter of her eyelids as she dreams, listening to her breathe. She will be left alone. Or not alone, because she will have Jon. Motherless. This is unthinkable.

I turn on the lights in the living room. I know I must start packing, but I don't know what to take. Clothes, toys for Sarah, her furry rabbit. It seems too difficult, so I go to bed. Jon is already in there, turned toward the wall. We have gone through a pretence of truce and reformation, straight into deadlock. I don't wake him up.

In the morning, after he leaves, I bundle Sarah into the

stroller and take some of my grant money out of the bank. I don't know where to go. All I can think of is away. I buy us tickets to Vancouver, which has the advantage of being warm, or so I suppose. I stuff our things into duffel bags, which I've bought at Army Surplus.

I want Jon to come back and stop me, because now that I'm in motion I can't believe I'm actually doing this. But he doesn't come.

I leave a note, I make a sandwich: peanut butter. I cut it in two and give half to Sarah, and a glass of milk. I call a taxi. We sit at the kitchen table with our coats on, eating our sandwiches and drinking our milk, and waiting.

This is when Jon comes back. I keep eating.

"Where the hell do you think you're going?" he says.

"Vancouver," I say.

He sits down at the table, stares at me. He looks as if he hasn't slept for weeks, although he's been sleeping a lot, oversleeping. "I can't stop you," he says. It's a statement of fact, not a maneuver: he will let us go without a fight. He too is exhausted.

"I think that's the taxi," I say. "I'll write."

I'm good at leaving. The trick is to close yourself off. Don't hear, don't see. Don't look back.

We don't have a sleeper, because I need to save the money. I sit up all night, Sarah sprawled and snuffling in my lap. She's done some crying, but she's too young to realize what I've done, what we're doing. The other passengers extend themselves into the aisles; baggage expands, smoke drifts in the stale air, food wrappings clog the washrooms. There's a card game going on up at the front of the car, with beer.

The train runs northwest, through hundreds of miles of scraggy forests and granite outcrops, hundreds of small blue anonymous lakes edged with swamp and bulrushes and dead spruce, old snow in the shadows. I peer out through the glass of the train window, which is streaked with rain and dust, and there is the landscape of my early childhood, smudged and scentless and untouchable and moving backward.

At long intervals the train crosses a road, gravel or thin

and paved, with a white line down the middle. This looks like emptiness and silence, but to me it is not empty, not silent. Instead it's filled with echoes.

Home, I think. But it's nowhere I can go back to.

It's worse than I thought it would be, and also better.

Some days I think I'm crazy to have done this; other times that it's the sanest move I've made in years.

It's cheaper in Vancouver. After a short spell in a Holiday Inn, I find a house I can rent, on the rise behind Kitsilano Beach, one of those toytown houses that are bigger inside than they look. It has a view of the bay, and the mountains across it, and, in the summer, endless light. I find a co-op preschool for Sarah. For a time I live on grant money. I freelance a little, then get a part-time job refinishing furniture for an antique dealer. I like this, because it's mindless and the furniture can't talk. I am thirsty for silence.

I lie on the floor, washed by nothing and hanging on. I cry at night. I am afraid of hearing voices, or a voice. I have come to the edge, of the land. I could get pushed over.

I think maybe I should go to see a shrink, because that is the accepted thing, now, for people who are not in balance, and I am not. Finally I do go. The shrink is a man, a nice man. He wants me to talk about everything that happened to me before I was six, nothing after. Once you are six, he implies, you are cast in bronze. What comes after is not important.

I have a good memory. I tell him about the war.

I tell him about the Exacto knife and the wrist, but not about the voice. I don't want him to think I'm a loony. I want him to think well of me.

I tell him about nothing.

He asks if I have orgasms. I say that isn't the problem.

He thinks I am hiding things.

After a while I stop going.

Gradually I grow back, into my hands. I take to getting up early in the morning, before Sarah is awake, to paint. I find I have a minor, ambiguous reputation, from the show in

Toronto, and I am invited to parties. At first there is a resentful edge, because I am from what is known as *back east*, which is supposed to confer unfair advantages; but after a time I've been here long enough so I can pass, and after that I can do the resentful act myself, to easterners, and get away with it.

I'm also invited to take part in several group showings, mostly by women: they've heard about the ink throwing, read the snotty reviews, all of which render me legitimate, although from the east. Women artists of many kinds, women of many kinds are in ferment here, they are boiling with the pressured energy of explosive forces confined in a small space, and with the fervor of all religious movements in their early, purist stages. It is not enough to give lip service and to believe in equal pay: there has to be a conversion, from the heart. Or so they imply.

Confession is popular, not of your flaws but of your sufferings, at the hands of men. Pain is important, but only certain kinds of it: the pain of women, but not the pain of men. Telling about your pain is called sharing. I don't want to share in this way; also I am insufficient in scars. I have lived a privileged life, I've never been beaten up, raped, gone hungry. There is the issue of money, of course, but Jon was as poor as me.

There is Jon. But I don't feel overmatched by him. Whatever he did to me, I did back, and maybe worse. He's twisting now, because he misses Sarah. He calls long distance, his voice on the phone fading in and out like a wartime broadcast, plaintive with defeat, with an archaic sadness that seems, more and more, to be that of men in general.

No mercy for him, the women would say. I am not merciful, but I am sorry.

A number of these women are lesbians, newly declared or changing over. This is at the same time courageous and demanded. According to some, it's the only equal relationship possible, for women. You are not genuine otherwise.

I am ashamed of my own reluctance, my lack of desire; but the truth is that I would be terrified to get into bed with a woman. Women collect grievances, hold grudges and change shape. They pass hard, legitimate judgments, unlike the purblind guesses of men, fogged with romanticism and ignorance and bias and wish. Women know too much, they can neither be deceived nor trusted. I can understand why men are afraid of them, as they are frequently accused of being.

At parties they start to ask leading questions that have the ring of inquisition; they are interested in my positions, my dogmas. I am guilty about having so few of these: I know I am unorthodox, hopelessly heterosexual, a mother, quisling and secret wimp. My heart is a dubious object at best, blotchy and treacherous. I still shave my legs.

I avoid gatherings of these women, walking as I do in fear of being sanctified, or else burned at the stake. I think they are talking about me, behind my back. They make me more nervous than ever, because they have a certain way they want me to be, and I am not that way. They want to improve me. At times I feel defiant: what right have they to tell me what to think? I am not Woman, and I'm damned if I'll be shoved into it. *Bitch*, I think silently. *Don't boss me around*.

But also I envy their conviction, their optimism, their carelessness, their fearlessness about men, their camaraderie. I am like someone watching from the sidelines, waving a cowardly handkerchief, as the troops go boyishly off to war, singing brave songs.

I have several women friends, not very close ones. Single mothers, as I am. I meet them at preschool. We trade kids for nights out and grumble harmlessly together. We avoid each other's deeper wounds. We're like Babs and Marjorie from my old Life Drawing class, with the same sense of rueful comedy. It's an older pattern, for women; but by now we are older.

Jon comes to visit, a tentative move toward reconciliation, which I think I want as well. It doesn't work, and we divorce, finally, by long distance.

My parents come as well. They miss Sarah, I think, more than they miss me. I have made excuses not to go east for Christmas. Against the backdrop of the mountains, they seem out of place, a little shrunken. They are more themselves in their letters. They are saddened by me and what they probably think of as my broken home, and don't know what to say about it. "Well, dear," says my mother, talking about Jon. "I always thought he was very intense." A bad word that spells trouble.

I take them to Stanley Park, where there are big trees. I show them the ocean, sloshing around in seaweed. I show them a giant slug.

My brother Stephen sends postcards. He sends a stuffed dinosaur for Sarah. He sends a water pistol. He sends a counting book, about an ant and a bee. He sends the solar system, in the form of a plastic mobile, and stars you can stick on the ceiling that light up at night.

After a time I find that, in the tiny world of art (tiny, because who knows about it really? It's not on television), swirls, squares, and giant hamburgers are out and other things are in, and I am suddenly at the front of a smallish wave. There's a flurry, as such things go. More of my pictures sell, for higher prices. I'm represented by two regular galleries now, one east, one west. I go to New York, briefly, leaving Sarah with one of my single mother friends, for a group show organized by the Canadian government which is attended by many people who work with the Trade Commission. I wear black. I walk on the streets, feeling sane in comparison with the other people there, who all seem to be talking to themselves. I come back.

I have men, at long intervals and in some desperation. These affairs are rushed and unsatisfactory: I don't have time for the finer points. Even these brief interludes are almost too strenuous for me.

None of these men rejects me. I don't give them the chance. I know what is dangerous for me, and keep away from the edges of things. From anything too bright, too sharp. From lack of sleep. When I start feeling shaky I lie down, expecting nothing, and it arrives, washing over me in a wave of black vacancy. I know I can wait it out.

. . .

After more time I meet Ben, who picks me up in the most ordinary way, in the supermarket. Actually he asks if he can carry my shopping bags, which look heavy and are, and I let him, feeling silly and archaic and looking first to make sure no women I know are watching.

Years before, I would have considered him too obvious, too dull, practically simple-minded. And for years after that, a chauvinist of the more amiable sort. He is all these things; but he is also like an apple, after a prolonged and gluttonous binge.

He comes over and fixes my back porch with his own saw and hammer, as in the women's magazines of long ago, and has a beer afterward, on the lawn, as in ads. He tells me jokes I haven't heard since high school. My gratitude for these mundane enjoyments amazes me. But I don't require him, he's no transfusion. Instead he pleases me. It's a happiness, to be so simply pleased.

He takes me to Mexico, as in drugstore romances. He's just bought his small travel business, more as a hobby than anything: he made his money earlier, in real estate. But he likes to take photographs and sit in the sun. To do what he likes and make money at the same time is what he's wanted all his life.

He is shy in bed, easily surprised, quickly delighted.

We combine households, in a third, larger house. After a while we get married. There is nothing dramatic about it. To him it seems appropriate, to me eccentric: it's a defiance of convention, but of a convention he's never heard of. He doesn't know how outlandish I think I'm being.

He's ten years older than I am. He has a divorce of his own, and a grown son. My daughter Sarah becomes the daughter he wanted, and soon we have Anne. I think of her as a second chance. She is less pensive than Sarah, more stubborn. Sarah knows, already, that you can't always have everything you want.

Ben considers me good, and I don't disturb this faith: he doesn't need my more unsavory truths. He considers me also a little fragile, because artistic: I need to be cared for, like a potted plant. A little pruning, a little watering, a little weeding and straightening up, to bring out the best in me. He makes up a set of books, for the business end of my painting: what has sold, and for how much. He tells me what I can deduct on my income tax return. He fills out the return. He arranges the spices in alphabetical order, on a special shelf in the kitchen. He builds the shelf.

I could live without this. I have before. But I like it, all the same.

My paintings themselves he regards with wonder, and also apprehension, like a small child looking at a candle. What he focuses on is how well I do hands. He knows these are hard. He once wanted to take up something like that himself, he says, but never got around to it because of having to earn a living. This is a lot like the kinds of things people have said to me at gallery openings, but in him I forgive it.

He goes away at judicious intervals, on business, giving me a chance to miss him.

I sit in front of the fireplace, with his arm around me solid as the back of a chair. I walk along the breakwater in the soothing Vancouver drizzle, the halftones of the seashore, the stroking of the small waves. In front of me is the Pacific, which sends up sunset after sunset, for nothing; at my back are the improbable mountains, and beyond them an enormous barricade of land.

Toronto lies behind it, at a great distance, burning in thought like Gomorrah. At which I dare not look.

PICOSECONDS

67

I wake late. I eat an orange, some toast, an egg, mushing it up in a teacup. The hole poked in the bottom of the eggshell was not to keep the witches from going to sea, as Cordelia said. It's to break the vacuum between shell and egg cup, so the shell can be extracted. Why did it take me forty years to figure that out?

I put on my other jogging suit, the cerise one, and do some desultory stretching exercises on Jon's floor. It's Jon's floor again, not mine. I feel I've returned it to him, along with whatever fragments of his own life, or of our life together, I've been keeping back till now. I remember all those mediaeval paintings, the hand raised, open to show there is no weapon: *Go in peace*. Dismissal, and blessing. My way of doing this was not exactly the way of the saints, but seems to have worked just as well. The peace was for the bestower of it, also.

I go down to get the morning paper. I leaf through it, without reading much. I know I'm killing time. I've almost forgotten what I'm supposed to be doing here, and I'm impatient to be gone, back to the west coast, back to the time zone where I live my life now. But I can't do that yet. I'm suspended, as in airports or dentists' waiting rooms, expecting yet another interlude that will be textureless and without desire, like a painkiller or the interiors of planes. This is how I think of the coming evening, the opening of the show: something to get through without disaster.

I should go to the gallery, check to see that everything's in order. I should perform at least that minimal courtesy. But instead I take the subway, get off near the main gate of the cemetery, wander south and east, scuffing through the fallen leaves, scanning the gutters; looking down at the sidewalk, for silver paper, nickels, windfalls. I still believe such things exist, and that I could find them.

With a slight push, a slip over some ill-defined edge, I could turn into a bag lady. It's the same instinct: rummaging in junk heaps, pawing through discards. Looking for something that's been thrown away as useless, but could still be dredged up and reclaimed. The collection of shreds, of space in her case, time in mine.

This is my old route home from school. I used to walk along this sidewalk, behind or in front of the others. Between these lampposts my shadow on the winter snow would stretch ahead of me, double, shrink again and disappear, the lamps casting their haloes around them like the moon in fog. Here is the lawn where Cordelia fell down backward, making a snow angel. Here is where she ran.

The houses are the same houses, though no longer trimmed in peeling white winter-grayed paint, no longer down-at-heels, postwar. The sandblasters have been here, the skylight people; inside, the benjamina trees and tropical climbers have taken over, ousting the mangy African violets once nurtured on kitchen windowsills. I can see through these houses, to what they used to be; I can see the colors that used to cover the walls, dusty rose, muddy green, mushroom, and the chintz curtains no longer there. What time do they really belong in, their own or mine?

I walk along the street, slightly uphill, against a scattered traffic of small children going home for lunch. Although the girls wear jeans, denoting freedom, they aren't as noisy as they used to be; there are no chants, no catcalls. They trudge along doggedly, or so it seems to me. Maybe that's because I'm not at their level any more: I'm higher, so the sound comes up to me filtered. Or maybe it's me, the presence among them of someone they think is an adult, and has power.

A few of them stare, many don't. What's to see? A middle-aged woman, hands in her coat pockets, the legs of her jogging suit bunching above her boot tops, no more bizarre than most and easily forgotten.

Some of the porches have pumpkins on them, carved with faces, happy or sad or threatening, waiting for tonight. All Souls' Eve, when the spirits of the dead will come back to the living, dressed as ballerinas and Coke bottles and spacemen and Mickey Mice, and the living will give them candy to keep them from turning vicious. I can still taste that festival: the tart air, caramel in the mouth, the hope at the door, the belief in something for nothing all children take for granted. They won't get homemade popcorn balls any more though, or apples: rumors of razor blades abound, and the possibility of poison. Even by the time of my own children, we worried about the apples. There's too much loose malice blowing around.

In Mexico they do this festival the right way, with no disguises. Bright candy skulls, family picnics on the graves, a plate set for each individual guest, a candle for the soul. Everyone goes away happy, including the dead. We've rejected that easy flow between dimensions: we want the dead unmentionable, we refuse to name them, we refuse to feed them. Our dead as a result are thinner, grayer, harder to hear, and hungrier.

68

*M*y brother Stephen died five years ago. I shouldn't say died: was killed. I try not to think of it as murder, although it was, but as some kind of accident, like an exploding train. Or else a natural catastrophe, like a landslide. What they call for insurance purposes an act of God.

He died of an eye for an eye, or someone's idea of it. He died of too much justice.

He was sitting on a plane. He had a window seat. This much is known.

In the nylon webbing pocket in front of him was an in-flight magazine with an article in it about camels, which he'd read, and another about upgrading your business wardrobe, which he hadn't. There was also a set of earphones and a vomit bag.

Under the seat in front, beyond his bare feet—he's taken off his shoes and socks—is his briefcase, with a paper in it written by himself, on the subject of the probable composition of the universe. The universe, he once thought, may well be made up of infinitesimal pieces of string, in thirty-two different colors. The pieces of string are so small that "colors" is only a manner of speaking. But he is having doubts: there are other theoretical possibilities, two of which he has outlined in his paper. The universe is hard to pin down; it changes when you look at it, as if it resists being known.

He was supposed to deliver his paper the day before yesterday, in Frankfurt. He would have heard other papers. He would have learned.

Stuffed under the seat along with the briefcase is his suit

jacket, one of the three he now owns. His shirtsleeves are rolled up, which doesn't solve much: the air-conditioning is on the fritz and the air on the plane is overheated. Also it smells bad: at least one washroom toilet is out of order, and people fart more on planes, as my brother has had occasion to observe before, having taken a lot of plane trips. This is now compounded by panic, which is bad for the digestion. Two seats over, a fat bald-headed man is snoring with his mouth open, releasing an invisible cloud of halitosis.

The shades on the windows are pulled down. My brother knows that if he were to raise his he would see a runway, shimmering with heat, and beyond that a dun landscape alien as the moon, with a blinding sea in the background; and some oblong brown buildings with flat roofs, from which reprieve will come, or not. He saw all this before the shades came down. He doesn't know what country the buildings are in.

He hasn't had anything to eat since this morning. Sandwiches arrived from outside, strange granular bread, the butter on it liquid, some sort of beige meat paste that hinted of ptomaine. Also a piece of pale sweaty cheese in plastic wrap. He ate this cheese and the sandwich, and now his hands smell of old picnics, the roadside lunches of wartime.

The last drink of water was doled out four hours ago. He has a roll of peppermint LifeSavers: he always takes them on trips, in case of bumpy rides. He gave one to the middle-aged woman in oversized glasses and a plaid pant suit who was sitting beside him. He is somewhat relieved she's gone: her voiceless, colorless weeping, snuffly and monotonous, was beginning to get to him. The women and the children have all been allowed off, but he is not a woman or a child. Everyone left on the plane is a man.

They have been spaced, two by two, with an empty seat between each pair. Their passports have been collected. Those who have done the collecting are standing at intervals in the aisles of the plane, six of them, three with small machine guns, three with visible grenades. They are all wearing airplane pillowcases over their heads, with holes cut for the eyes and for the mouths, which show in the dim light as white glints, pink glistenings. Below these pillow-cases, which are red, their clothes are ordinary: a leisure outfit, a pair of gray flannel slacks with a white shirt tucked in, the bottom of a conservative navy-blue suit.

Naturally they came on board in the guise of passengers; though how they got the weapons past Security is anyone's guess. They must have had help, someone at the airport, so that they could jump up, the way they did, somewhere over the English Channel, and start shouting orders and waving around the firearms. Either that or the things were already on the plane, in pre-arranged hiding places, because nothing metal gets through the X-rays these days.

There are two or possibly three other men up in the cockpit, negotiating with the control tower over the radio. They haven't yet told the passengers who they are or what they want; all they've said, in heavily accented but under-standable English, is that everyone on the plane will live together or else die together. The rest has been monosyl-lables and pointing: *You, here*. It's hard to tell how many of them there are altogether, because of the identical pillow-cases. They're like those characters in old comic books, the ones with two identities. These men have been caught half-way through their transformation: ordinary bodies but with powerful, supernatural heads, deformed in the direction of heroism, or villainy.

I don't know whether or not this is what my brother thought. But it's what I think for him, now.

Unlike the open-mouthed man beside him, my brother can't sleep. So he occupies himself with theoretical strata-gems: what would he do if he were in their place, the place of the men with pillowcase heads? It's their tension, their hair-trigger excitement and blocked adrenaline that fills the plane, despite the lax bodies of the passengers, their fatigue and resignation.

If he were them, he would of course be ready to die. Without that as a given, such an operation would be point-less and unthinkable. But die for what? There's probably a religious motif, though in the foreground something more immediate: money, the release of others jailed in some sink-hole for doing more or less the same thing these men are doing. Blowing something up, or threatening to. Or shooting someone.

In a way this is all familiar. It's as if he's lived through it before, a long time ago; and despite the unpleasantness, the irritation of it, the combination of boredom and fear, he has a certain fellow feeling. He hopes these men can keep their heads and carry it off, whatever it is. He hopes there will be no sniveling and pants wetting among the passengers, that no one will go berserk and start screaming, and trigger a jittery massacre. A cool hand and a steady eye is what he wishes for them.

A man has entered from the front of the plane and is talking with two of the others. It seems to be an argument: there are gestures of the hands, a raised word. The other standing men tense, their square red heads scanning the passengers like odd radars. My brother knows he should avoid eye contact, keep his head down. He looks at the nylon webbing pocket in front of him, furtively peels off a LifeSaver.

The new man starts to walk down the aisle of the plane, his oblong, three-holed head turning from side to side. A second man walks behind him. Eerily, the taped music comes on over the intercom, saccharine, soporific. The man pauses; his oversized head moves ponderously left, like the head of some shortsighted, dull-witted monster. He extends an arm, gestures with the hand: *Up*. It's my brother he points to.

Here I stop inventing. I've spoken with the witnesses, the survivors, so I know that my brother stands up, eases himself past the man in the aisle seat, saying, "Excuse me." The expression on his face is one of bemused curiosity: these people are unfathomable, but then so are most. Perhaps they have mistaken him for someone else. Or they may want him to help negotiate, because they're walking toward the front of the plane, where another pillowhead stands waiting.

It's this one who swings open the door for him, like a polite hotel doorman, letting in the full glare of day. After the semidarkness it's ferociously bright, and my brother stands blinking as the image clears to sand and sea, a happy vacation postcard. Then he is falling, faster than the speed of light.

This is how my brother enters the past.

I was on planes and in airports for fifteen hours, getting there. I saw the buildings after that, the sea, the stretch of runway; the plane itself was gone. All they got in the end was safe conduct.

I didn't want to identify the body, or see it at all. If you don't see the body, it's easier to believe nobody's dead. But I did want to know whether they shot him before throwing him out, or after. I wanted it to be after, so he could have had that brief moment of escape, of sunlight, of pretended flight.

I did not stay up at night, on that trip. I did not want to look at the stars.

The body has its own defenses, its way of blocking things out. The government people said I was wonderful, by which they meant not a nuisance. I didn't collapse or make a spectacle of myself; I spoke with reporters, signed the forms, made the decisions. There was a great deal I didn't see or think about until much later.

What I thought about then was the space twin, the one who went on an interplanetary journey and returned in a week to find his brother ten years older.

Now I will get older, I thought. And he will not.

69
.

$\mathcal{M}$y parents never understood Stephen's death, because it had no reason; or no reason that had anything to do with him. Nor did they get over it. Before it, they were active, alert, vigorous; after it they faded.

"It doesn't matter how old they are," my mother said. "They're always your children." She tells me this as something I will need to know, later on.

My father became shorter and thinner, visibly shriveled; he sat for long periods, without doing anything. Unlike himself. This is what my mother told me, over the telephone, long distance.

Sons should not die before their fathers. It's not natural, it's the wrong order. Because who will carry on?

My parents themselves died in the usual way, of the things elderly people die of, that I myself will die of sooner than I think: my father instantly, my mother a year later, of a slower and more painful disease. "It's a good thing your father went the way he did," she said. "He would have hated this." She didn't say anything about hating it herself.

The girls came for a week, early on, at the end of summer, when my mother was still in her house in the Soo and we could all pretend this was just another visit. I stayed on after them, digging weeds out of the garden, helping with the dishes because my mother had never got a dishwasher, doing the laundry downstairs in the automatic washer but hanging it out on the line because she thought driers used up too much electricity. Greasing the muffin tins. Impersonating a child.

. . .

My mother is tired, but restless. She won't take naps in the afternoon, insists on walking to the corner store. "I can manage," she says. She doesn't want me to cook for her. "You'll never find anything in this kitchen," she says, meaning she thinks she'll never find anything herself if I start messing around in there. I smuggle frozen TV dinners into the refrigerator and con her into eating them by saying they'll go to waste if she doesn't. Waste is still a bugaboo for her. I take her to a movie, checking it first for violence, sex and death, and to a Chinese restaurant. In the north, in the old days, the Chinese restaurants were the only ones that could be depended on. The others went in for white bread and gravy mix sandwiches, lukewarm baked beans, pies made from cardboard and glue.

She is on painkillers, then stronger painkillers. She lies down more. "I'm glad I don't have to have an operation, in a hospital," she says. "The only time I was ever in a hospital was with you kids. With Stephen they gave me ether. I went out like a light, and when I woke up, there he was."

A lot of what she says is about Stephen. "Remember those smells he used to make, with that chemistry set of his? That was the day I was having a bridge party! We had to open the doors, and it was the middle of winter." Or else: "Remember all those comic books he had stowed away under his bed? There were too many to save. I chucked them out, after he went away. I didn't think there was any use for them. But people collect them, I read about it; now they'd be worth a fortune. We always thought they were just trash." She tells this like a joke on herself.

When she talks about Stephen, he is never more than twelve years old. After that he got beyond her. I come to realize that she was, or is, in awe of him, slightly afraid of him. She didn't intend to give birth to such a person.

"Those girls gave you a bad time," she says one day. I've made both of us a cup of tea—she's permitted this—and we sit at the kitchen table, drinking it. She's still surprised to catch me drinking tea, and has asked several times whether I wouldn't prefer milk.

"What girls?" I say. My fingers are a wreck; I shred

them quietly, out of sight beneath the tabletop, as I do in times of stress; an old bad habit I cannot seem to break.

"Those girls. Cordelia and Grace, and the other one. Carol Campbell." She looks at me, a little slyly, as if testing.

"Carol?" I say. I remember a stubby girl, turning a skipping rope.

"Of course, Cordelia was your best friend, in high school," she says. "I never thought she was behind it. It was that Grace, not Cordelia. Grace put her up to it, I always thought. What became of her?"

"I have no idea," I say. I don't want to talk about Cordelia. I still feel guilty, about walking away from her and not helping.

"I didn't know what to do," she says. "They came to me that day and said you'd been kept in at school, for being rude to the teacher. It was that Carol who said it. I didn't think they were telling the truth." She avoids the word *lie*, if possible.

"What day?" I say carefully. I don't know what day she means. She's begun to get things mixed up, because of the drugs.

"That day you almost froze. If I'd believed them I wouldn't have gone to look for you. I went down the road, along by the cemetery, but you weren't there." She regards me anxiously, as if wondering what I will say.

"Oh yes," I say, pretending I know what she's talking about. I don't want to confuse her. But I am growing confused myself. My memory is tremulous, like water breathed on. For an instant I see Cordelia and Grace, and Carol, walking toward me through the astonishing whiteness of the snow, their faces in shadow.

"I was so worried," she says. What she wants from me is forgiveness, but for what?

On some days she is stronger, and gives the illusion of improvement. Today she wants me to help her sort through the things in the cellar. "So you won't have to go through a lot of that old junk, later on," she says delicately. She won't say *death*; she wants to spare my feelings.

I don't like cellars. This one is unfinished: gray cement,

rafters above. I make sure the upstairs door is left open. "You should have a railing put on these stairs," I say. They are narrow, undependable.

"I can manage," says my mother. From the days when managing was enough.

We sort through the old magazines, the stash of different-sized cardboard boxes, the shelves of clean jars. She threw out a lot less than she could have, when they moved; or else she's accumulated more. I carry things up the stairs and stow them in the garage. In there they seem disposed of.

There's a whole shelf of my father's shoes and boots, lined-up pairs: city shoes with perforated toecaps, overshoes, rubber boots, wading boots for fishing, heavy-soled boots for walking in the woods, with a bacon grease patina and leather laces. Some of them must be fifty years old, or more. My mother will not throw them out, I know; but neither does she mention them. I can sense what she expects of me, in the way of control. I did my mourning at the funeral. She doesn't need to deal with a tearful child, not now.

I remember the old Zoology Building where we used to go on Saturdays, the creaking, overheated corridors, the bottles of eyeballs, the comforting smells of formaldehyde and mice. I remember sitting at the dinner table, with Cordelia, his warnings washing over our heads, the ruined water, the poisoned trees, species after species snuffed out like stepped-on ants. We did not think such things were prophecies. We thought they were boring then, a form of adult gossip that did not concern us. Now it's all come true, except worse. I live in his nightmare, no less real for being invisible. You can still breathe the air, but for how long?

Against his bleak forecasting is set my mother's cheerfulness, in retrospect profoundly willed.

We start on the steamer trunk. It's the one I remember from our Toronto house; I still think of it as mysterious, the repository of treasure. My mother too views this as an adventure: she says she hasn't looked into that trunk for years, she has no idea what's in there. She is no less alive because dying.

I open the trunk, and the smell of mothballs blossoms upward. Out come the baby clothes folded in tissue paper, the pieces of flowery silver, yellowy-black. "Keep these for the girls," she says. "You have this one." The wedding dress, the wedding pictures, the sepia-colored relatives. A packet of feathers. Some bridge tallies with tassels on them, two pairs of white kid gloves. "Your father was a wonderful dancer," she says. "Before we were married." I have never known this.

We go down through the layers, unearthing discoveries: my high school pictures, my lipsticked mouth unsmiling, somebody's hair in an envelope, a single knitted baby sock. Old mittens, old neckties. An apron. Some things are to be kept, others thrown out or given away. Some things I will take back with me. We have several piles.

My mother is excited, and I catch some of this excitement from her: it's like a Christmas stocking. Although not pure joy.

Stephen's packets of airplane trading cards, held together with rotting elastic bands. His scrapbooks, his drawings of explosions, his old report cards. These she sets aside.

My own drawings and scrapbooks. There are the pictures of little girls I now remember, with their puffed sleeves and pink skirts and hairbows. Then, in the scrapbooks, some unfamiliar pictures cut from magazines: women's bodies, in clothes of the forties, with other women's heads glued onto them. *This is a Watchbird watching YOU.*

"You loved those magazines," says my mother. "You used to pore over them for hours, when you were sick in bed."

Underneath the scrapbooks is my old photo album, the black pages held together with the tie like a shoelace. Now I can remember putting it into the trunk, before I went to high school.

"We gave you that," says my mother. "For Christmas, to go with your camera." Inside is my brother, poised with a snowball, and Grace Smeath crowned with flowers. A couple of large boulders, with names printed underneath them in white pencil. Myself, in a jacket with the sleeves too short, standing against a motel cabin door. The number 9.

"I wonder what happened to that camera?" says my mother. "I must've given it away. You lost interest in it, after a while."

I'm aware of a barrier between us. It's been there for a long time. Something I have resented. I want to put my arms around her. But I am held back.

"What's that?" she says.

"My old purse," I say. "I used to take it to church." I did. I can see the church now, the onion on the spire, the pews, the stained-glass windows. THE·KINGDOM·OF·GOD· IS·WITHIN·YOU.

"Well, what do you know. I don't know why I saved that," says my mother, with a little laugh. "Put it on the throw-out pile." It's squashed flat; the red plastic is split at the sides, where the sewing is. I pick it up, push at it to make it go back into shape. Something rattles. I open it up and take out my blue cat's eye.

"A marble!" says my mother, with a child's delight. "Remember all those marbles Stephen used to collect?"

"Yes," I say. But this one was mine.

I look into it, and see my life entire.

70

*D*own this street is where the store was. We bought red licorice whips, bubble gum, orange Popsicles, black jaw-breakers that faded to a seed. Things you could buy for a penny, with the King's head on it. *Georgius VI Dei Gratia.*

I've never got used to the Queen being grown up. Whenever I see her cut-off head on the money, I think of her as fourteen years old, in her Girl Guide uniform, her back as straight as ours were supposed to be, looking down at me from the yellowing newspaper clippings on Miss Lumley's Grade Four blackboard; standing in front of the clumsy diamond of a radio microphone, frowning with earnestness and well-concealed fear, rallying the forces as the bombs fell on London, as we sang "There'll Always Be an England" to the waving of Miss Lumley's life-threatening wooden pointer, in a time warp eight years later.

The Queen has had grandchildren since, discarded thousands of hats, grown a bosom and (heresy to think it) the beginning of a double chin. None of this fools me. She's in there somewhere, that other one.

I walk the next blocks, turn the corner, expecting to see the familiar dingy oblong of the school, in weathered red brick the color of dried liver. The cindered schoolyard, the tall thin windows with orange paper pumpkins and black cats stuck onto them for Hallowe'en, the graven lettering over the doors, BOYS and GIRLS, like the inscriptions on mausoleums of the late nineteenth century.

But the school has disappeared. In its place a new school has risen instantly, like a mirage: light-colored, block-shaped, glossy and modern.

I feel hit, in the pit of the stomach. The old school has

been erased, wiped from space. It's as if it was never there at all. I lean against a telephone pole, bewildered, as if something has been cut out of my brain. Suddenly I'm bone tired. I would like to go to sleep.

After a while I approach the new school, go toward it through the gate, walk slowly around it. BOYS and GIRLS have been abolished, that much is clear; though there's still a chain-link fence. The schoolyard is dotted with swings, with climbing bars and slides, in bright primary colors; a few children have come back early from lunch and are clambering about.

It's all so clean-cut, so open. Surely behind those glassy, candid doors there are no more long wooden pointers, no black rubber strap, no hard wooden desks in rows; no King and Queen in their stiff regalia, no inkwells; no sniggering about underpants; no bitter, whiskery old women. No cruel secrets. Everything like that is gone.

I come around the back corner, and there is the eroded hill, with its few sparse trees. That much is still the same, then.

No one's up there.

I climb up the wooden steps, stand where I used to stand. Where I am still standing, never having been away. The voices of the children from the playground below could be any children's voices, from any time, the light under the trees thickens, turns malevolent. Ill will surrounds me. It's hard to breathe. I feel as if I'm pushing against something, a pressure on me, like opening the door against a snowstorm.

Get me out of this, Cordelia. I'm locked in.

I don't want to be nine years old forever.

The air is soft, autumnal, the sun shines. I am standing still. And yet I walk head down, into the unmoving wind.

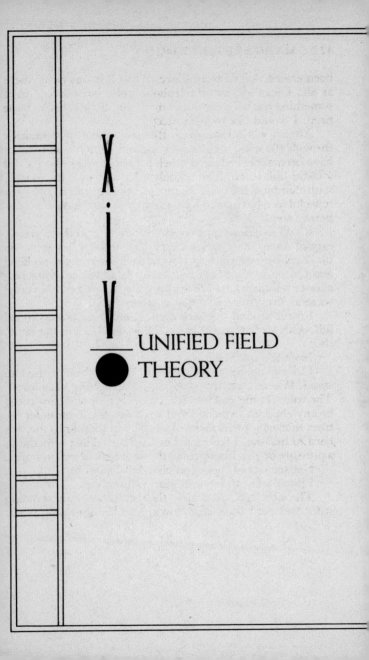

XIV

UNIFIED FIELD THEORY

?!

I put on my new dress, cutting off the price tag with Jon's wire cutters. I ended up with black, after all. Then I go into the bathroom to squint at myself in the inadequate, greasy mirror: now that I've got the thing on, it looks much the same as all the other black dresses I've ever owned. I check it for lint, apply my pink lipstick, and end up looking nice, as far as I can tell. Nice, and negligible.

I could jazz myself up somehow. I ought to have some dangly earrings, some bangles, a silver bow tie on a little chain, an outsize Isadora Duncan strangle-yourself-by-mistake scarf, a rhinestone brooch of the thirties, in sly bad taste. But I don't have any of these things, and it's too late to go out and buy any. I will have to do. Come-as-you-are parties, they used to have. I will come as I am.

I'm at the gallery an hour early. Charna is not here, or the others; they may have gone out to eat, or more likely to change. Everything is set up though, the rented thick-stemmed wineglasses, the bottles of mediocre hooch, the mineral water for teetotalers, because who would serve unadulterated chlorine from the tap? The cheeses hardening at the edges, the sulfur-drenched grapes, luscious and shiny as wax, plumped with blood from the dying field workers of California. It doesn't pay to know too many of these things; eventually there's nothing you can put into your mouth without tasting the death in it.

The bartender, a severe-eyed young woman in gelled hair and unstructured black, is polishing glasses behind the long table that serves as the bar. I extract a glass of wine from her. She's doing the bartending for money, her nonchalance implies: her true ambitions lie elsewhere. She tight-

ens her lips while doling out my drink: she doesn't approv
of me. Possibly she wants to be a painter, and thinks I hav
compromised my principles, knuckled under to succes
How I used to revel in such bitter little snobberies mysel
how easy they were, once.

I walk slowly around the gallery, sipping at my glass o
wine, permitting myself to look at the show, for the fir
time really. What is here, and what is not. There's a cata
logue, put together by Charna, a professional-looking com
puter-and-laser-printer affair. I remember the catalogu
from the first show, done on a mimeo machine, smeare
and illegible, its poverty a badge of authenticity. I rememb
the sound of the roller turning, the tang of the ink, the pai
in my arm.

Chronology won out after all: the early things are on th
east wall, what Charna calls the middle period on the en
wall, and on the west wall are five recent pictures whic
I've never shown before. They're all I've been able to do i
the past year. I work more slowly, these days.

Here are the still lifes. "Early forays by Risley into th
realm of female symbolism and the charismatic nature c
domestic objects," says Charna. In other words, the toaste
the coffee percolator, my mother's wringer washer. Th
three sofas. The silver paper.

Farther along are Jon and Josef. I look at them with som
fondness, them and their muscles and their cloudy-heade
notions about women. Their youngness is terrifying. Ho
could I have put myself into the hands of such inexperience

Next to them is Mrs. Smeath; many of her. Mrs. Smeat
sitting, standing, lying down with her holy rubber plant
flying, with Mr. Smeath stuck to her back, being screwe
like a beetle; Mrs. Smeath in the dark-blue bloomers of Mis
Lumley, who somehow combines with her in a frightenin
symbiosis. Mrs. Smeath unwrapped from white tissue pa
per, layer by layer. Mrs. Smeath bigger than life, bigge
than she ever was. Blotting out God.

I put a lot of work into that imagined body, white as
burdock root, flabby as pork fat. Hairy as the inside of a

ear. I labored on it, with, I now see, considerable malice. But these pictures are not only mockery, not only desecration. I put light into them too. Each pallid leg, each steel-rimmed eye, is there as it was, as plain as bread. I have said, *Look*. I have said, *I see*.

It's the eyes I look at now. I used to think these were self-righteous eyes, piggy and smug inside their wire frames; and they are. But they are also defeated eyes, uncertain and melancholy, heavy with unloved duty. The eyes of someone for whom God was a sadistic old man; the eyes of a small town threadbare decency. Mrs. Smeath was a transplant to the city, from somewhere a lot smaller. A displaced person; as I was.

Now I can see myself, through these painted eyes of Mrs. Smeath: a frazzle-headed ragamuffin from heaven knows where, a gypsy practically, with a heathen father and a feckless mother who traipsed around in slacks and gathered weeds. I was unbaptized, a nest for demons: how could she know what germs of blasphemy and unfaith were breeding in me? And yet she took me in.

Some of this must be true. I have not done it justice, or rather mercy. Instead I went for vengeance.

An eye for an eye leads only to more blindness.

I move to the west wall, where the new paintings are. They are larger than my usual format, and space out the wall nicely.

The first one is called *Picoseconds*. "A *jeu d'esprit*," says Charna, "which takes on the Group of Seven and reconstructs their vision of landscape in the light of contemporary experiment and postmodern pastiche."

It is in fact a landscape, done in oils, with the blue water, the purple underpainting, the craggy rocks and windswept raggedy trees and heavy impasto of the twenties and thirties. This landscape takes up much of the painting. In the lower right-hand corner, in much the same out-of-the-way position as the disappearing legs of Icarus in the painting by Bruegel, my parents are making lunch. They have their fire going, the billy tin suspended over it. My mother in her plaid jacket bends over, stirring, my father adds a stick of wood to the fire. Our Studebaker is parked in the background.

They are painted in another style: smooth, finely modulated, realistic as a snapshot. It's as if a different light falls on them; as if they are being seen through a window which has opened in the landscape itself, to show what lies behind or within it.

Underneath them, like a subterranean platform, holding them up, is a row of iconic-looking symbols painted in the flat style of Egyptian tomb frescoes, each one enclosed in a white sphere: a red rose, an orange maple leaf, a shell. They are in fact the logos from old gas pumps of the forties. By their obvious artificiality, they call into question the reality of landscape and figures alike.

The second painting is called *Three Muses*. Charna has had some trouble with this one. "Risley continues her disconcerting deconstruction of perceived gender and its relationship to perceived power, especially in respect to numinous imagery," she says. If I hold my breath and squint, I can see where she gets that: all Muses are supposed to be female, and one of these is not. Maybe I should have called it *Dancers*, and put her out of her misery. But they are not dancers.

To the right is a short woman, dressed in a flowered housecoat and mules with real fur. On her head is a red pillbox hat with cherries. She has black hair and large golden earrings, and is carrying a round object the size of a beach ball, which is in fact an orange.

To the left is an older woman with blue-gray hair, wearing a waltz-length lavender silk gown. In her sleeve is tucked a lace handkerchief, over her nose and mouth is a gauze nurse's mask. Above the mask her bright blue eyes look out crinkly at the edges and sharp as tacks. In her hands she holds a globe of the world.

In the middle is a thin man with medium-brown skin and white teeth, smiling an uncertain smile. He is wearing a richly worked gold and red oriental costume reminiscent of Balthazar's in Jan Gossaert's *Adoration of the Magi*, but without the crown and scarf. He too holds out a round object: it's flat like a disc and appears to be made of purple stained glass. On its surface are arranged, seemingly at random, several bright pink objects not unlike those to be found in abstract paintings. They are in fact spruce budworm eggs, in section; though I would not expect anyone but a biologist to recognize them.

The arrangement of the figures recalls that of classical Graces, or else of the different-colored children wreathed around Jesus on the front of my old Sunday school paper. But those were facing in, and these are facing out. They hold their gifts forward, as if presenting them to someone who sits or stands outside the painting.

Mrs. Finestein, Miss Stuart from school, Mr. Banerji. Not as they were, to themselves: God knows what they really saw in their own lives, or thought about. Who knows what death camp ashes blew daily through the head of Mrs. Finestein, in those years right after the war? Mr. Banerji probably could not walk down a street here without dread, of a shove or some word whispered or shouted. Miss Stuart was in exile, from plundered Scotland still declining, three thousand miles away. To them I was incidental, their kindness to me casual and minor; I'm sure they didn't give it a second thought, or have any idea of what it meant. But why shouldn't I reward them, if I feel like it? Play God, translate them into glory, in the afterlife of paint. Not that they'll ever know. They must be dead by now, or elderly. Elsewhere.

The third picture is called *One Wing*. I painted it for my brother, after his death.

It's a triptych. There are two smaller, flanking side panels. In one is a World War Two airplane, in the style of a cigarette card; in the other is a large pale-green luna moth.

In the larger, central panel, a man is falling from the sky. That he is falling and not flying is clear from his position, which is almost upside-down, slantwise to the few clouds; nevertheless he appears calm. He is wearing a World War Two RCAF uniform. He has no parachute. In his hand is a child's wooden sword.

This is the kind of thing we do, to assuage pain.

Charna thinks it's a statement about men, and the juvenile nature of war.

The fourth painting is called *Cat's Eye*. It's a self-portrait, of sorts. My head is in the right foreground, though it's shown only from the middle of the nose up: just the upper half of the nose, the eyes looking outward, the forehead and the topping of hair. I've put in the incipient wrinkles, the little chicken feet at the corners of the lids. A few gray hairs. This is cheating, as in reality I pull them out.

Behind my half-head, in the center of the picture, in the empty sky, a pier glass is hanging, convex and encircled by an ornate frame. In it, a section of the back of my head is visible; but the hair is different, younger.

At a distance, and condensed by the curved space of the mirror, there are three small figures, dressed in the winter clothing of the girls of forty years ago. They walk forward, their faces shadowed, against a field of snow.

The last painting is *Unified Field Theory*. It's a vertical oblong, larger than the other paintings. Cutting across it a little over a third up is a wooden bridge. To either side of the bridge are the tops of trees, bare of leaves, with a covering of snow on them, as after a heavy moist snowfall. This snow is also on the railing and struts of the bridge.

Positioned above the top railing of the bridge, but so her feet are not quite touching it, is a woman dressed in black, with a black hood or veil covering her hair. Here and there on the black of her dress or cloak there are pinpoints of light. The sky behind her is the sky after sunset; at the top of it is the lower half of the moon. Her face is partly in shadow.

She is the Virgin of Lost Things. Between her hands, at the level of her heart, she holds a glass object: an oversized cat's eye marble, with a blue center.

Underneath the bridge is the night sky, as seen through a telescope. Star upon star, red, blue, yellow, and white, swirling nebulae, galaxy upon galaxy: the universe, in its incandescence and darkness. Or so you think. But there are also stones down there, beetles and small roots, because this is the underside of the ground.

At the lower edge of the painting the darkness pales and merges to a lighter tone, the clear blue of water, because the creek flows there, underneath the earth, underneath the bridge, down from the cemetery. The land of the dead people.

I go to the bar, ask for another glass of wine. It's better quality than the rotgut we used to buy for such affairs.

I walk the room, surrounded by the time I've made; which is not a place, which is only a blur, the moving edge we live in; which is fluid, which turns back upon itself, like a wave. I may have thought I was preserving something from time, salvaging something; like all those painters, centuries ago, who thought they were bringing Heaven to earth, the revelations of God, the eternal stars, only to have their slabs of wood and plaster stolen, mislaid, burnt, hacked to pieces, destroyed by rot and mildew.

A leaky ceiling, a match and some kerosene would finish all this off. Why does this thought present itself to me, not as a fear but as a temptation?

Because I can no longer control these paintings, or tell them what to mean. Whatever energy they have came out of me. I'm what's left over.

$\mathcal{N}$ow Charna hustles toward me in mauve leather, clanking with ersatz gold. She whisks me into the back office: she doesn't want me dangling around in the empty gallery, at loose ends while the first revelers trickle in, she doesn't want me looking unsuccessful and too eager. She will make an entrance with me, later, when the noise level is high enough.

"You can relax here," she says; which is unlikely. In her office I drink my second drink, pacing the empty space. This is like birthday parties, with streamers and balloons at the ready and the hot dogs waiting in the kitchen, but what if nobody comes? Which will be worse: if they don't come, or if they do? Soon the door will open, and in will crowd a horde of snide and treacherous little girls, whispering and pointing, and I will be servile, grateful.

My hands begin to sweat. I think another drink will calm me down, which is a bad sign. I will go out there and flirt, just for the hell of it, to see if I can still interest anyone. But there may not be anyone to flirt with. In which case I will get drunk. Maybe I will throw up in the toilet, with or without the excess alcohol.

I'm not like this in other places, not this bad. I shouldn't have come back here, to this city that has it in for me. I thought I could stare it down. But it still has power; like a mirror that shows you only the ruined half of your face.

I think about escaping, out the back way. I could send a telegram later, claiming illness. That would start a good rumor: a lingering, invisible illness, which would get me out of such things forever.

But Charna reappears through the door in time, flushed with excitement. "There's lots of people here already," she says. "They're dying to meet you. We're all very proud of you." This is so much like what a family would say, a mother or an aunt, that I'm thrown off guard. Who is this family, and whose family is it? I've been framed: the recalcitrant child before the piano recital, or, more like it, the bullet-scarred war horse, veteran of early, barely remembered battles, about to be presented with a gold watch and a handshake and a heartfelt vote of thanks. A fading halo of blue ink clings round me.

Suddenly Charna reaches over to me, gives me a quick metallic hug. Maybe that warmth is genuine, maybe I should be ashamed of my dour, cynical thoughts. Maybe she really does like me, wish me well. I can almost believe it.

I stand in the main gallery, black from neck to toes, with my third glass of red wine. Charna is off now, rummaging in the crowd for people who are dying to meet me. I am at her disposal. I crane my neck, peering through the crowd, which has blotted out the paintings; only a few tops of heads are visible, a few skies, a few backgrounds and clouds. I keep expecting, or fearing, that people I should know, have known, will appear, and I will only half recognize them. They will stride forward, hands outstretched, girls from high school bloated or diminished, skins crinkled, frowns permanent, smooth-fleshed boyfriends of thirty years ago who've gone bald or grown mustaches or shrunk. *Elaine! What the heck! Good to see you!* They will have the advantage of me, it's my face on the poster. My smile will be welcoming, my mind frantic as I scramble through the past, trying to locate their names.

Really it's Cordelia I expect, Cordelia I want to see. There are things I need to ask her. Not what happened, back then in the time I lost, because now I know that. I need to ask her why.

If she remembers. Perhaps she's forgotten the bad things, what she said to me, what she did. Or she does remember them, but in a minor way, as if remembering a game, or a single prank, a single trivial secret, of the kind girls tell and then forget.

She will have her own version. I am not the center of her story, because she herself is that. But I could give her something you can never have, except from another person: what you look like from outside. A reflection. This is the part of herself I could give back to her.

We are like the twins in old fables, each of whom has been given half a key.

Cordelia will walk toward me through the opening crowd, a woman of wavering age, dressed in Irish tweed of a muted green, mother-of-pearl earrings circled with gold, beautiful shoes; well-groomed, soignée as they used to say. Taking care of herself, as I am. Her hair will be gently frosted, her smile quizzical. I won't know who she is.

There are a lot of women in this room, several other painters, some rich people. Mostly it's the rich people Charna drags over. I shake their hands, watch their mouths move. Elsewhere I have more stamina for these things, these acts of self-exposure; I could brazen them out. But here I feel scraped naked.

In the gap between rich people, a young girl pushes her way through. She's a painter, it goes without saying, but she says it anyway. She's in a miniskirt and tight leggings and flat clumpy black shoes with laces, her hair is shaved up the back the way my brother's used to be, a late forties squareboy's cut. She is post everything, she is what will come after *post*. She is what will come after me.

"I loved your early work," she says. "*Falling Women*, I loved that. I mean, it sort of summed up an era, didn't it?" She doesn't mean to be cruel, she doesn't know she's just relegated me to the dust heap along with crank telephones and whalebone stays. In former days I would have said something annihilating to her, some scabby, scalding remark, but I can't think of anything right off the bat. I'm out of training, I'm losing my nerve. In any case, what purpose would it serve? Her past-tense admiration is sincere. I should be gracious. I stand there, my grin turning to stone, institutionalized. Eminence creeps like gangrene up my legs.

"I'm glad," I manage. When in doubt, lie through your teeth. I'm lucky I still have teeth to lie through.

I'm standing back to the wall, with a new, full glass of wine. I crane my neck, peering through the crowd, over the well-arranged heads: it's time for Cordelia to appear, but she has not appeared. Disappointment is building in me, and impatience; and then anxiety. She must have set out, in this direction. Something must have happened to her on the way here.

This goes on while I shake more hands and say more things, and the room gradually clears.

"That went off very well," says Charna, with a sigh, of relief I think. "You were wonderful." She's happy because I haven't bitten anyone or spilled my drink down their legs, or otherwise acted like an artist. "How about dinner, with all of us?"

"No," I say. "No thank you. I'm bone-tired. I think I'll just go back." I look around once more: Cordelia is not here.

Bone-tired, an old phrase, of my mother's. Though bones as such do not get tired. They're strong, they have a lot of stamina; they can go on for years and years, after the rest of the body has quit.

I'm headed for a future in which I sprawl propped in a wheelchair, shedding hair and drooling, while some young stranger spoons mushed food into my mouth and I stand in the snow under the bridge, and stand and stand. While Cordelia vanishes and vanishes.

I go out, into the sidewalk twilight, outside the gallery. I want to take a taxi, but I can barely lift my hand.

I've been prepared for almost anything; except absence, except silence.

73

I take a taxi back to the studio, climb the four flights of stairs, dimly lit as they are at night, resting on the landings. I listen to my heart, going dull and fast in there under the layers of cloth. A flawed heart, in decline. I shouldn't have drunk all that wine. It's cold here, they've been skimping on the heat. The sound of my breath comes to me, a disembodied gasping, as if it's someone else breathing.

Cordelia has a tendency to exist.

I fumble the key into the keyhole, grope for the light switch. I could do without all the fake body parts around here. I make my way to the kitchenette, shambling a little, keeping my coat on because of the cold.

Coffee is what I need. I make some, wrap my hands around the warm cup, carry it to the workbench, clearing a space for my elbows among the wire and sharp-edged tools. Tomorrow I'm out of this city, and not a moment too soon. There's too much old time here.

So, Cordelia. Got you back.

Never pray for justice, because you might get some.

I drink my coffee, holding the shaking cup, hot liquid slopping down my chin. It's a good thing I'm not in a restaurant. It isn't chic for women to be drunk. Men drunks are more excusable, more easily absolved, but why? It must be thought they have better reasons.

I wipe my arm with the coat sleeve across my face, which is wet because I'm crying. This is the kind of thing I should look out for: crying without reason, making a spectacle of myself. I feel it's a spectacle, even though no one's watching.

You're dead, Cordelia.
No I'm not.
Yes you are. You're dead.
Lie down.

XV

● BRIDGE

74

I'm light-headed, as if convalescing. I slept rolled up in the duvet, still in my black dress, which I did not have the energy to take off. I woke at noon, with a large, cottony skull, pulsing with hangover, to discover I'd missed my plane. It's a long time since I've drunk that much of anything. As with many things, I should know better.

Now it's late afternoon. The sky is soft and gray, low, damp and blurred like wet blotting paper. The day feels vacant, as if everyone has moved out of it; as if there's nothing more to come.

I pace along the sidewalk, away from the demolished school. My old direction, I could still do it blindfolded. As always on these streets I feel disliked.

Down below me is the bridge. From here it looks neutral. I stand at the top of the hill, take a breath. Then I start down.

It's surprising how little has changed. The houses on either side are the same, although the muddy path is gone: in its place is a neat little hand railing, a trim cement walk. The smell of the fallen leaves is still here, the burning smell of their slow decay, but the nightshade vines with their purple flowers and red blood-drop berries and the weeds and random debris have been cleared away, and everything is pruned and civic.

Nevertheless there's a rustling, a rank undertone of cats and their huntings and furtive scratchings, still going on behind the deceptive tidiness. Another, wilder and more tangled landscape rising up, from beneath the surface of this one.

We remember through smells, as dogs do.

The willow trees overhanging the path are the same.

Although they've grown, I've grown also, so the distance between us remains constant. The bridge itself is different, of course; it's made of concrete and lighted up at night, not wooden and falling apart and rotten-smelling. Nevertheless it's the same bridge.

Stephen's jar of light is buried down there somewhere.

At this time of year the day darkens early. It's silent, no voices of children; only the monotonous cawing of a crow, and behind it the sea sound of distant traffic. I rest my arms on the concrete wall and look down through the bare branches that are like dry coral. I used to think that if I jumped over, it would not be like falling, it would be more like diving; that if I died that way it would be soft, like drowning. Though far below, on the ground, there's a pumpkin, tossed over and smashed open, looking unpleasantly like a head.

The ravine is more filled in with bushes and trees than it used to be. In among them is the creek, running with clear water unsafe to drink. They've cleaned up the junk, the rusted car parts and discarded tires; this is no longer an unofficial garbage dump but a joggers' route. The neatly graveled runners' path beneath me leads uphill to the distant road and to the cemetery, where the dead people wait, forgetting themselves atom by atom, melting away like icicles, flowing downhill into the river.

That was where I fell into the water, there is the bank where I scrambled up. That's where I stood, with the snow falling on me, unable to summon the will to move. That's where I heard the voice.

There was no voice. No one came walking on air down from the bridge, there was no lady in a dark cloak bending over me. Although she has come back to me now in absolute clarity, acute in every detail, the outline of her hooded shape against the lights from the bridge, the red of her heart from within the cloak, I know this didn't happen. There was only darkness and silence. Nobody and nothing.

There's a sound: a shoe against loose rock.

It's time to go back. I push away from the cement wall, and the sky moves sideways.

I know that if I turn, right now, and look ahead of me along the path, someone will be standing there. At first I think it will be myself, in my old jacket, my blue knitted hat. But then I see that it's Cordelia. She's standing halfway up the hill, gazing back over her shoulder. She's wearing her gray snowsuit jacket but the hood is back, her head is bare. She has the same green wool knee socks, sloppily down around her ankles, the brown school brogues scuffed at the toes, one lace broken and knotted, the yellowish-brown hair with the bangs falling into her eyes, the eyes gray-green.

It's cold, colder. I can hear the rustle of the sleet, the water moving under the ice.

I know she's looking at me, the lopsided mouth smiling a little, the face closed and defiant. There is the same shame, the sick feeling in my body, the same knowledge of my own wrongness, awkwardness, weakness; the same wish to be loved; the same loneliness; the same fear. But these are not my own emotions any more. They are Cordelia's; as they always were.

I am the older one now, I'm the stronger. If she stays here any longer she will freeze to death; she will be left behind, in the wrong time. It's almost too late.

I reach out my arms to her, bend down, hands open to show I have no weapon. *It's all right*, I say to her. *You can go home now.*

The snow in my eyes withdraws like smoke.

When I turn, finally, Cordelia is no longer there. Only a middle-aged woman, pink-cheeked and bareheaded, coming down the hill toward me, in jeans and a heavy white pull-over, with a dog on a green leash, a terrier. She passes me smiling, a civil, neutral smile.

There's nothing more for me to see. The bridge is only a bridge, the river a river, the sky is a sky. This landscape is empty now, a place for Sunday runners. Or not empty: filled with whatever it is by itself, when I'm not looking.

75

I'm on the plane, flying or being flown, westward toward the watery coast, the postcard mountains. Ahead of me, out the window, the sun sinks in a murderous, vulgar, unpaintable and glorious display of red and purple and orange; behind me the ordinary night rolls forward. Down on the ground the prairies unscroll, vast and mundane and plausible as hallucinations, dusted already with snow and scrawled with sinuous rivers.

I have the window seat. In the two seats beside me are two old ladies, old women, each with a knitted cardigan, each with yellowy-white hair and thick-lensed glasses with a chain for around the neck, each with a desiccated mouth lipsticked bright red with bravado. They have their trays lowered and are drinking tea and playing Snap, fumbling the slippery cards, laughing like cars on gravel when they cheat or make mistakes. From time to time they get up, unbuckling themselves laboriously, and hobble to the back of the plane, to smoke cigarettes and line up for the washroom. When they return they make bathroom jokes, quips about wetting your pants and running out of toilet paper, eyeing me cunningly while they do so. I wonder how old they think they are, underneath the disguise of their bodies; or how old they think I am. Perhaps, to them, I look like their mother.

They seem to me amazingly carefree. They have saved up for this trip and they are damn well going to enjoy it, despite the arthritis of one, the swollen legs of the other. They're rambunctious, they're full of beans; they're tough as thirteen, they're innocent and dirty, they don't give a hoot. Responsibilities have fallen away from them, obligations, old hates and grievances; now for a short while they can play again like children, but this time without the pain.

This is what I miss, Cordelia: not something that's gone, but something that will never happen. Two old women giggling over their tea.

Now it's full night, clear, moonless and filled with stars, which are not eternal as was once thought, which are not where we think they are. If they were sounds, they would be echoes, of something that happened millions of years ago: a word made of numbers. Echoes of light, shining out of the midst of nothing.

It's old light, and there's not much of it. But it's enough to see by.

About the Author

Margaret Atwood, the author of more than twenty volumes of poetry, fiction, and nonfiction, is best known for her seven novels: *The Edible Woman, Surfacing, Lady Oracle, Life Before Man, Bodily Harm, The Handmaid's Tale,* and *The Robber Bride.* Born in Ottawa, she grew up in northern Ontario, Quebec, and Toronto. Throughout her life, Ms. Atwood has traveled extensively, living for periods of time in England, Germany, France, Italy, and the United States. She now resides in Toronto with novelist Graeme Gibson and their daughter, Jess.